ATLAS OF HUMAN ANATOMY

Mark Nielsen
University of Utah

Shawn Miller
University of Utah

WILEY

JOHN WILEY & SONS, INC.

Vice President & Executive Publisher	Kaye Pace
Acquisitions Editor	Bonnie Roesch
Project Editor	Lorraina Raccuia
Production Manager	Dorothy Sinclair
Senior Production Editor	Anna Melhorn
Marketing Manager	Clay Stone
Creative Director	Harry Nolan
Senior Designer	Madelyn Lesure
Media Editor	Linda Muriello
Cover Photo Credit	Mark Nielsen

This book was set in Minion Pro by Mark Nielsen & Aptara®, Inc. and printed and bound by Quad/Graphics Versailles. The cover was printed by Quad/Graphics Versailles.

Founded in 1807, John Wiley & Sons, Inc. has been a valued source of knowledge and understanding for more than 200 years, helping people around the world meet their needs and fulfill their aspirations. Our company is built on a foundation of principles that include responsibility to the communities we serve and where we live and work. In 2008, we launched a Corporate Citizenship Initiative, a global effort to address the environmental, social, economic, and ethical challenges we face in our business. Among the issues we are addressing are carbon impact, paper specifications and procurement, ethical conduct within our business and among our vendors, and community and charitable support. For more information, please visit our website: *www.wiley.com/go/citizenship*.

Printed in the United States of America

SKY10022037_102820

Preface

Anatomy is a visual science, and in no other subject does the age-old saying ring so true — "a picture is worth a thousand words." With this in mind we created this book to teach anatomy with the real thing — photographs of cadaver dissections and the bones of the skeleton, and micrographs of the body's tissues. We believe that every word that has ever been written about anatomy is the result of someone describing what they observed in a dissection (or as is the case of many authors today, the words are paraphrased from somebody else's knowledge and writings about dissection). In this book we provide you with the *images* of *real* anatomy, with the hope that this will help you better visualize the *words* of anatomy.

We often hear that photographs can never clarify and teach anatomy as well as art. While it is true that the artist has much more creative license than the dissector, it is also true that a lot of anatomical art does not always accurately depict what is actually observed by a dissector; or for that matter, a surgeon in a clinical setting. We believe that *good* dissection and photography can be instructive, especially when creatively coupled with teaching concepts. With this in mind, another objective of this book is to present images that teach, and not just showcase a plethora of anatomy. Each dissection was made with an instructive purpose and reference images are used to highlight and focus on the patterns or concepts depicted by the dissections. There are many simple patterns of design that organize and clarify the structure of the vertebrate body. We attempt to show these patterns in our presentation of anatomical structure throughout the chapters of this book. The few words that accompany the images in the book draw attention to the patterns and the basic structure-function relationships of the dissections and micrographs.

It has also been our goal to create a book that will benefit students at all levels of anatomy education. The chapters are constructed with a systematic approach to anatomy to meet the needs of the typical undergraduate anatomy course. Each chapter illustrates the concepts and features of a body system and depicts those features with clear dissections and reference images of the dissections. On the other hand, because it is dissection based the book is also an excellent reference for the medical student, physical therapy student, or other graduate student who is studying cadaver anatomy from a regional approach. Even the layperson who wants to learn more about their amazing body can benefit from the beautiful anatomy images throughout the book. Students can continue their exploration of anatomy using Real Anatomy, 3-D imaging software that enables students to dissect through layers of the real human body.

To learn more about Real Anatomy, visit http://www.wiley.com/college/sc/realanatomy

In conclusion we would like to thank a few individuals for their help with the dissections that were photographed for this book. Good dissection is a time consuming task that requires a strong knowledge of anatomy, skill and dexterity, and above all a lot of patience. Nathan Mortensen played a major role in helping with the dissections throughout the pages of this book. Also, the following individuals each contributed one or two dissections, and we want to thank them for their contribution: Richard Homer, Torrence Meyer, Jordan Barker, Jon Groot, and John Dimitropoulos. We also want to thank Alexa Doig who took a few of the cadaver photographs.

We hope this book expands your vista of the amazing machine we call the human body. We would love to have any feedback you have on how we might improve the book for future editions.

Mark Nielsen, University of Utah
marknielsen@bioscience.utah.edu
Shawn Miller, University of Utah
smiller@biology.utah.edu

Content

1 | Introduction

Human anatomy is the science that deals with the structure and design of the human body. A knowledge of anatomy is not only important for the anatomist, but is an essential tool for all the professionals who deal with the human body in any of a variety of ways. Furthermore, everyone can benefit from a knowledge of anatomy because it is what we are, and understanding our bodies can be invaluable.

Anatomy is an ancient science. The principal methods anatomists used, and still use, to reveal what is known about anatomy are dissection and microscopy. Dissection involves the cutting apart of a body to reveal its gross structure. This was the first technique used to discover the structure of the body and is still the best way to truly understand the design and relationship of anatomical detail. The best drawings, photos, and virtual images can never reveal what the dissector experiences during a dissection. The advent of the microscope expanded anatomical knowledge by revealing microscopic perspectives that were not available to the unaided eye. This understanding of microscopic structure opened the door to an increased knowledge of the functional aspects of anatomy.

In this atlas we attempt to teach the elegant structure and design of the human body using the tools and methods of the anatomist — dissection and microscopy. While there are numerous excellent visual resources that depict anatomy, we believe that, with the exception of personal dissection study, excellent photographs based on excellent dissections and microscopy are the truest form of anatomical imagery. Nothing depicts the actual thing as well as the actual thing. Our goal is to create images that teach, and to use that imagery to highlight the patterns and design features of anatomy.

This atlas approaches the body from a systemic perspective; that is, it covers each body system and the organs associated with that system. Each system is highlighted in the dissection photos. However, the dissections of the systemic anatomy often reveal regional perspectives and relationships, and the structural details of regional anatomy are labeled on every image. Have fun exploring what we think might be the next best thing to dissection.

Find more information about anatomy in REAL ANATOMY

1

Design of the Book

The design features of the *Atlas of Human Anatomy* are illustrated on this page using a sample page from the book. Each page will begin with a short introduction to the featured anatomy of the page. This brief narrative will occupy this text space. Below this narrative, the majority of the page will focus on the images of anatomy and the appropriate labels for the images. The design elements used to teach and illustrate the anatomy are highlighted in the boxes below.

Featured Structure
The page heading will list the anatomical structure or feature that is the focal point of the page

Descriptive Narrative
A brief description of the structure and function of the anatomical structures on the page

Reference Image
The reference image helps to quickly identify the featured anatomy and see its relationships

Stomach

The stomach is a J-shaped organ of variable size and shape and has the greatest diameter of any part of the gut tube. It occupies the upper left quadrant of the abdominal cavity, where it is anchored to the posterior abdominal wall by a mesentery. The stomach performs several functions, the most important of which is to store ingested food until it can be emptied into the small intestine at a rate that allows for optimal digestion and absorption.

Structure List
Numbered list of all the structures visible on the anatomical images

1 Stomach	7 Pylorus	13 Surface mucous cell
2 Cardia of stomach	8 Pyloric sphincter	14 Lamina propria
3 Fundus of stomach	9 Gastric rugae	15 Mucous neck cell
4 Body of stomach	10 Greater curvature	16 Gastric glands
5 Pyloric antrum	11 Lesser curvature	17 Liver
6 Pyloric canal	12 Gastric pit	18 Gallbladder
		19 Spleen
		20 Greater omentum

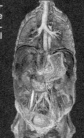

Dissection Images
Beautiful dissections illustrate the anatomy of the body system

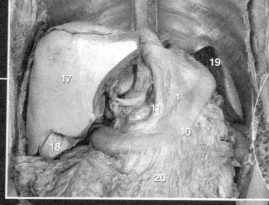

Abdominal dissection revealing stomach
Anterior view

Microscope Images
Crisp histology photomicrographs illustrate the contextual microscopic structure of the anatomy

Numbered Structures
Unobtrusive numbered structures without the clutter and distraction of leader lines

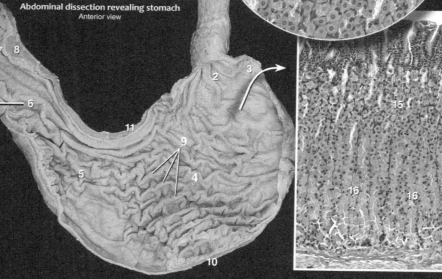

Frontal section of stomach
Anterior view

Photomicrograph of stomach mucosa with callout above
40x and 100x

Captions
Captions describe the image and the view or magnification of the anatomy or histology

2 | Histology

Histology is the study of tissues, and tissues are the building materials of the body. Like the materials we use to make the clothing we wear, tissues are the materials that form the various layers and structures of all the body's organs. For example, you might wear a light undershirt of cotton beneath a silk long-sleeved shirt and wear a wool sweater over the top of the two shirts. Each layer of clothing is made of a different material, and the material is organized into a unique structure that has its own functional qualities. The same is true of the organs of the body. Each organ consists of distinct structural layers, and each layer is a specific type of tissue. For example, the stomach has an inner lining of simple columnar epithelium that is in contact with the food we eat and secretes enzymes to help digest the food. This epithelial layer is surrounded by a vascular layer of loose connective tissue that contains the blood vessels that transport the absorbed molecules from the stomach. Smooth muscle tissue surrounds the two inner layers and helps toss and turn the food within the stomach and move it toward the small intestine. The smooth muscle tissue is covered by a slippery, thin layer of simple squamous epithelium that forms the outer surface of the stomach and allows it to move against neighboring organs while reducing the damaging friction. And just as the layers of clothing have names — undershirt, long-sleeved shirt, sweater — so also do the structural layers of an organ such as the stomach — mucosa, submucosa, muscularis, and serosa.

All the tissues of the body can be organized into four basic tissue categories — epithelial tissue, connective and supporting tissue, muscle tissue, and nervous tissue. Each tissue category has unique structural features that are shared by the tissues of that category. Epithelial tissues are surface tissues that consist of numerous cells tightly packed together. Connective and supporting tissues share the common feature of having relatively few cells that are scattered within a surrounding fibrous extracellular matrix. Muscle tissue consists of elongated cells with specialized protein arrangements that are designed to shorten. Nervous tissue cells are branching, wire-like cells with a great variety of shapes and lengths. In this chapter you will explore these four tissue categories and the specific tissue types that comprise each category. In the chapters that follow, the different tissues will be observed in the context of the organs and organ systems they form.

Find more information about histology in

REAL ANATOMY

Tissues

The facing pages show photomicrograph collages of the four principal tissue categories—epithelial tissue, connective and supporting tissue, muscle tissue, and nervous tissue. The photomicrographs illustrate the key structural features shared by the tissues in each category. Note the numerous closely packed cells of the epithelial tissues and contrast them with the scattered cells and the fibrous surrounding matrix of the connective and supporting tissues. In the muscle tissue observe the long, slender specialized cells that are designed to shorten, and in the nerve tissue the branched, wire-like cells. We will explore each of the principal tissue categories in more detail on the pages that follow.

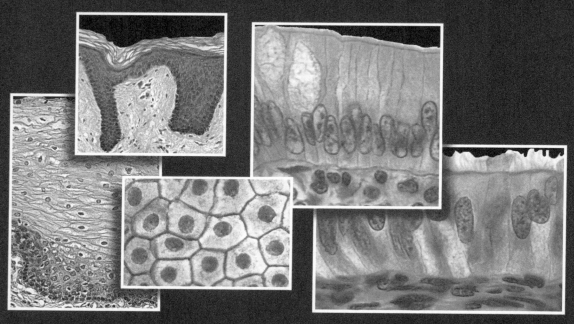

Epithelial Tissues

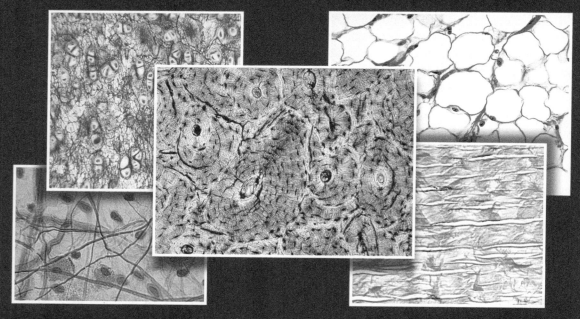

Connective and Supporting Tissues

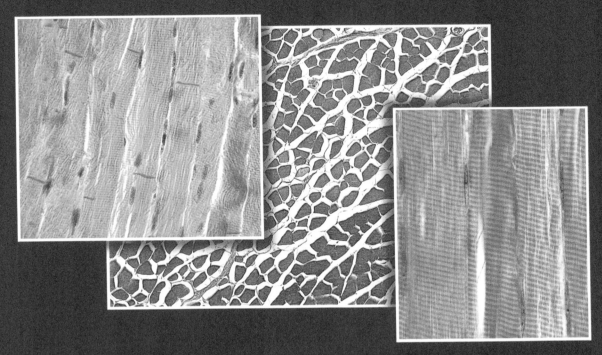

Muscle Tissues

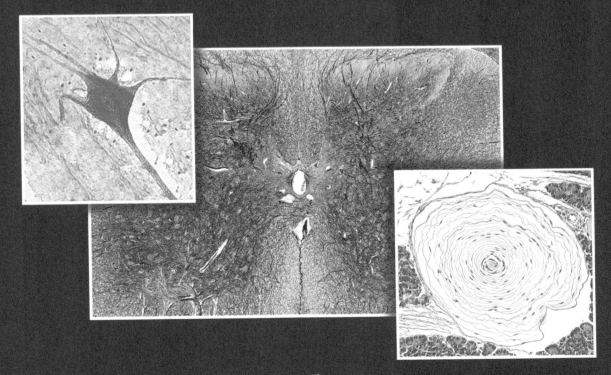

Nerve Tissues

Epithelial Tissue

Epithelial tissues are surface tissues that consist of numerous cells, with each cell forming membrane to membrane contact with its neighbors. As a general rule, descriptions of epithelial tissues are based on the shape of their cells and on the number of cell layers present. By combining the shape names — squamous (flat cells), cuboidal, and columnar — with the term simple if there is a single layer of cells or the term stratified if there is more than one layer of cells, almost all of the epithelial tissues can be described and named. The photomicrographs on this page and the facing page represent the simple (single cell layer) epithelial tissues.

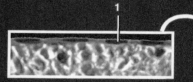

Simple squamous epithelium, mesothelium
Section of mesentery, 400x

1	Nucleus	7	Mucous in goblet cell
2	Cytoplasm	8	Microvilli
3	Cell membrane	9	Basement membrane
4	Capillary lumen	10	Blood vessel with red blood cells
5	Glandular lumen	11	Cilia
6	Connective tissue	12	Basal cell

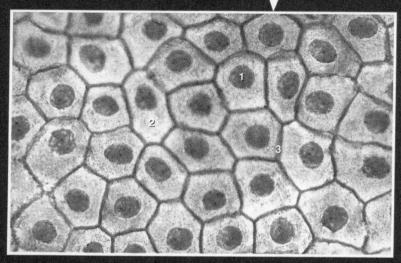

Simple squamous epithelium, mesothelium
Surface view of mesentery, 400x

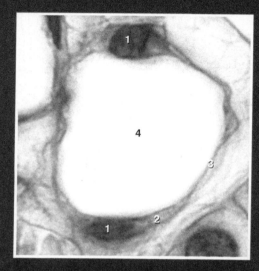

Simple squamous epithelium, endothelium
Section of capillary, 630x

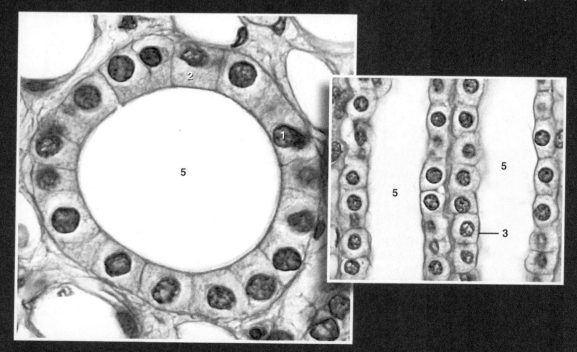

Simple cuboidal epithelium
Urinary tubes in kidney - transverse section, 630x (left); longitudinal section, 400x (right)

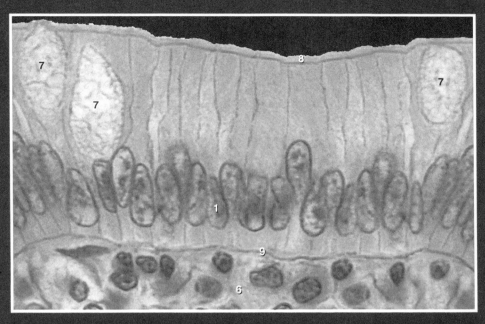

Simple columnar epithelium
Section of mucosa of small intestine, 630x

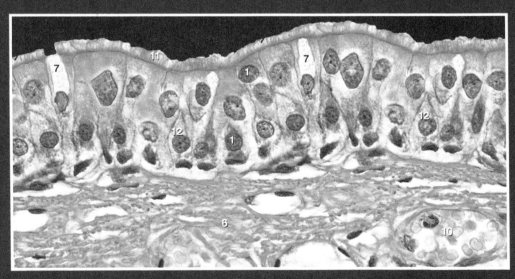

Pseudostratified columnar epithelium
Section of mucosa of larynx, 400x

Epithelial Tissue

The photomicrographs on this and the facing page illustrate the stratified (more than one layer of cells) epithelial tissues. Note that the tissues range from two layers to numerous layers and the cell shape used for the tissue name is the shape of the cells found in the surface layer.

1 Nucleus
2 Cytoplasm
3 Basal cell layer
4 Intermediate cell layer
5 Superficial cell layer
6 Stratum basale
7 Stratum spinosum
8 Stratum granulosum
9 Stratum lucidum
10 Stratum corneum
11 Connective tissue
12 Basement membrane
13 Glandular lumen

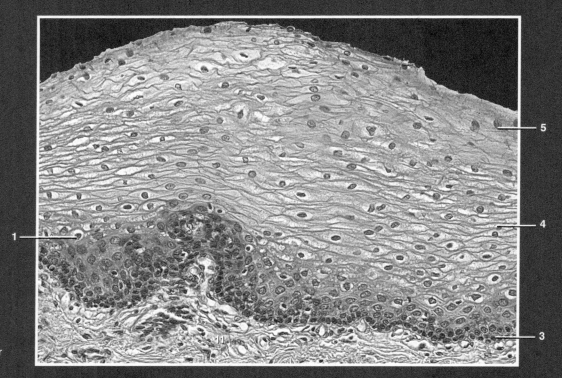

Nonkeratinized stratified squamous epithelium
Section of esophageal mucosa, 200x

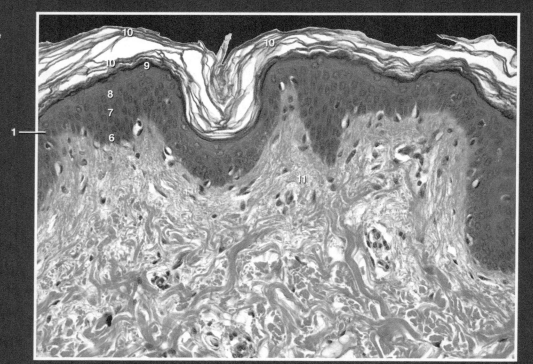

Keratinized stratified squamous epithelium
Section of skin, 200x

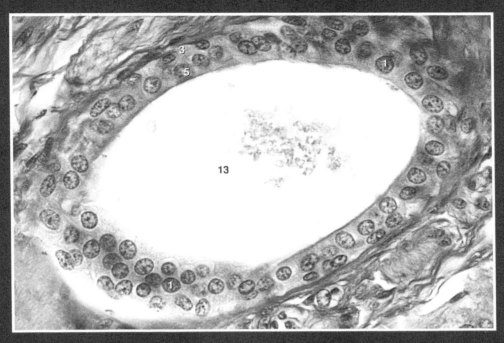

Stratified cuboidal epithelium
Section of duct of esophageal gland, 400x

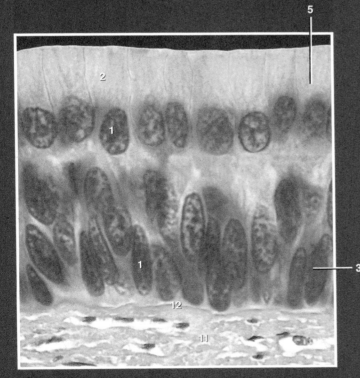

Stratified columnar epithelium
Section of pharyngeal mucosa, 400x

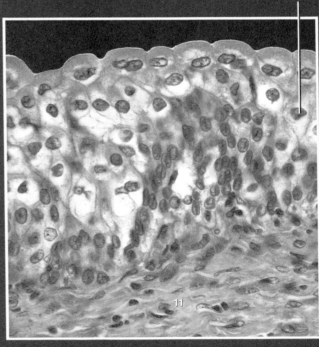

Transitional epithelium
Section of urinary bladder mucosa, 400x

Connective Tissue

Connective tissues have relatively few cells and the cells are surrounded by an extracellular matrix of fibers, which the cells secrete. The classification and names of connective tissues arise from the type and arrangement of the fibers produced by the cells and secreted into the surrounding matrix. There are three named fibers in the matrix — collagen fibers, reticular fibers (actually a thin form of collagen), and elastic fibers. The fibers are deposited in varying degrees of density and are arranged in different patterns. The tissue names are based on the different fiber types and patterns in the matrix.

1	Mast cell	7	Cytoplasm
2	Fibroblast	8	Plasma membrane
3	Collagen fiber	9	Lipid storage area
4	Elastic fiber	10	Nucleus of reticular cell
5	Reticular fiber	11	Nucleus of fibroblast
6	Nucleus of adipose cell	12	Elastic lamella

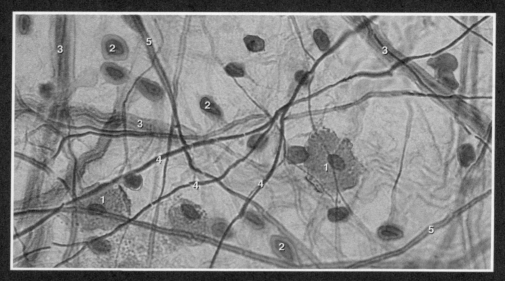

Loose (areolar) connective tissue
Section of subcutaneous layer of integument, 400x

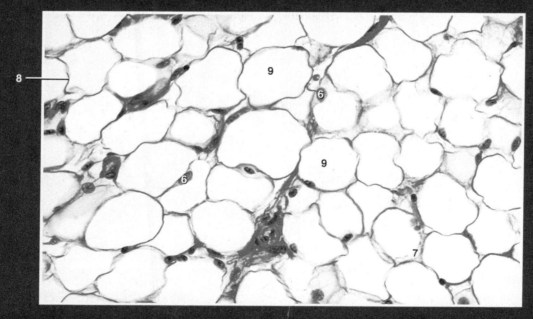

Adipose tissue
Section of epicardial fat, 200x

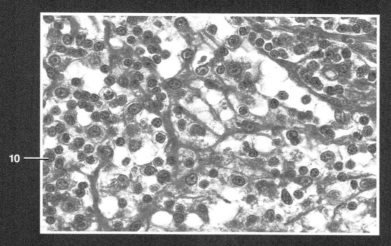

Reticular tissue
Section of lymph node, 400x

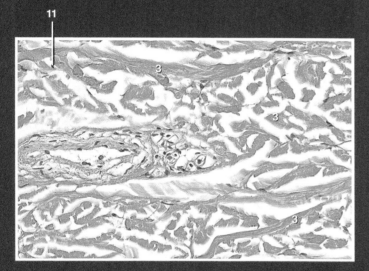

Dense irregular connective tissue
Section of dermis, 200x

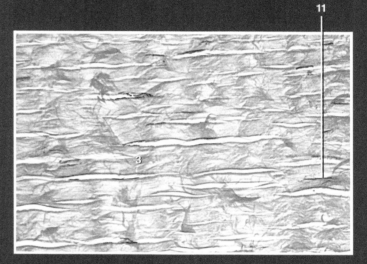

Dense regular (collagenous) connective tissue
Section of tendon, 200x

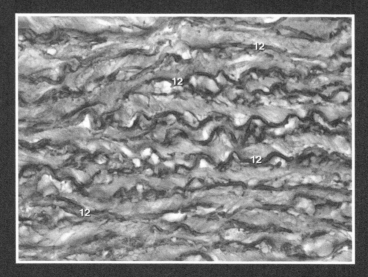

Dense regular (elastic) connective tissue
Section of tunica media of aorta, 400x

Supporting Tissue

The supporting tissue category consists of the skeletal tissues— cartilage and bone. Like the connective tissues, the supporting tissues have relatively few cells surrounded by a significant amount of extracellular matrix, which for the most part the cells produce. However, unlike the soft matrix of the connective tissues, the extracellular matrix of the supporting tissues is firm and rubber-like in cartilage and hard in bone tissue.

1 Hyaline ground substance
2 Collagen fibers in ground substance
3 Elastic fibers in ground substance
4 Chondrocyte nucleus
5 Chondrocyte in lacuna
6 Perichondrium
7 Bone trabecula
8 Osteocyte
9 Red bone marrow
10 Canaliculi
11 Lacuna
12 Lamella
13 Central canal

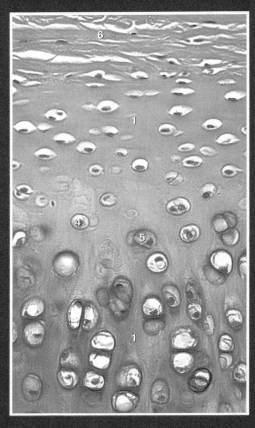

Hyaline cartilage
Section of cartilage in developing fetal bone, 200x

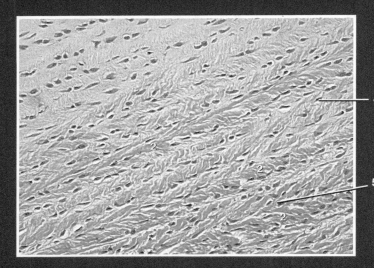

Fibrocartilage
Section of intervertebral disc, 200x

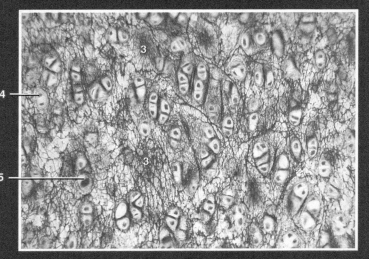

Elastic cartilage
Section of cartilage from auricle of ear, 400x

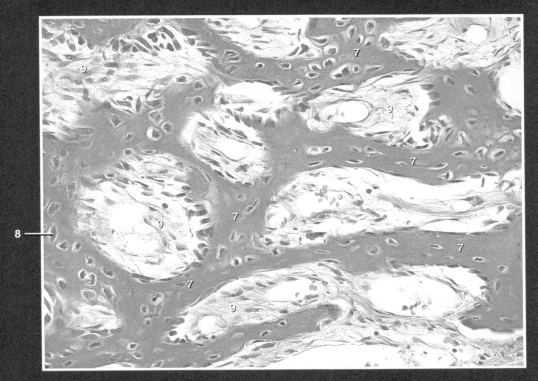

Spongy bone
Section of epiphysis of metacarpal bone, 200x

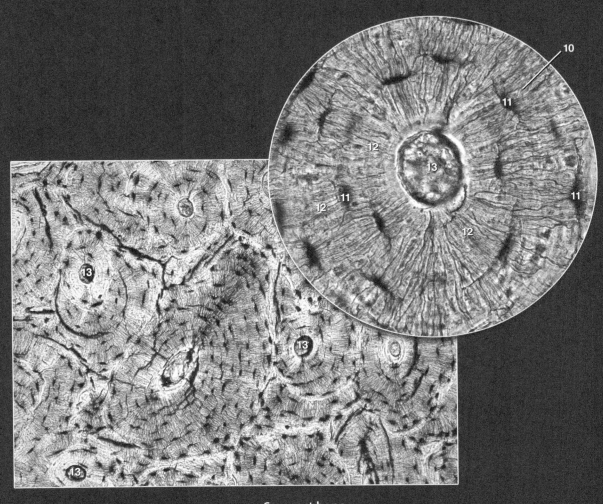

Compact bone
Section of diaphysis of fibula, 100x; callout of osteon, 400x

Hematolymphoid Complex

The tissues blood and lymph traditionally were classified as connective tissues because, like all connective tissues, the extracellular matrix is a greater percentage of the tissue then are the cells. However, the extracellular matrix of blood and lymph is a liquid matrix called plasma, rather than the soft, firm matrix of connective tissues. The most recent *Terminologia Histologica* places blood and lymph in their own subcategory called the hematolymphoid complex.

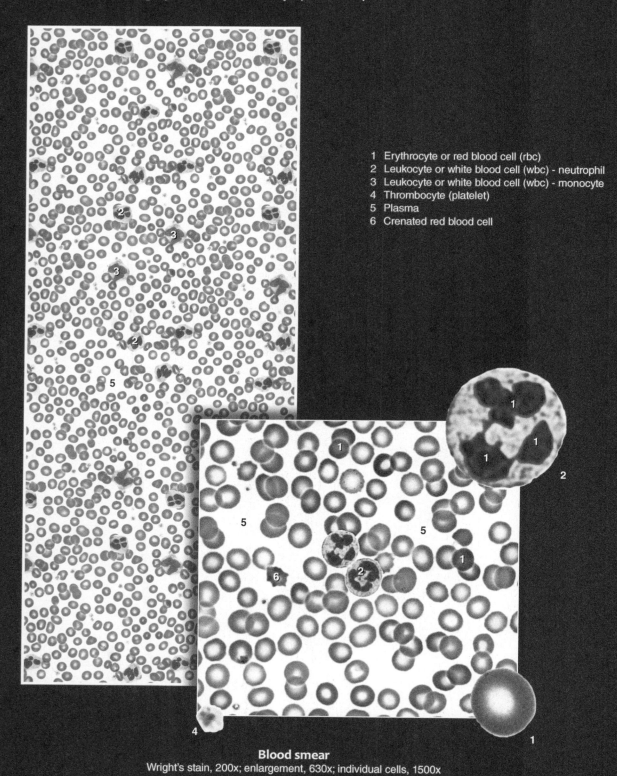

1 Erythrocyte or red blood cell (rbc)
2 Leukocyte or white blood cell (wbc) - neutrophil
3 Leukocyte or white blood cell (wbc) - monocyte
4 Thrombocyte (platelet)
5 Plasma
6 Crenated red blood cell

Blood smear
Wright's stain, 200x; enlargement, 630x; individual cells, 1500x

Muscle Tissue

Muscle cells are long, slender cells that have special arrangements of the proteins actin and myosin within the cytoplasm. The architectural design of these proteins forms the muscle cell "machinery" that allows the cell to specialize at contracting (shortening). The names of the different types of muscle tissues arise from the arrangement of the contractile proteins within their cells. In some tissues the protein arrangement gives the cell a striated, or striped, appearance (striated muscle), while in other tissues the striped appearance is not evident (non-striated or smooth muscle).

1 Nucleus
2 Sarcoplasm
3 Smooth muscle cell
4 Cardiac muscle cell
5 Skeletal muscle cell
6 Intercalated disc

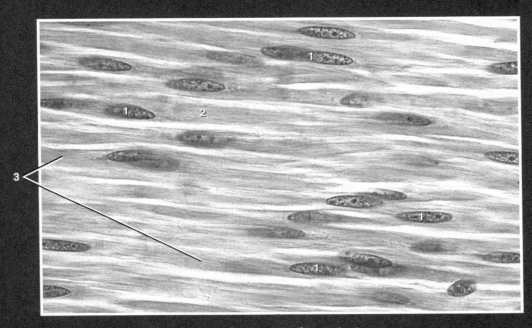

Smooth (nonstriated) muscle tissue
Longitudinal section of muscular wall of intestine, 500x

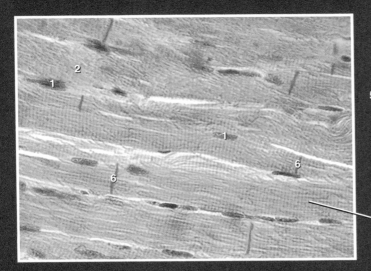

Cardiac striated muscle tissue
Section of ventricle of heart, 500x

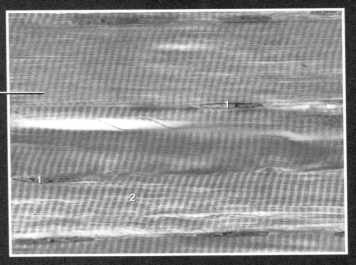

Skeletal striated muscle tissue
Section of vastus lateralis muscle, 400x

Nerve Tissue

Nervous tissue forms the complex electrical computing system of the body. The cells that characterize nervous tissue are the branched, wire-like cells called neurons. Surrounding the neurons of the nervous tissue are the smaller, more numerous glial cells that are involved in protecting, insulating, and nourishing the neurons. The neurons can be grouped together in long slender structures called nerves, or they can form the complex circuit boards we call the spinal cord and brain.

1 Nucleus of multipolar neuron
2 Cell body of multipolar neuron
3 Nucleus of glial cell
4 Axon
5 Dendrite

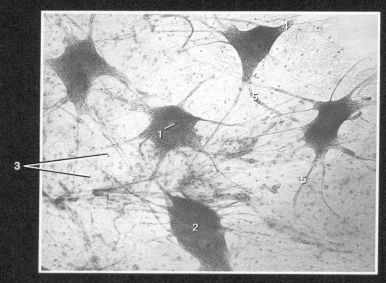

Nerve tissue
Multipolar neuron smear, 400x

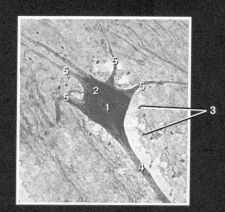

Neuron
400x

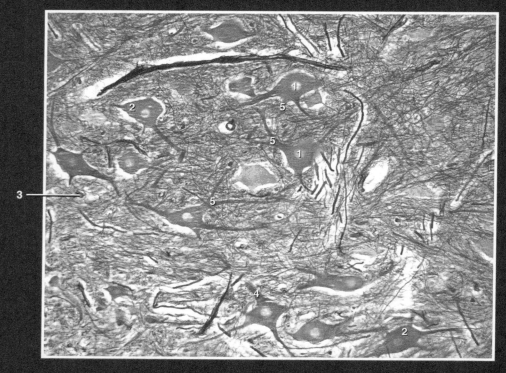

Nerve tissue
Section of ventral horn of spinal cord, 200x

3 | Integument

The integument forms the organ system that covers the body. From the Latin meaning to cover inward, the integument is an important system that performs a variety of functions that are essential to life. The outer layers of the integument called the epidermis and dermis form the skin, which is an important protective layer. The skin protects the body in a number of ways. Its tough, outer-covering of dead cells protects the more delicate deeper layers from friction and abrasion. The pigment cells in the epidermis produce melanin, a protective pigment that absorbs damaging ultra-violet radiation from the sun, to protect the rapidly dividing keratinocytes that make up the majority of the epidermal layer of the skin. The structure of the epidermal layer of the skin and its secretions also protect the body from excessive water loss or gain. The large network of blood vessels and numerous sweat glands form an evaporative cooling system that help to protect the body from overheating in warm conditions or during exercise. Additionally, the impenetrable skin and some of its special cells form a first line of defense against bacterial invasion.

These are just some of the functions of the integument. Other important functions are the following: it is a major surface for sensory perception to receive input or stimuli from the environment, it is an excretory surface to help rid the body of metabolic wastes, it plays an important role in energy storage and metabolism, it provides an important site for the production of vitamin D and various growth factors, and it plays a major role in sociosexual communication and identification. This chapter will depict the structural features of the integument that account for this wide variety of important functions.

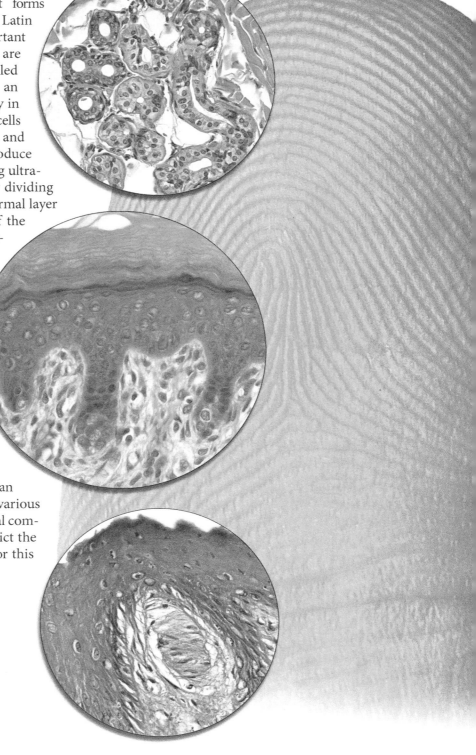

Subdivisions of the Integument

The integument consists of two major parts or layers of anatomy, the skin and the subcutaneous layer, or hypodermis. The cadaver and histology images on this and the facing page illustrate these two layers of anatomy. The skin, consisting of the superficial epidermis and the deeper dermis, structurally combines an epithelial tissue and connective tissue to form the body's covering organ. The skin is an organ that produces hairs, various glands, finger and toe nails, and accounts for the majority of the functions of the integument. The subcutaneous layer is a variable layer that can consist of fat, fibrous connective tissue, loose connective tissue, and smooth muscle.

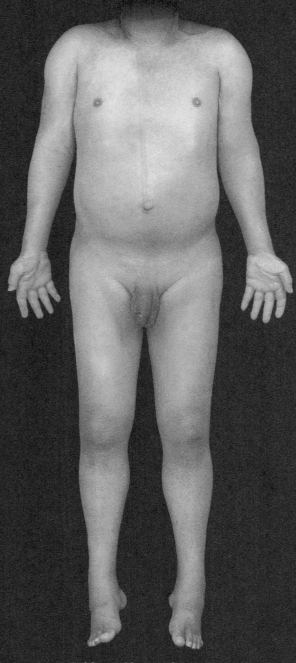

Epidermal layer of the skin
Anterior view

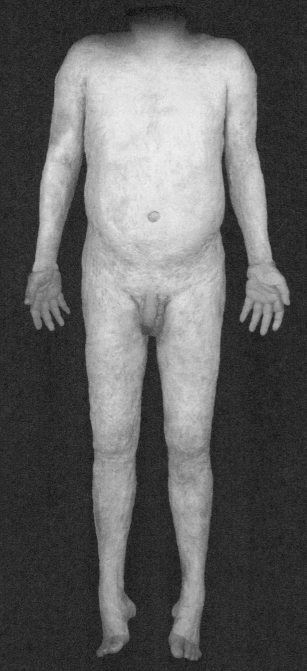

Subcutaneous layer of the integument
Anterior view

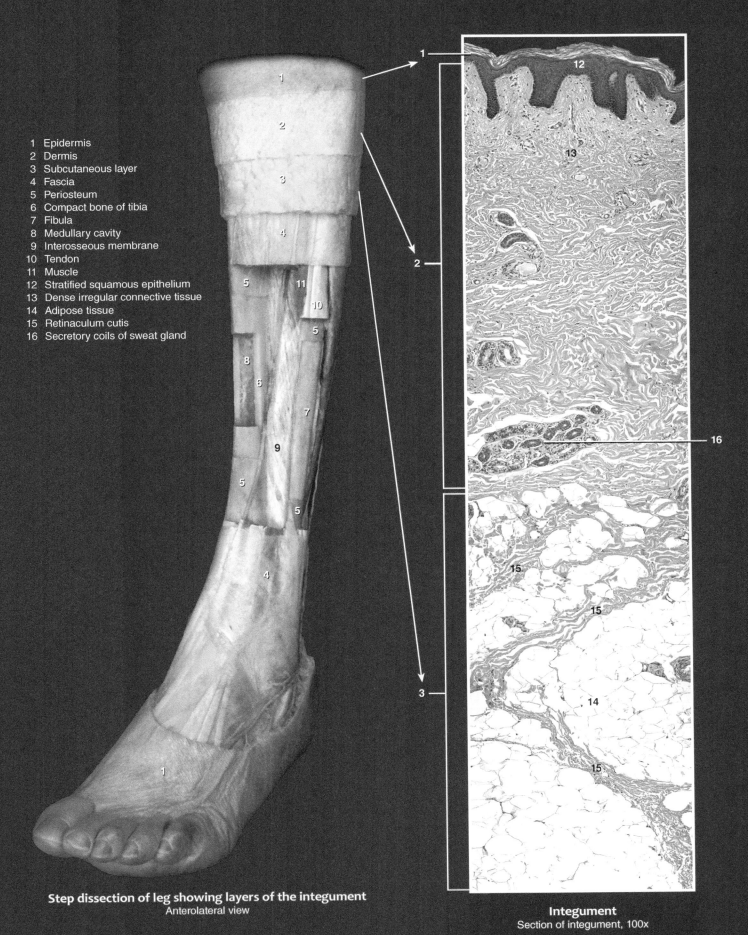

1 Epidermis
2 Dermis
3 Subcutaneous layer
4 Fascia
5 Periosteum
6 Compact bone of tibia
7 Fibula
8 Medullary cavity
9 Interosseous membrane
10 Tendon
11 Muscle
12 Stratified squamous epithelium
13 Dense irregular connective tissue
14 Adipose tissue
15 Retinaculum cutis
16 Secretory coils of sweat gland

Step dissection of leg showing layers of the integument
Anterolateral view

Integument
Section of integument, 100x

Skin - Epidermis

The stratified squamous epithelial epidermis is the superficial layer of the skin. This cellular layer and its derivatives — hairs, nails, and glands — is the most recognizable part of our anatomy. It can range in thickness from a .10 mm (0.0039 in) on the eyelids to 1.5 mm (0.059 in) on the palms and soles. Keratinocytes are the primary cells of the epidermis. They proliferate from the stratum basale and differentiate as they push toward the surface, where they eventually form dead cells filled with the protein keratin. Also present in the basal layer are melanocytes that produce the brown pigment melanin to protect the skin from the ultraviolet radiation from the sun.

1 Stratum basale
2 Stratum spinosum
3 Stratum granulosum
4 Stratum lucidum
5 Stratum corneum
6 Connective tissue of dermis

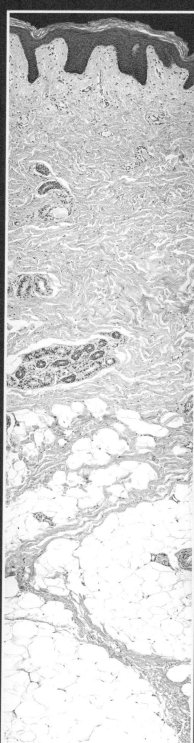

Epidermis of integument
100x

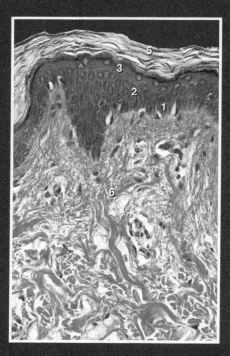

Epidermis of skin of a Caucasian
Section of thin skin, 200x

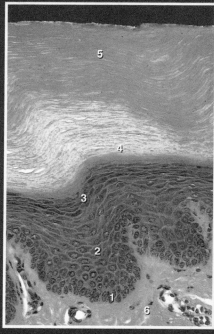

Epidermis of skin of a Caucasian
Section of thick palmar skin, 200x

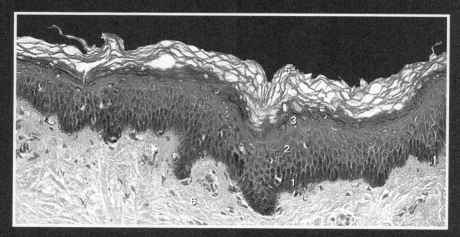

Epidermis of skin of a black
Section of thin skin, 200x

Skin - Dermis

The connective tissue dermis sits deep to the epidermis where it forms the strong binding layer of the skin. The zone of interface between the dermis and epidermis is an intricate peg and socket-like arrangement between the two layers. The dermal pegs are called dermal papillae. This arrangement has multiple functions. It assures that the two layers are strongly united, it increases the surface area to improve the blood supply to the avascular epidermis, and it increases the contact surface for sensory receptors. On the palms and soles the arrangement of the dermal papillae creates the friction ridges we call fingerprints.

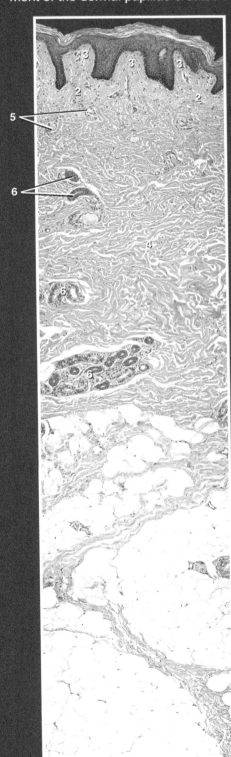

Dermis of integument
100x

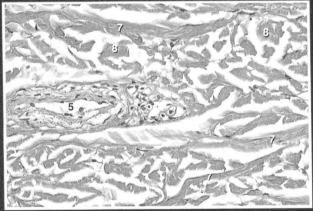

Loose connective tissue of stratum papillare
Section of dermis, 200x

Dense irregular connective tissue of stratum reticulare
Section of dermis, 200x

1 Epidermis
2 Loose connective tissue of stratum papillare
3 Dermal papilla of the stratum papillare
4 Dense connective tissue of stratum reticulare
5 Blood vessel in dermis
6 Sweat glands in dermis
7 Longitudinal collagen fibers
8 Transverse collagen fibers
9 Friction ridges formed by dermal papillae
10 Flexion crease line

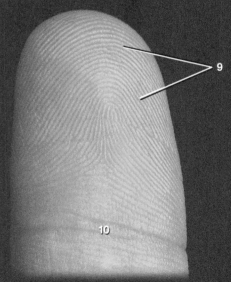

Friction ridges (fingerprints) of right index finger
Anterior view

Skin - Hairs and Nails

During embryonic and fetal development, the epithelial cells of the epidermis push down (invaginate) into the connective tissue dermis. This developmental process creates a hair follicle, a baglike extension of the epidermis that projects into the dermis and is responsible for producing the hair. The hair is a column of dead keratinocytes that arise from the basal keratinocytes at the bottom of the hair follicle. A sebaceous gland, also derived from the epidermal epithelium, empties into the hair follicle, and a small band of dermal smooth muscle, the arrector pili muscle, attaches to the base of the follicle. When the muscle shortens it produces "goose bumps" on the surface of the skin and causes the hair to "stand up." Nails also arise from invaginations that produce the shallow nail fold and root. A plate of strongly keratinized tissue emerges from the nail root to cover the dorsal ends of the fingers and toes.

1 Epidermis
2 Dermis
3 Follicle wall
4 Hair
5 Papilla
6 Root of nail
7 Nail
8 Nail bed
9 Lunula
10 Eponychium (cuticle)
11 Hyponychium
12 Eccrine sweat glands
13 Cartilage
14 Bone

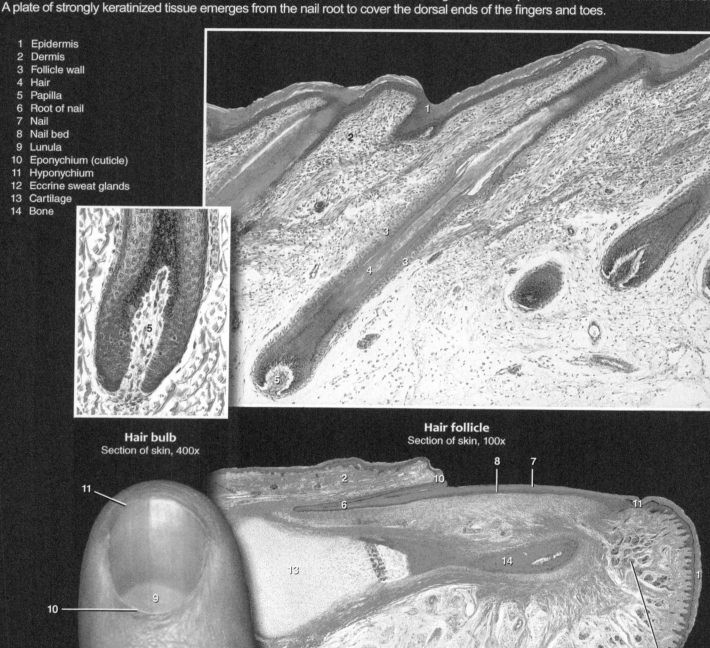

Hair bulb
Section of skin, 400x

Hair follicle
Section of skin, 100x

Fingernail of an adult
Dorsal view

Finger of a child
Longitudinal section, 50x

22

Skin - Glands

Like hairs, glands arise as invaginations of the epidermis into the dermis during embryonic and fetal life. The three prominent glands of the skin are the sebaceous gland, the eccrine sweat gland, and the apocrine sweat gland. The sebaceous and apocrine sweat glands typically empty into a hair follicle, whereas the eccrine sweat gland empties onto the surface of the epidermis.

1 Sebaceous secretory cells
2 Eccrine secretory cell
3 Eccrine duct cell
4 Apocrine secretory cell
5 Hair
6 Hair follicle
7 Arrector pili muscle

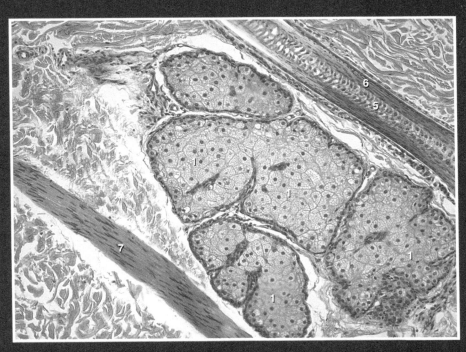

Sebaceous gland
Section of dermis, 200x

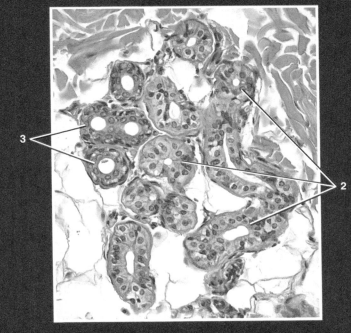

Eccrine sweat gland
Section of dermis, 200x

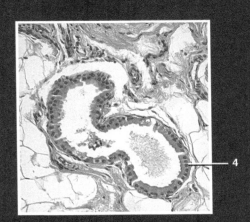

Apocrine sweat gland
Section of thin skin, 200x

Subcutaneous Layer

The subcutaneous layer, also called the hypo-dermis, is a layer of variable thickness that ranges from a thin layer of loose connective tissue to a thick fibroadipose layer. This layer is a prominent location of fat storage in the body. In addition, it functions as an insulative layer and is the site of distribution of the main venous drainage channels of the integument and the cutaneous nerves that supply the skin.

1 Epidermis of skin
2 Subcutaneous layer
3 Fascia
4 Superficial veins
5 Cutaneous nerve
6 Tendon
7 Muscle
8 Retinaculum cutis
9 Adipose cell membrane
10 Nucleus of adipose cell
11 Fat storage vacuole of adipose cell
12 Blood vessel

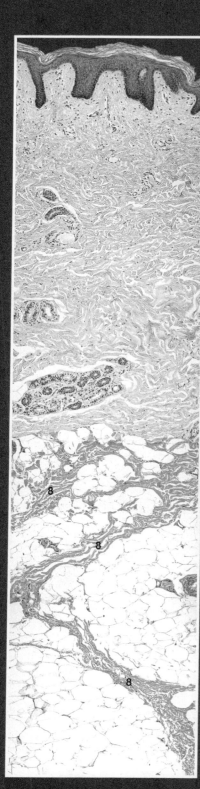

Subcutaneous layer of integument
100x

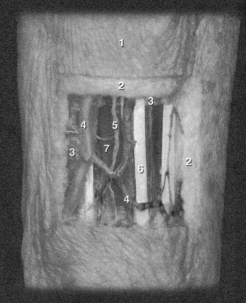

Superficial veins and cutaneous nerves in the subcutaneous layer
Step dissection of antebracial integument, anterior view

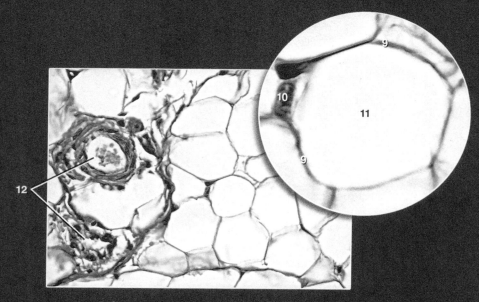

Subcutaneous adipose tissue (left), adipose cell (callout)
Section of subcutaneous layer, 200x and 640x

4 | Skeletal System

The skeletal system forms the internal framework for the soft tissues of the body. This is not a static framework, but a highly dynamic internal scaffolding. It is dynamic in many ways. On one hand, because of its jointed design, it shows extreme flexibility of movement when acted upon by muscles. At another extreme, the cells of skeletal tissue are constantly monitoring and changing the micro-structure of this amazing tissue called bone, providing it with maximal strength, toughness, and resilience. In addition to its dynamic role of support, it also serves a protective role for many organs of the body. This dynamic framework also exhibits a tremendous capacity for growth and repair. It is a storehouse of calcium ions, ions that play a significant role in many of the body's functions.

The skeleton consists of 206 separate bones, ignoring various sesamoid bones and the fact that some bones represent the fusion of multiple bones. These bones range in size from the small ear ossicles measuring a few millimeters in length to the large femur measuring up to fifty centimeters. The skeleton is divisible into two portions, the axial skeleton and the appendicular skeleton. The axial skeleton includes the cranium, vertebral column, ribs, and sternum. The appendicular skeleton consists of the bones of the limbs and their girdles. The individual bones of the skeleton come in a variety of shapes. Some are long and tubular, while others have the spread-winged appearance of a butterfly. Bones can be grouped into four shape categories. Although not that meaningful, the four categories descriptively group the bones. The four shape categories are: long bones, short bones, flat bones, and irregular bones. Long bones are unique in having a diaphysis or shaft with a medullary cavity. The other bone types lack this hollow tubular region. The short, flat, and irregular bones are similar in having outer plates of compact bone surrounding internal centers of spongy bone. In general, long bones and short bones are found in the appendicular skeleton, while flat bones and irregular bones occur in the axial skeleton. In the right hands, the skeleton can be a library of information. Its markings, foramina, landmarks, and canals each tell a story about the soft tissues of the body. A strong foundation of skeletal anatomy is an important starting point in understanding anatomy.

This chapter covers bone tissue and the general structure of bones and the skeleton. In the two chapters that follow you will explore the two subdivisions of the skeleton — the axial skeleton and the appendicular skeleton.

Find more information about the skeletal system in

REAL ANATOMY

Bone Tissue

The tissue bone has two general forms — trabecular or spongy bone and compact bone. Trabecular bone is an internal bone that always resides deep to the more dense compact bone. Like its name implies, trabecular bone has many small beams of bone tissue connected together in complex array around obvious spaces in the tissue. To the unaided eye this gives the bone a spongy appearance. Bone marrow fills the spaces in the trabecular bone. The second type of bone tissue, compact bone, is very dense and solid looking to the unaided eye. Compact bone forms the outer surface of all bones and can range in thickness from paper thin to many centimeters thick. Microscopic analysis of this dense bone reveals that it has many microscopic spaces containing cells and blood vessels in circular arrangements called osteons.

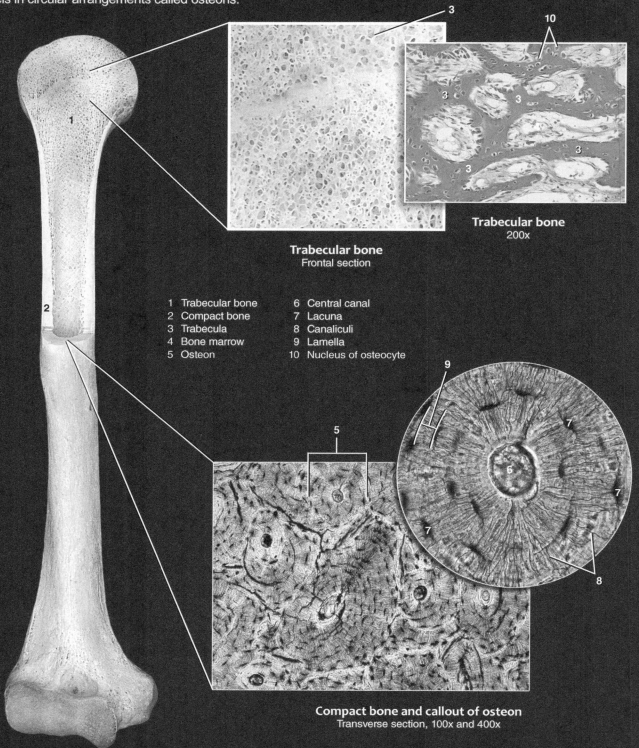

Trabecular bone
Frontal section

Trabecular bone
200x

1 Trabecular bone
2 Compact bone
3 Trabecula
4 Bone marrow
5 Osteon

6 Central canal
7 Lacuna
8 Canaliculi
9 Lamella
10 Nucleus of osteocyte

Compact bone and callout of osteon
Transverse section, 100x and 400x

Sectioned humerus
Anterior view, proximal half frontal section

Cartilage Growth Plate

Bone tissue forms during development by either replacing cartilage tissue precursors (endochondral ossification) or by developing within mesenchymal connective tissue (intramembranous ossification). In endochondral ossification cartilaginous growth plates remain between developing bone centers to allow a bone to increase in length and size. During an individual's young life, the growth plates are evident on a radiograph and are a clear indication that the individual is still growing.

1 Radial diaphysis
2 Radial epiphysis
3 Ulnar diaphysis
4 Ulnar epiphysis
5 Growth plate
6 Carpal bones
7 Metacarpal bones
8 Developing diaphysial bone
9 Zone of calcified cartilage
10 Zone of hypertrophied cartilage
11 Zone of proliferating cartilage
12 Zone of resting cartilage

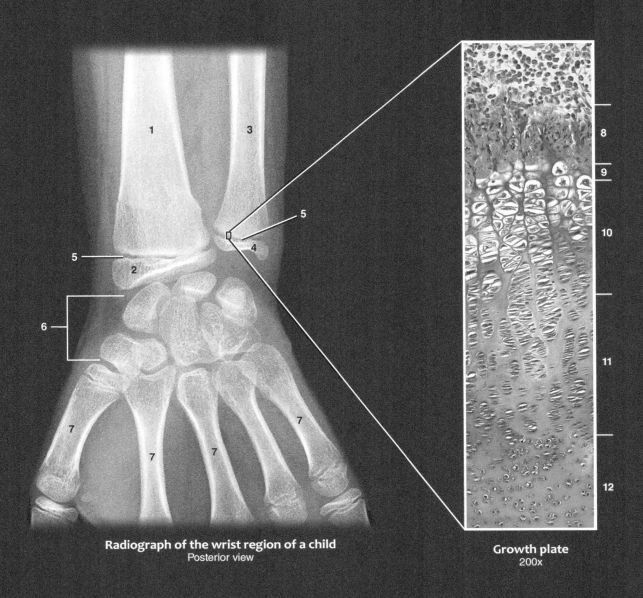

Radiograph of the wrist region of a child
Posterior view

Growth plate
200x

Bone Types

The bones of the skeleton come in a variety of sizes and shapes. The form of each bone emerges from its position and functional role in the skeletal system. In an effort to classify the different bones of the body anatomists define four general categories of bones based on their size and shape. Long bones, as their name suggests, are longer in one dimension than any other dimension. The long bones range in size from the short phalanges of the digits to the long proximal humerus and femur of the limb skeletons. Conversely, short bones are small, block-like bones. Like the long bones, short bones occur in the limb skeletons where they form the bones of the wrist and ankle. Flat bones are plate-like bones and are common in the cranium. The final category, irregular bones, is a mixed group of bones that have a variety of shapes and locations within the skeleton.

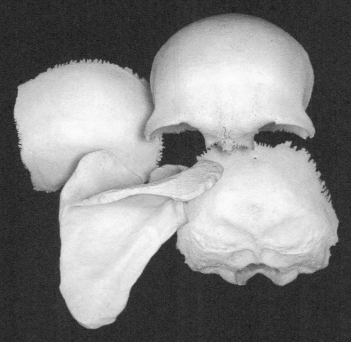

Flat bones

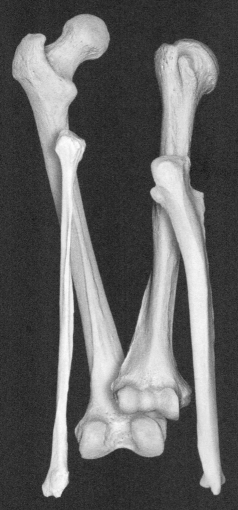

Long bones

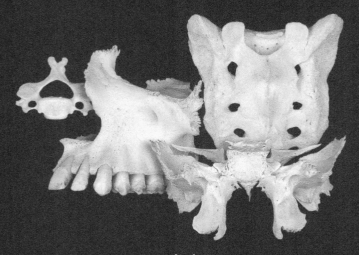

Irregular bones

Short bones

Anatomy of a Bone

All bones share basic features in common. Compact bone tissue forms all the visible outer surface of the bone and can vary from a paper-thin covering to a thick wall of bone. Trabecular bone tissue occupies the core of the bone beneath the compact bone. Areas of compact bone covered by articular cartilage form smooth subchondral compact bone surfaces. These subchondral bone surfaces mark the joint surfaces of bones. The photos below illustrate the basic parts and features of a long bone.

1 Epiphyses
2 Diaphysis
3 Metaphysis
4 Compact bone
5 Subchondral bone
6 Trabecular bone
7 Medullary cavity
8 Epiphysial line

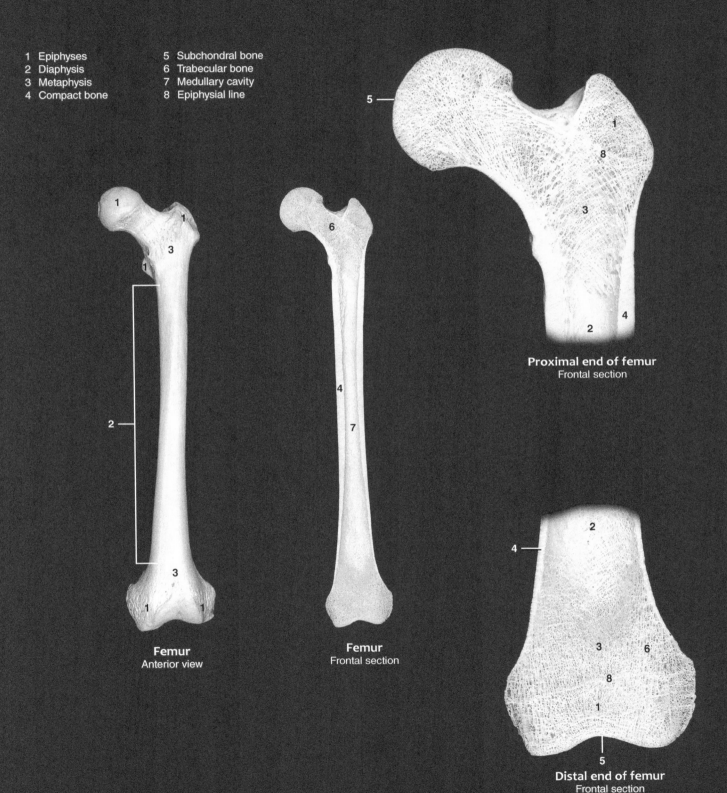

Proximal end of femur
Frontal section

Femur
Anterior view

Femur
Frontal section

Distal end of femur
Frontal section

Skeleton

The first appearance of the skeletal elements arises during the second month of embryonic life when connective tissue and cartilage precursors to the bones arise. Slowly through fetal life, childhood, puberty, and the teenage years the bones mature into their adult forms. This developmental process combines more than 500 bone-forming centers into the final 206 bones of the skeleton. This page, the facing page, and the page that follows depict changes in the skeleton from a newborn to an adult.

1 Cranial bones
2 Vertebral column
3 Ribs
4 Clavicle
5 Scapula
6 Humerus
7 Ulna
8 Radius
9 Carpals
10 Metacarpals
11 Phalanges
12 Os coxae
13 Femur
14 Patella
15 Tibia
16 Fibula
17 Tarsals
18 Metatarsals

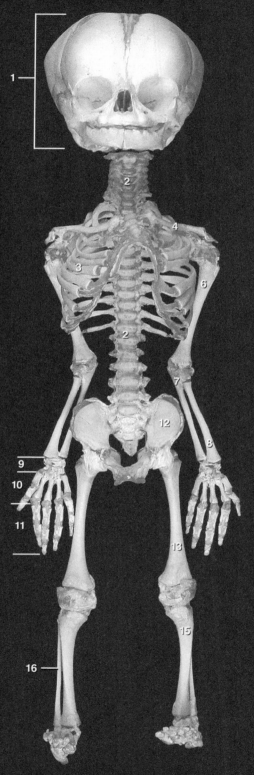

Newborn skeleton
Anterior view

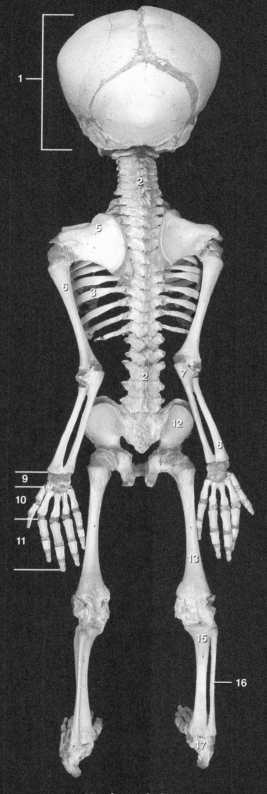

Newborn skeleton
Posterior view

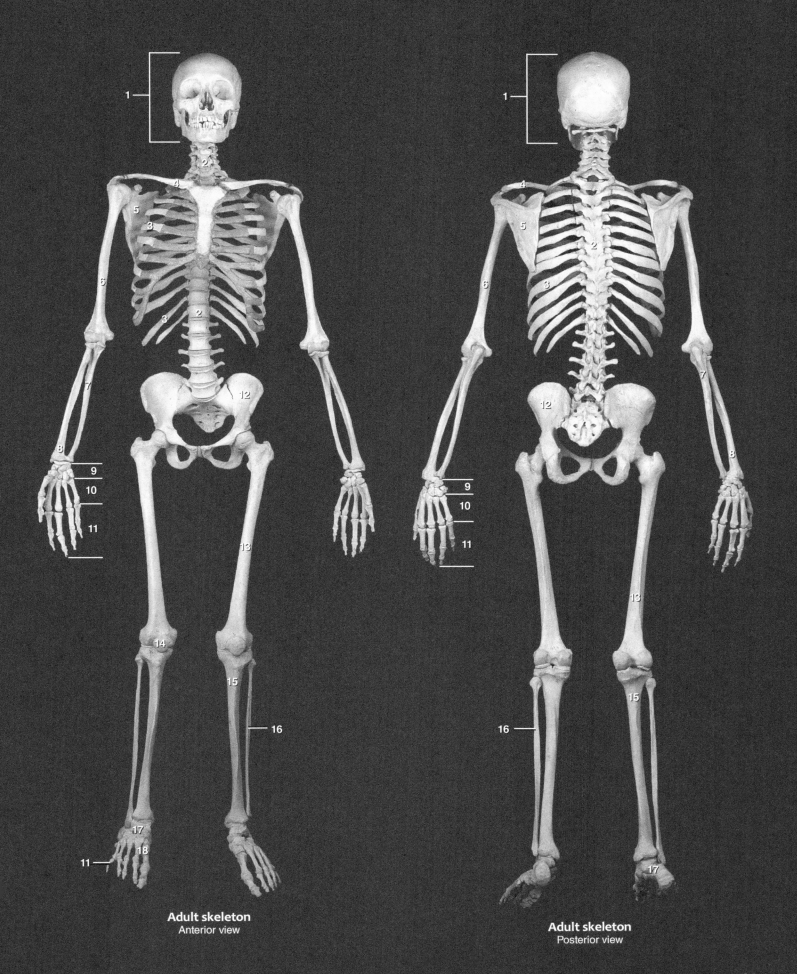

Adult skeleton
Anterior view

Adult skeleton
Posterior view

1	Cranial bones	7	Ulna	13	Femur
2	Vertebral column	8	Radius	14	Patella
3	Ribs	9	Carpals	15	Tibia
4	Clavicle	10	Metacarpals	16	Fibula
5	Scapula	11	Phalanges	17	Tarsals
6	Humerus	12	Os coxae	18	Metatarsals

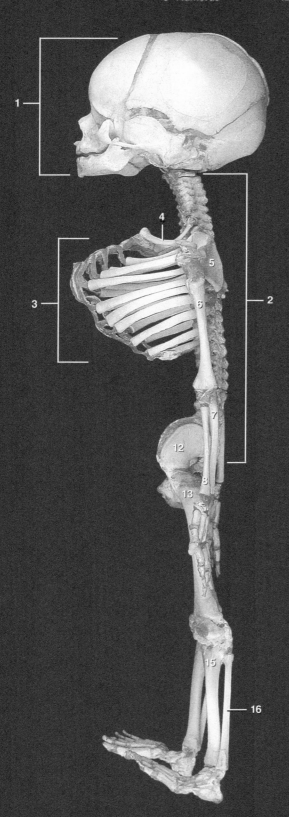

Newborn skeleton
Lateral view

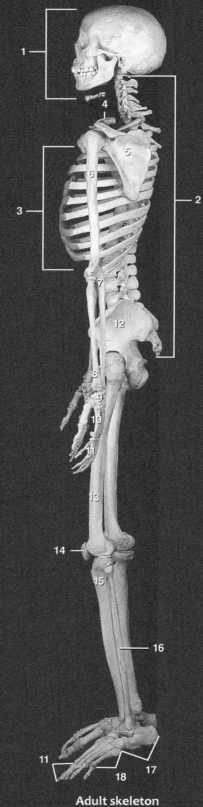

Adult skeleton
Lateral view

5 | Axial Skeleton

The axial skeleton, comprised of the skull, vertebral column, ribs, and sternum, forms the central axis of the body. This sturdy central core is the most primitive portion of the vertebrate skeletal system. It evolved as the initial skeleton of the first vertebrate animals, to which the limb bones (the subject of the next chapter) were much later additions. The majority of the axial skeleton's bony elements, from the bones at the base of the skull through all the vertebrae and ribs, form as serial homologues from the segmental embryonic somites. Because of this shared developmental similarity each body segment, from the base of the skull to the end of the coccyx, has the same basic skeletal design. This is clearly evident in the structure of the vertebrae and ribs. As you study these skeletal elements in the photos of this chapter, notice their similarities.

The elements of the axial skeleton have many functional roles in the body. Both the cranial skeleton and the vertebral column form a strong protective case around the delicate tissues of the central nervous system. Additionally, the cranium fixes in space important nervous structures, such as the internal ear and eye, both of which would not function properly in an unstable environment. The cranium also plays an important role in the acquisition and processing of food, respiratory gases, and sensory input such as sound. In addition to protecting the spinal cord, the vertebrae form a strong, flexible rod. This strong, flexible column not only forms the central support axis of the body from which the limbs are suspended, but is also capable of a varied range of joint movements that are essential to our daily functions.

*Find more information
about the axial skeleton in*

REAL ANATOMY

33

Axial Skeleton

The axial skeleton is clearly depicted in the photos below. Note that this portion of the skeleton consists of three principal skeletal regions — the cranium, the vertebral column, and the rib cage. There are 29 cranial bones, 26 vertebral bones, and 25 bones in the rib cage. On the pages that follow, each of the axial skeletal regions and the respective bones will be explored in greater detail.

1 Cranium
2 Hyoid bone
3 Cervical vertebral column
4 Cervical vertebra 1 - Atlas
5 Cervical vertebra 2 - Axis
6 Cervical vertebra 7
7 Thoracic vertebral column
8 Thoracic vertebra 1
9 Thoracic vertebra 12
10 Lumbar vertebral column
11 Lumbar vertebra 1
12 Lumbar vertebra 5
13 Sacrum
14 Sternum
15 Ribs

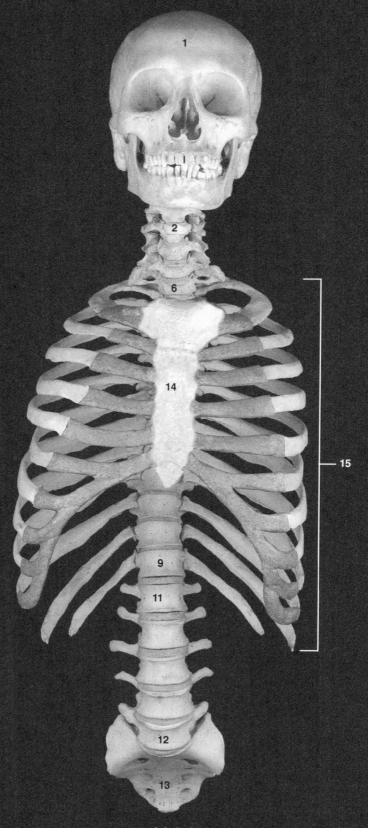

Axial skeleton
Anterior view

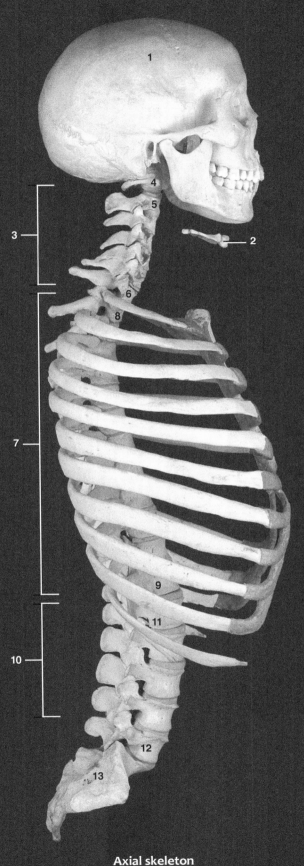

Axial skeleton
Lateral view

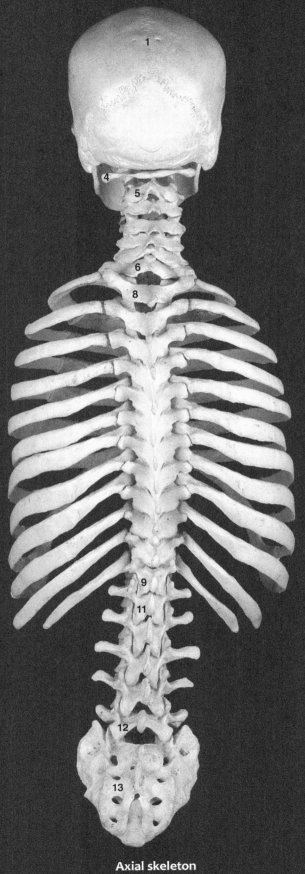

Axial skeleton
Posterior view

Cranium

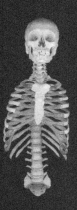

The cranium is the composite skeleton of the head and is composed of 29 bones. The bones of the cranium range from simple, non-descript plates of bone to the most intricate bones of the skeleton. The cranial bones have a range of important functions, that include protecting the delicate brain tissue, fixing the vestibular apparatus of the inner ear in three-dimensional space, maintaining open air passageways for respiration, and acquiring and processing food, to name a few. There are two main subdivisions of the cranium — the neurocranium or brain box is the region that surrounds and encases the brain, and the viscerocranium or facial skeleton is the area contributing to the orbits, nasal cavity, and oral cavity. This page and the facing page, and the four page spreads that follow, depict the five normas, or views, of the cranium in both articulated and disarticulated cranial images. The bones of the skull are labeled on these views, along with key landmarks that can only be labeled on the articulated cranium. Individual landmarks of the bones are labeled on the individual pictures of the cranial bones on the pages that follow. This spread is of the norma facialis or facial aspect of the cranium.

1 Frontal bone
2 Parietal bone
3 Occipital bone
4 Sphenoid bone
5 Temporal bone
6 Ethmoid bone
7 Inferior nasal concha
8 Lacrimal bone
9 Nasal bone
10 Vomer
11 Maxilla
12 Palatine bone
13 Zygomatic bone
14 Mandible
15 Bony nasal cavity
16 Piriform aperture
17 Inferior nasal meatus
18 Middle nasal meatu
19 Orbit

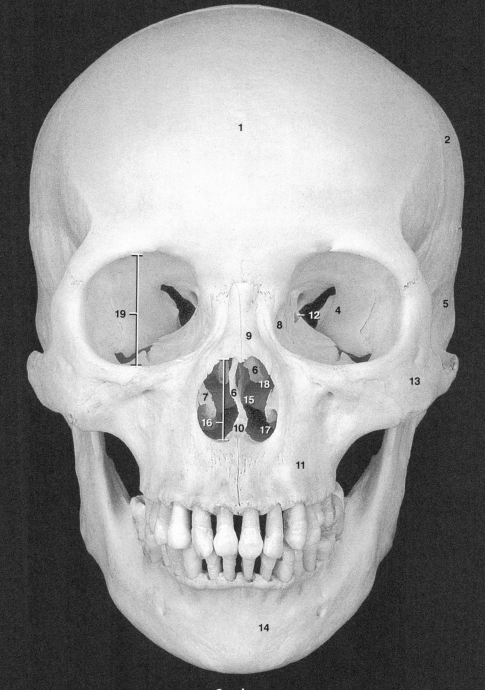

Cranium
Anterior view

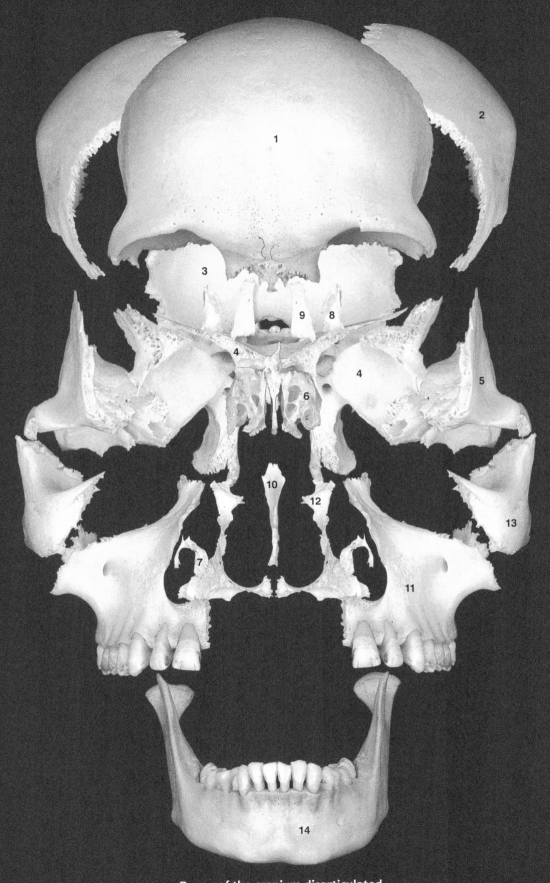

Bones of the cranium disarticulated
Anterior view

Cranium

This page spread depicts the norma lateralis, or lateral aspect of the cranium. In this view both the brain box and facial skeleton are clearly visible and the relative proportions of the two cranial regions are evident. In the disarticulated view, only those bones that are visible in the lateral aspect are shown.

1 Frontal bone
2 Parietal bone
3 Occipital bone
4 Sphenoid bone
5 Temporal bone
6 Ethmoid bone
7 Lacrimal bone
8 Nasal bone
9 Maxilla
10 Zygomatic bone
11 Mandible
12 Zygomatic arch
13 Pterygopalatine fossa

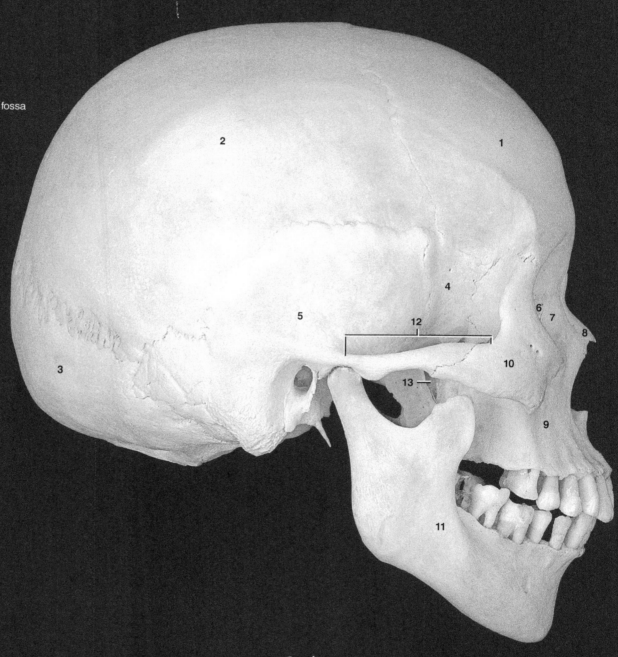

Cranium
Lateral view

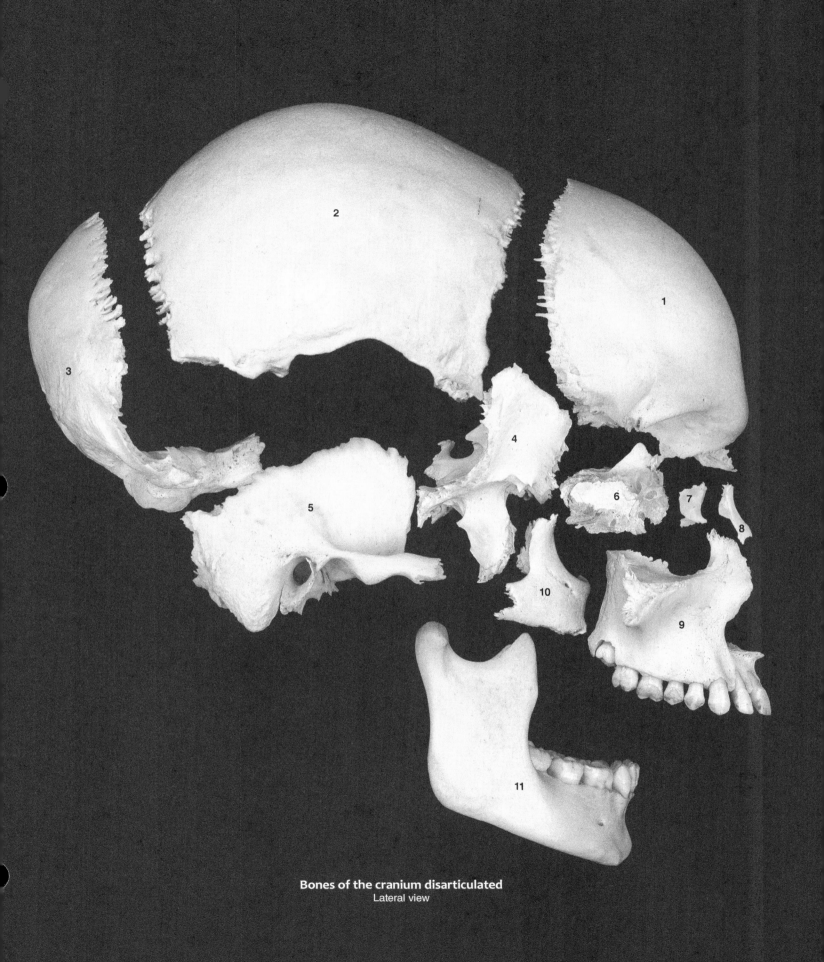

Bones of the cranium disarticulated
Lateral view

Cranium

This page spread depicts the norma occipitalis, or occipital aspect of the cranium. From this posterior view the internal aspects of the bones of the oral and nasal cavities are clearly visible. In the disarticulated view only those bones that are visible in the occipital aspect of the cranium are depicted.

1 Parietal bone
2 Occipital bone
3 Sphenoid bone
4 Temporal bone
5 Ethmoid bone
6 Inferior nasal concha
7 Vomer
8 Maxilla
9 Palatine bone
10 Zygomatic bone
11 Mandible
12 Choana or posterior nasal aperture
13 Inferior orbital fissure
14 Bony nasal cavity
15 Middle nasal meatus
16 Inferior nasal meatus
17 Bony palate
18 Sutural bone

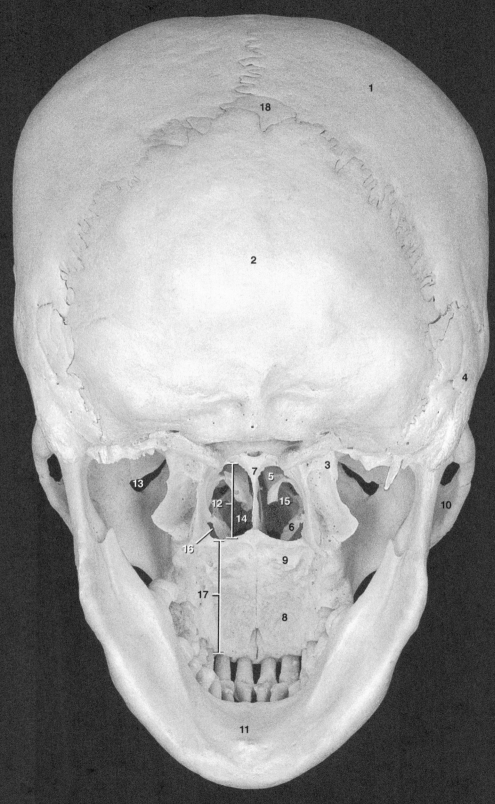

Cranium
Posterior view

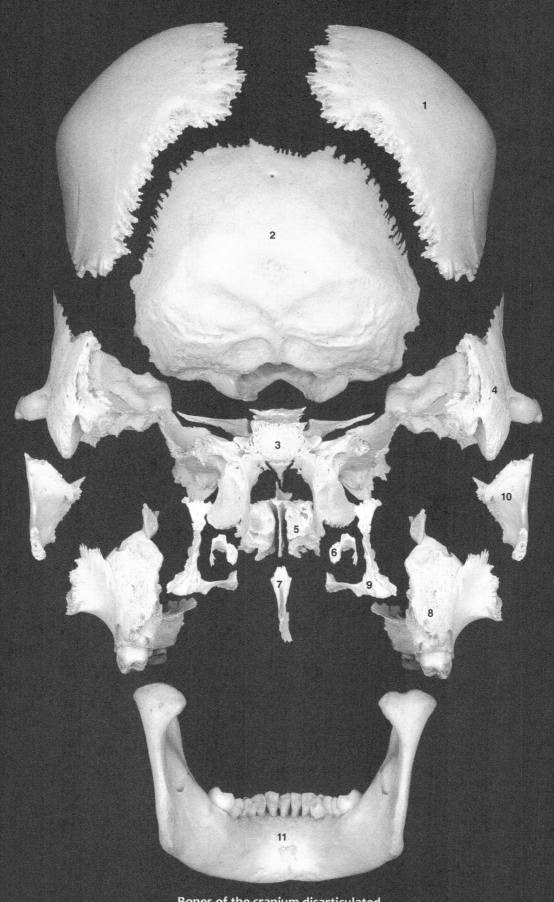

Bones of the cranium disarticulated
Posterior view

Cranium

This page spread depicts the norma superior, or superior aspect of the cranium. This view clearly depicts the neurocranium or brain box, while the facial skeleton is almost completely hidden from view. In the disarticulated view only those bones that are visible in the superior aspect of the cranium are depicted.

1 Frontal bone
2 Parietal bone
3 Occipital bone
4 Temporal bone
5 Nasal bone
6 Maxilla
7 Zygomatic bone

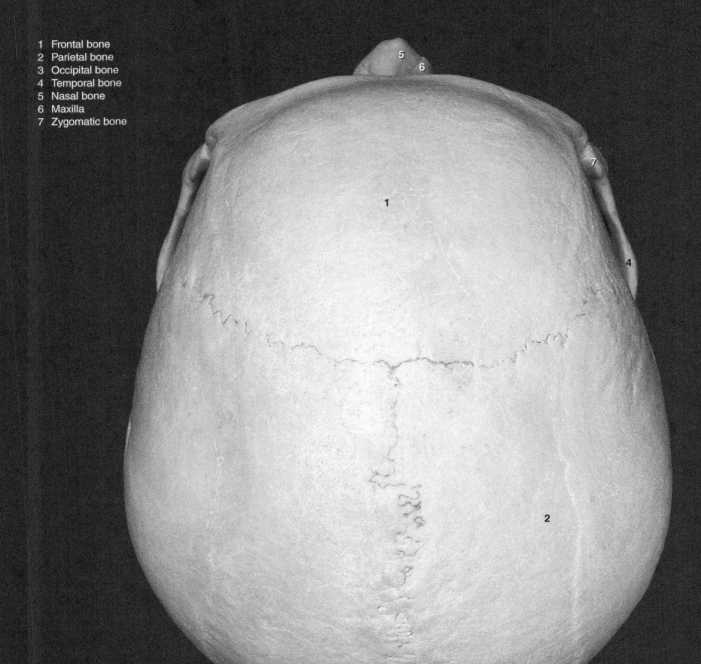

Cranium
Superior view

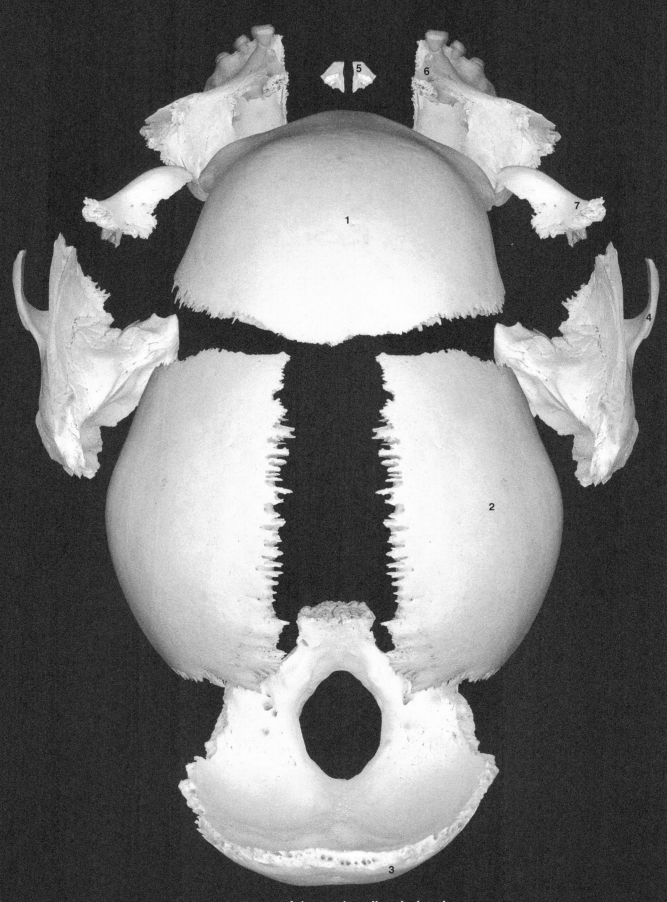

Bones of the cranium disarticulated
Superior view

Cranium

This page spread depicts the norma inferior (basalis), or inferior aspect of the cranium. The mandible has been removed to more clearly reveal the basicranium. This view clearly depicts the floor of the brain box, the bony palate forming the roof of the oral cavity, and mandibular tooth row. In the disarticulated view only those bones that are visible in the inferior aspect of the cranium are depicted.

1 Occipital bone
2 Sphenoid bone
3 Temporal bone
4 Vomer
5 Maxilla
6 Palatine bone
7 Zygomatic bone
8 Bony palate
9 Choana or posterior nasal aperture
10 Zygomatic arch
11 Jugular foramen
12 Foramen lacerum
13 Greater palatine foramen
14 Incisive fossa

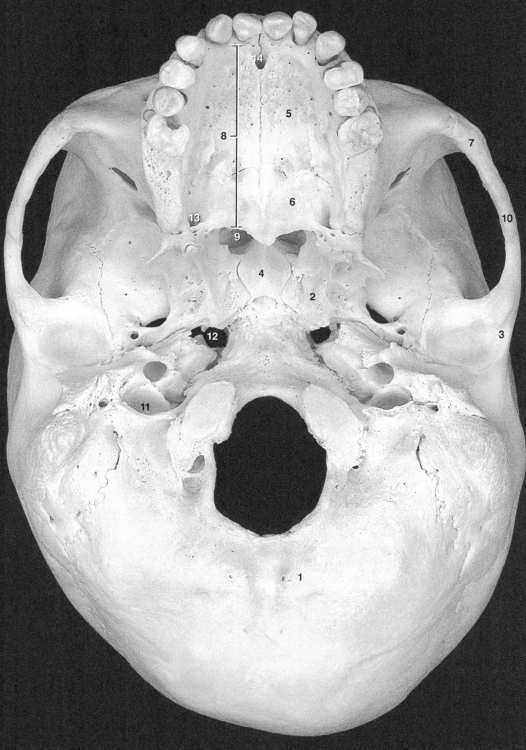

Cranium
Inferior view

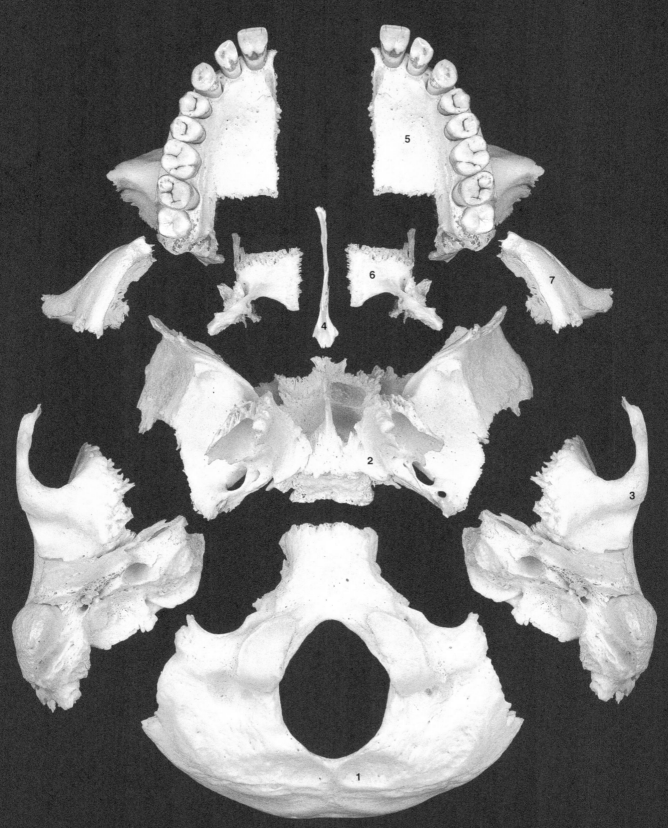

Bones of the cranium disarticulated
Inferior view

This page spread depicts the cranium sectioned in a parasagittal plane through the right side of the nasal cavity just lateral to the bony nasal septum. The section below depicts the lateral wall of the right nasal cavity, and the section on the opposite page depicts the medial (septal) wall of the right nasal cavity. The osseous sinuses that communicate with the nasal cavity are all visible in these sections.

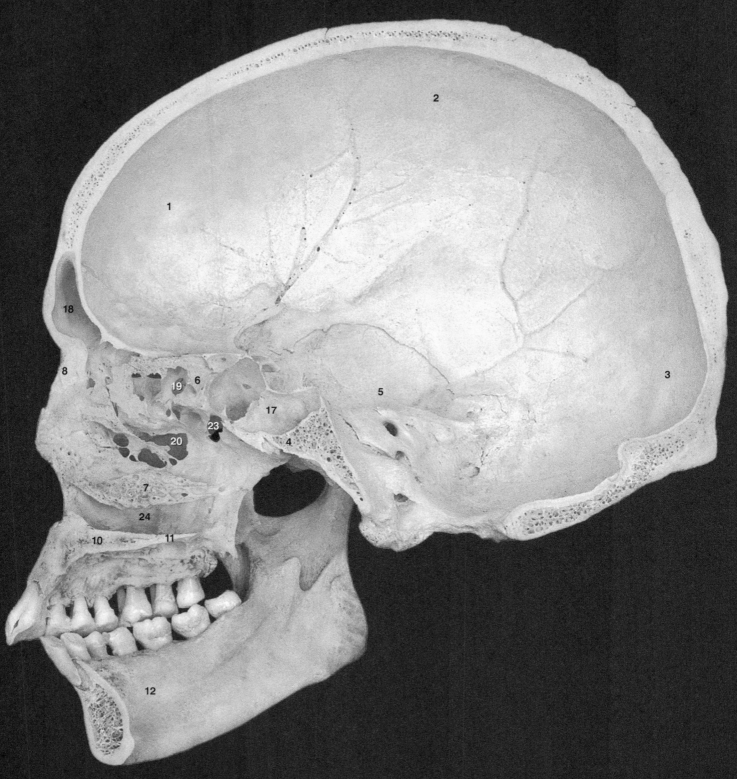

Parasagittal section of the cranium
Medial view of the right side

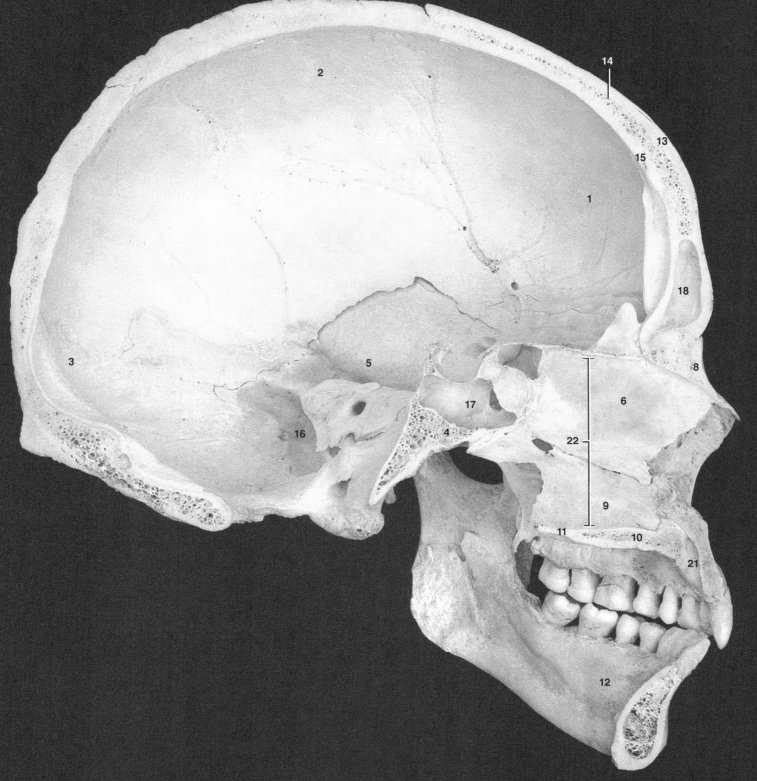

1	Frontal bone	7	Inferior nasal concha	13	External table of calvaria	19	Ethmoidal air cells (sinuses)
2	Parietal bone	8	Nasal bone	14	Diploë	20	Maxillary sinus
3	Occipital bone	9	Vomer	15	Internal table of calvaria	21	Incisive canal
4	Sphenoid bone	10	Maxilla	16	Groove for sigmoid sinus	22	Bony nasal septum
5	Temporal bone	11	Palatine bone	17	Sphenoidal sinus	23	Sphenopalatine foramen
6	Ethmoid bone	12	Mandible	18	Frontal sinus	24	Inferior nasal meatus

Parasagittal section of the cranium
Medial view of the left side

Cranium

This page spread depicts the cranium sectioned in a horizontal plane through the neurocranium, or brain box, revealing the internal aspects of the floor and roof of the sectioned cranial cavity. On this page the floor of the neurocranium is visible, while on the opposing page the roof of the neurocranium is visible. The superior portion of the cranium, depicted on the opposite page, is called the calvaria.

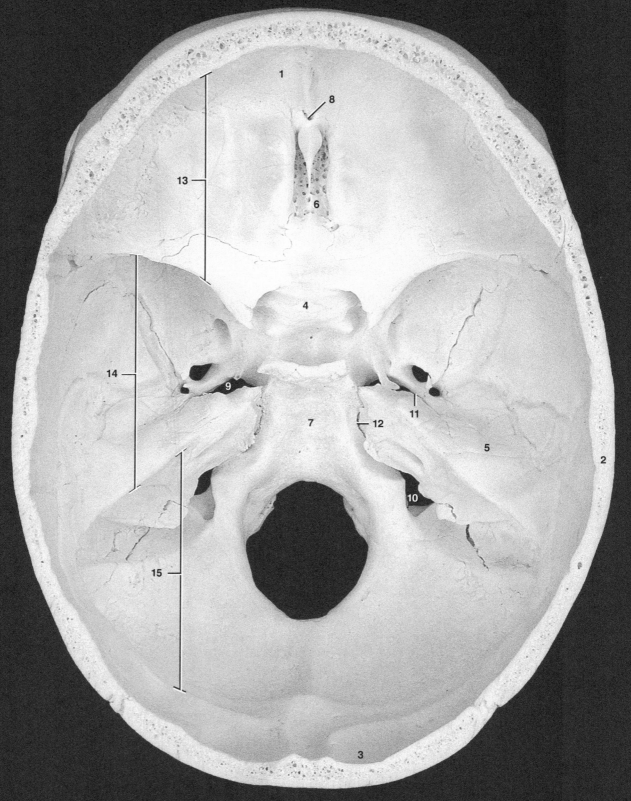

Cranium with calvaria removed
Superior or internal view of the cranial base

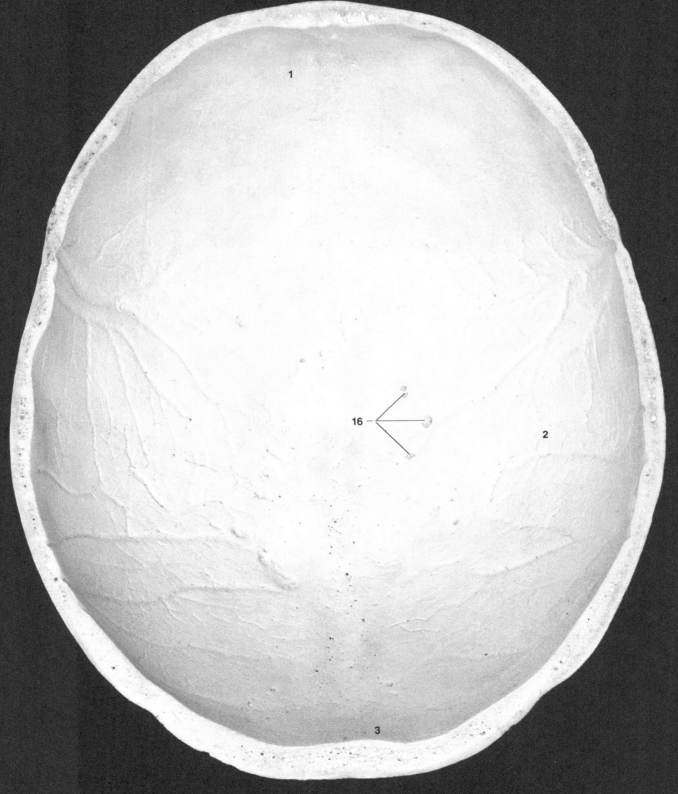

1 Frontal bone	5 Temporal bone	9 Foramen lacerum	13 Anterior cranial fossa
2 Parietal bone	6 Ethmoid bone	10 Jugular foramen	14 Middle cranial fossa
3 Occipital bone	7 Clivus	11 Petrosphenoidal fissure	15 Posterior cranial fossa
4 Sphenoid bone	8 Foramen caecum	12 Petro-occipital fissure	16 Granular foveolae

Removed calvaria
Inferior or internal view

Cranial Bones – Frontal

The unpaired frontal bone has a bowl-like shape that consists of two parts, an internally concave vertical portion termed the squama and a horizontal plate that forms the superior walls of the orbits. The bone has a smooth external surface, while its internal surface consists of impressions made by the meningeal vessels and scattered foramina that transmit diploic vessels. The squamous portion of the bone is thick. It consists of internal and external laminae of compact bone sandwiching a layer of trabecular bone called diploë. Near the anterior, inferior midline the spongy bone is absent between the external and internal laminae and in its place are variably sized spaces — the frontal sinuses. The orbital plate consists of a thin plate of compact bone, which is often so thin that it is translucent. The frontal bone articulates with twelve bones.

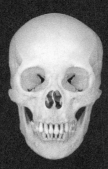

 1 Squamous part
 2 Frontal tuber
 3 Glabella
 4 Superciliary arch
 5 Supra-orbital notch or foramen
 6 Frontal notch or foramen
 7 Temporal surface
 8 Zygomatic process
 9 Frontal crest
10 Groove for superior sagittal sinus
11 Nasal spine
12 Orbital surface
13 Trochlear spine
14 Lacrimal fossa
15 Ethmoidal notch
16 Frontal sinus

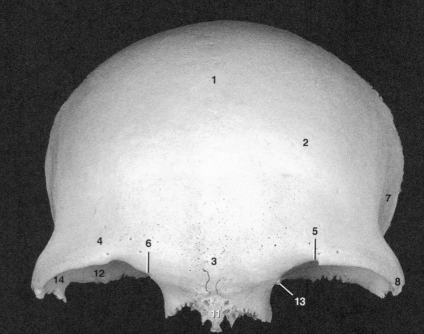

Frontal bone
Anterior view

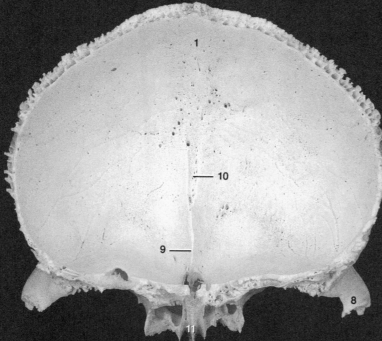

Frontal bone
Posterior view

50

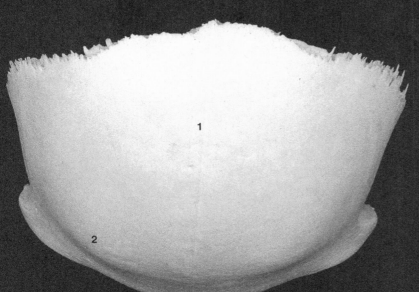

Frontal bone
Superior view, anterior to bottom

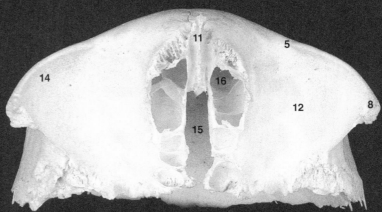

Frontal bone
Inferior view, anterior to top

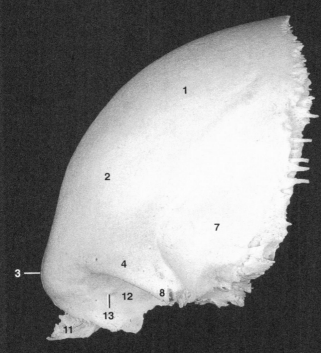

Frontal bone
Lateral view, anterior to left

Cranial Bones – Parietal

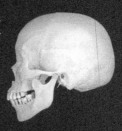

The parietal bones are large quadrilateral bones forming the greater part of the roof and sides of the cranium. The external surface of each parietal bone is slightly convex while the internal surface is concave and marked with impressions from meningeal vessels. The inferior border forms a beveled articular surface, while the superior, anterior, and posterior borders form deeply denticulate articular surfaces. The bone consists of inner and outer laminae of compact bone sandwiching a layer of trabecular bone, the diploë. Each parietal bone articulates with five bones.

1 Groove for sigmoid sinus
2 Groove for superior sagittal sinus
3 Grooves for middle meningeal artery
4 Superior temporal line
5 Inferior temporal line
6 Parietal tuber
7 Squamosal border
8 Occipital border
9 Frontal border
10 Sagittal border
11 Frontal angle
12 Occipital angle
13 Sphenoid angle
14 Mastoid angle
15 Parietal foramen

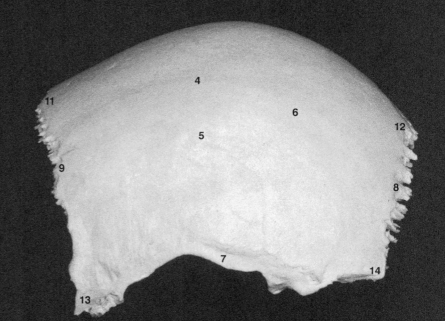

Left parietal bone
Lateral view, anterior to right

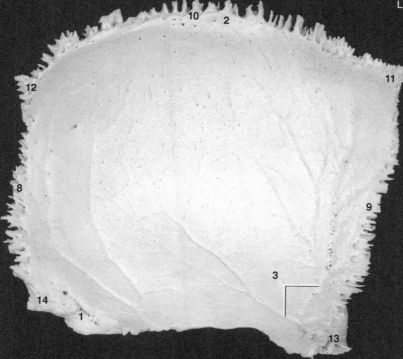

Left parietal bone
Medial view, anterior to right

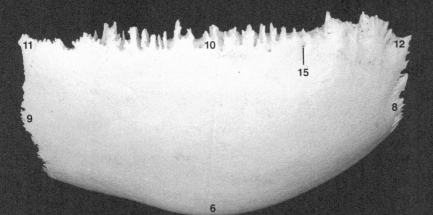

Left parietal bone
Superior view, anterior to left

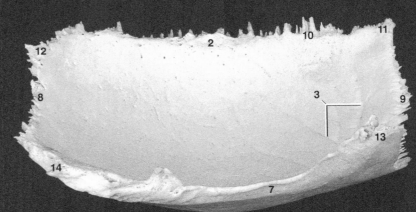

Left parietal bone
Inferior view, anterior to right

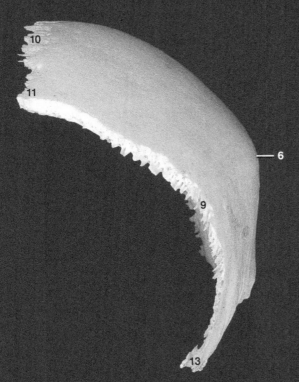

Left parietal bone
Anterior view

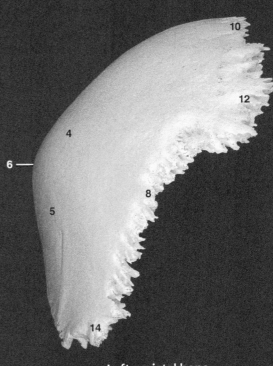

Left parietal bone
Posterior view

Cranial Bones – Occipital

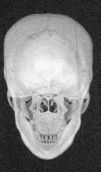

The occipital bone forms the greater part of the posterior and inferior cranium. Viewed from behind it has an oval to round shape. The bone has four distinct regions. The squamous portion is the internally concave posterosuperior plate and forms the greater part of the bone. The thick quadrilateral basi-occipital, or basilar part, contributes to the base of the cranium anterior to the foramen magnum. Lateral to this and converging with the squama are the two condylar parts or exoccipitals. Together the four regions of the bone form the borders to the large circular opening, the foramen magnum, which provides passage for the spinal cord between the cranial vault and the spinal canal. The occipital bone articulates with six bones.

1 Foramen magnum
2 Clivus
3 Pharyngeal tubercle
4 Squamous part
5 Mastoid border
6 Lambdoid border
7 Occipital condyle
8 Condylar canal
9 Hypoglossal canal
10 Condylar fossa
11 Jugular tubercle
12 Jugular notch
13 Jugular process
14 External occipital protuberance
15 Superior nuchal line
16 Inferior nuchal line
17 Internal occipital protuberance
18 Groove for transverse sinus
19 Groove for occipital sinus
20 Groove for superior sagittal sinus
21 Cerebral fossa
22 Cerebellar fossa

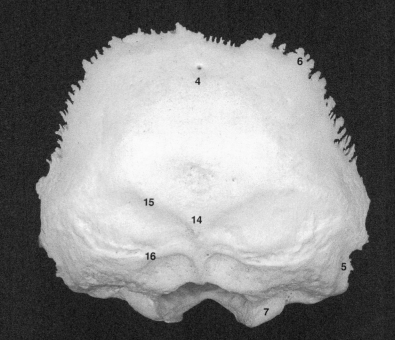

Occipital bone
Posterior view

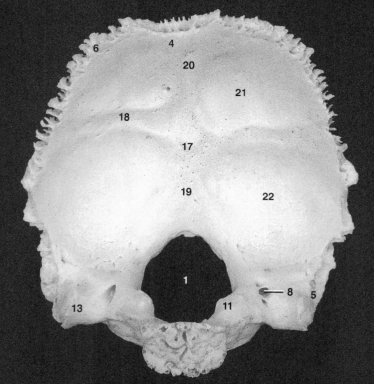

Occipital bone
Anterior view

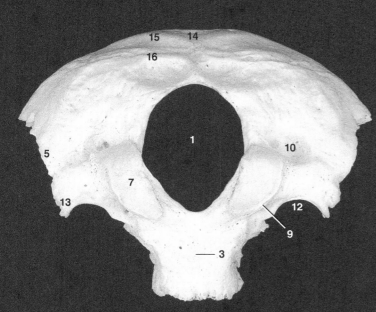

Occipital bone
Inferior view, anterior to bottom

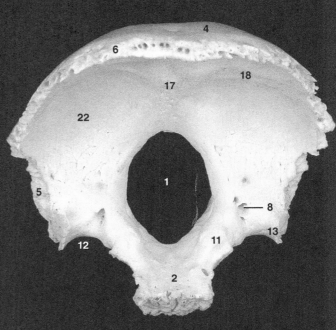

Occipital bone
Superior view, anterior to bottom

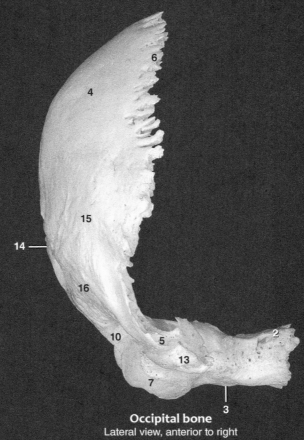

Occipital bone
Lateral view, anterior to right

Cranial Bones – Temporal

The temporal bone is a complex bone with five distinct parts. The squamous part of the bone is the thin lateral plate that contributes to the lateral wall of the cranium. It projects anteriorly as the zygomatic process and forms the mandibular fossa for the temporomandibular joint. The styloid part is represented by the styloid process. This projection of bone arises from the upper elements of the second pharyngeal arch. The petrous part forms the thick pyramidal base of the bone. It begins posterior to the external acoustic meatus as the mastoid process and ends where it forms a junction with the basi-occipital and greater wing of the sphenoid. The name petrous describes its rock-like appearance. This is the thickest part of the temporal bone. It arises from the otic capsules that stabilize the delicate internal ear structures. The mastoid is the posterolateral protuberance of the petrous portion that is easily palpable just posterior to the ear. The tympanic part of the temporal bone is the ring-like plate that forms the walls of the external acoustic meatus. Each temporal bone articulates with five bones.

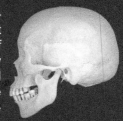

1 Petrous part
2 Mastoid process
3 Mastoid notch
4 Groove for sigmoid sinus
5 Carotid canal
6 Apex of petrous part
7 Musculotubal canal
8 Tegmen tympani
9 Hiatus for greater petrosal nerve
10 Hiatus for lesser petrosal nerve
11 Trigeminal impression
12 Internal acoustic meatus
13 Mastoid canaliculus
14 Tympanic canaliculus
15 Styloid process (broken)
16 Stylomastoid foramen
17 Jugular notch
18 Tympanic ring
19 External acoustic meatus
20 Greater tympanic spine
21 Lesser tympanic spine
22 Squamous part
23 Zygomatic process
24 Mandibular fossa
25 Articular tubercle
26 Petrotympanic fissure
27 Tympanomastoid fissure

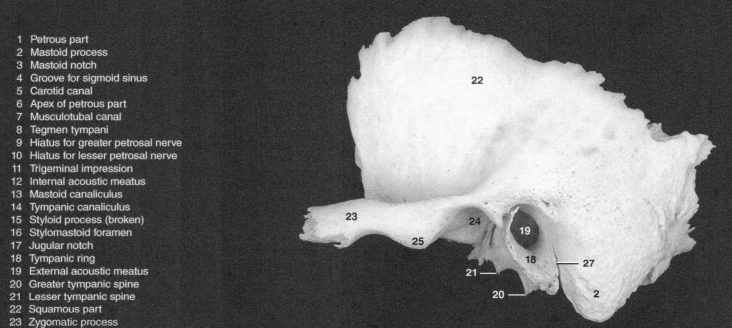

Left temporal bone
Lateral view, anterior to left

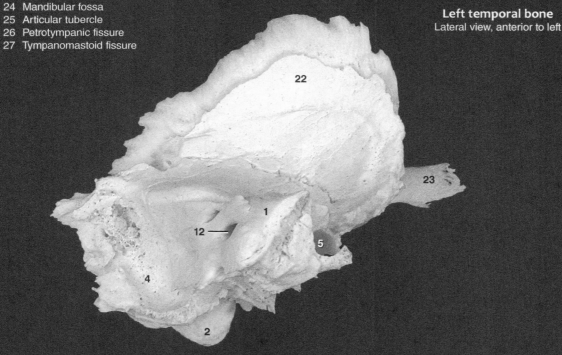

Left temporal bone
Medial view, anterior to right

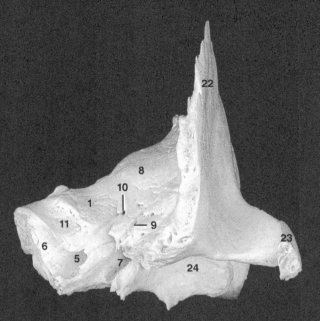

Left temporal bone
Anterior view

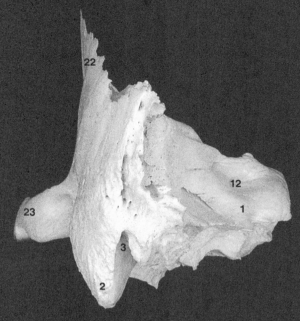

Left temporal bone
Posterior view

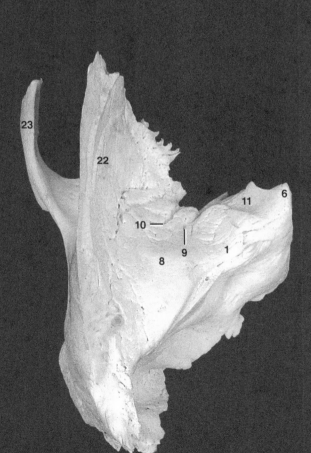

Left temporal bone
Superior view, anterior at top

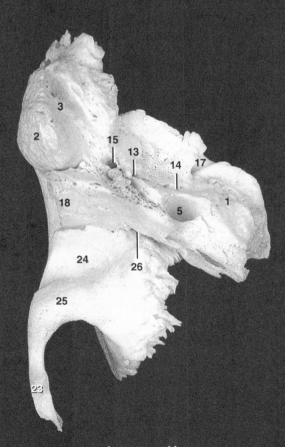

Left temporal bone
Inferior view, anterior at bottom

Cranial Bones – Sphenoid

The sphenoid bone is a complex bone that has the spread-winged appearance of a butterfly. Like it name suggests, it is wedged into the center of the cranium where it articulates with twelve neighboring bones and contributes to much of the cranial base. It is divisible into four principal components — the body, greater wings, lesser wings, and pterygoid processes. With the calvaria removed the bone is visible from any view. This bone plays a prominent role at the base of the skull. It supports the brain, serves to protect the optic stalks and capsules, provides passage for many vessels and nerves entering and leaving the skull, and forms a sinus cavity that communicates with the nasal cavity.

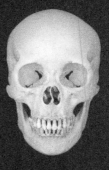

1 Jugum
2 Sella turcica
3 Tuberculum sellae
4 Hypophysial fossa
5 Dorsum sellae
6 Posterior clinoid process
7 Middle clinoid process
8 Carotid sulcus
9 Sphenoidal crest
10 Sphenoidal rostrum
11 Sphenoidal sinus
12 Sphenoidal concha
13 Lesser wing
14 Optic canal
15 Anterior clinoid process
16 Superior orbital fissure
17 Greater wing
18 Infratemporal crest
19 Orbital surface
20 Foramen rotundum
21 Foramen ovale
22 Foramen spinosum
23 Spine of sphenoid bone
24 Lateral plate of pterygoid process
25 Medial plate of pterygoid process
26 Pterygoid notch
27 Pterygoid fossa
28 Scaphoid fossa
29 Vaginal process
30 Pterygoid hamulus
31 Pterygoid canal

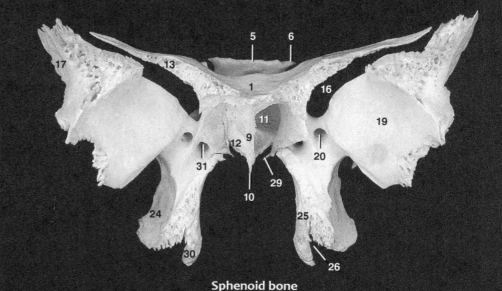

Sphenoid bone
Anterior view

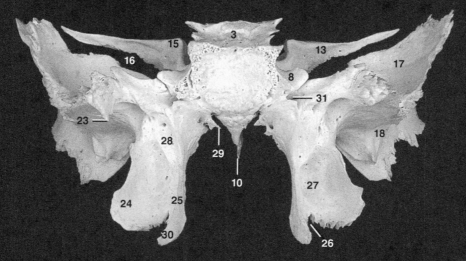

Sphenoid bone
Posterior view

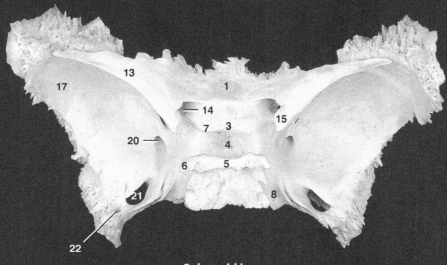

Sphenoid bone
Superior view, anterior at top

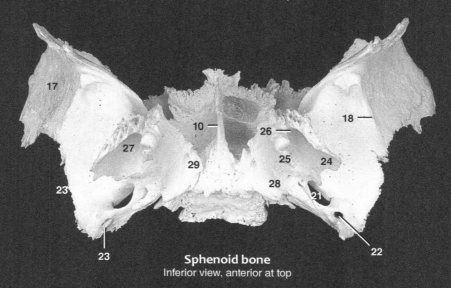

Sphenoid bone
Inferior view, anterior at top

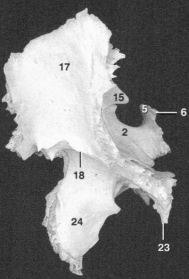

Sphenoid bone
Lateral view, anterior to left

Cranial Bones – Maxilla

The maxillae are large, paired bones that unite to form the upper jaw. They also contribute to the walls of the nasal cavity, orbit, oral cavity, and maxillary sinus. The maxillary sinus is the hollow central cavity within the large body of the maxilla. Four variable-shaped processes project from the maxillary body. The processes are the posterolateral zygomatic process, the medial projecting palatine process, the arched inferior process called the alveolar, and the superiorly projecting frontal process. Each maxilla articulates with nine bones.

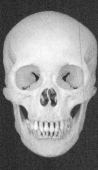

1 Orbital surface
2 Infra-orbital groove
3 Infra-orbital foramen
4 Anterior nasal spine
5 Canine fossa
6 Maxillary tuberosity
7 Lacrimal groove
8 Maxillary sinus
9 Greater palatine groove
10 Frontal process
11 Zygomatic process
12 Palatine process
13 Incisive canal
14 Alveolar process
15 Interalveolar septum

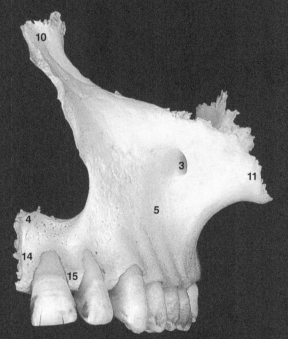

Left maxilla
Anterior view

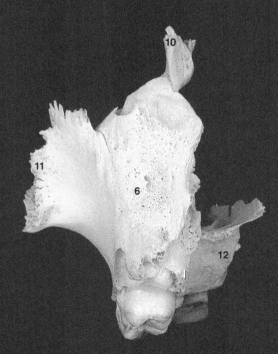

Left maxilla
Posterior view

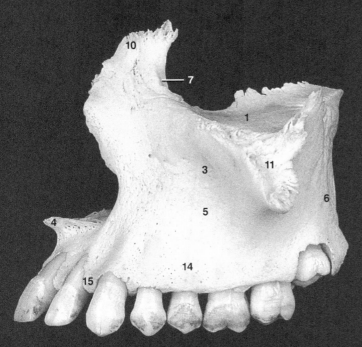

Left maxilla
Lateral view, anterior to left

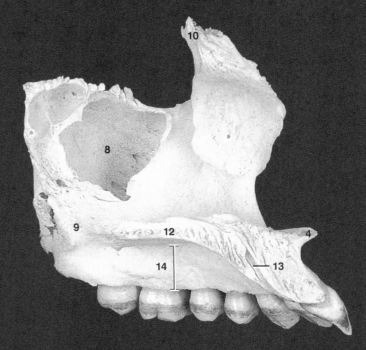

Left maxilla
Medial view, anterior to right

Left maxilla
Superior view, anterior at top

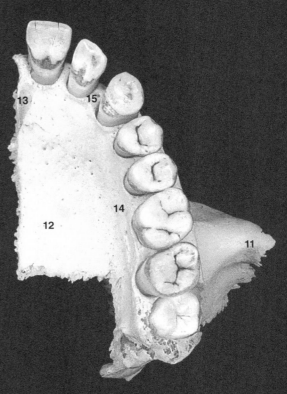

Left maxilla
Inferior view, anterior at top

Cranial Bones – Mandible

The mandible, the largest of the facial bones, forms the lower jaw. The bone has an arched body with a tooth-bearing alveolar process. Posteriorly each side of the arched body joins the vertically directed rami at the mandibular angle. The superior aspects of the two rami articulate with the temporal bones at the base of the cranium. The mandible is a strong bone composed predominantly of compact bone. It houses the lower tooth row in its alveolar arch. The strong masticatory muscles act on this bone to move it in the temporomandibular joint. Its shape can vary exceedingly with age. If the teeth are lost, bone gets resorbed on the alveolar surface leading to the thinning of the dental arch. The mandible articulates with two bones.

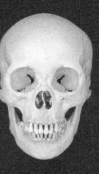

1   Body of mandible
2   Mental protuberance
3   Mental foramen
4   Mental tubercle
5   Oblique line
6   Digastric fossa
7   Mental spines
8   Mylohyoid line
9   Submandibular fossa
10 Alveolar part
11 Retromolar triangle
12 Ramus of mandible
13 Angle of mandible
14 Mandibular foramen
15 Coronoid process
16 Mandibular notch
17 Condylar process
18 Head of mandible
19 Pterygoid fovea
20 Masseteric tuberosity
21 Pterygoid tuberosity

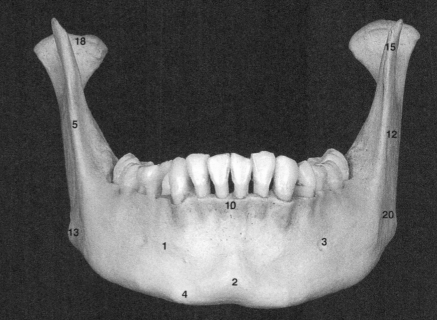

Mandible
Anterior view

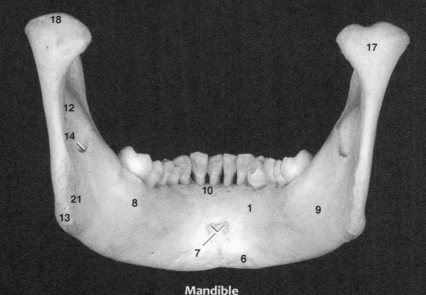

Mandible
Posterior view

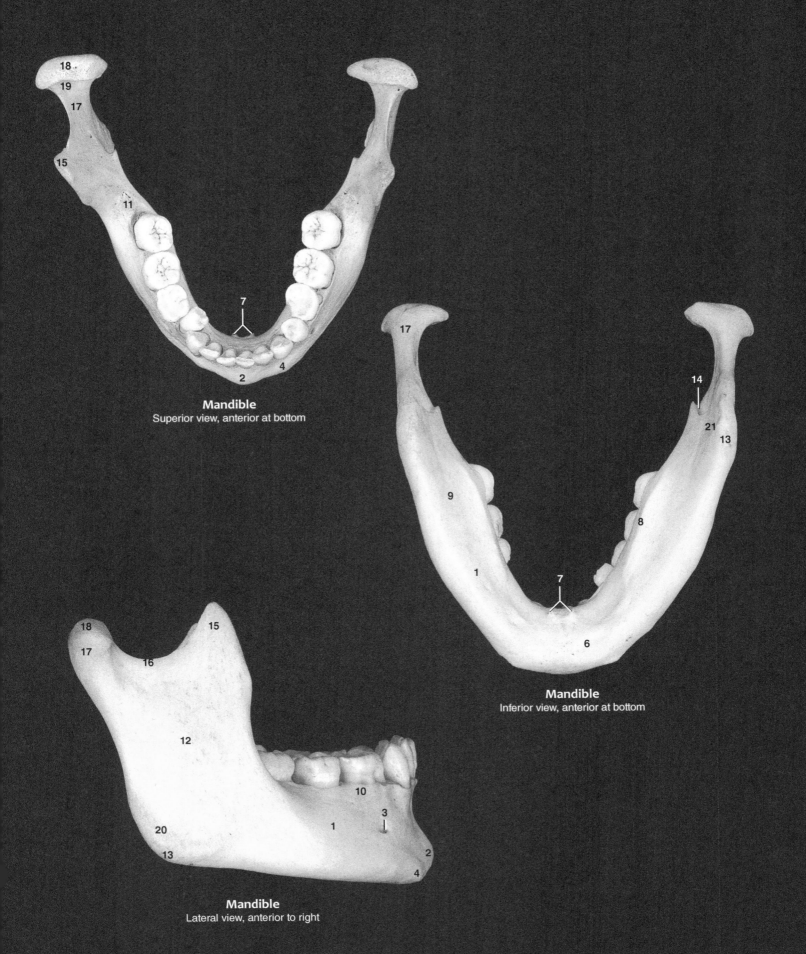

Mandible
Superior view, anterior at bottom

Mandible
Inferior view, anterior at bottom

Mandible
Lateral view, anterior to right

Cranial Bones – Ethmoid

The term ethmoid comes from the Greek term ethmos meaning sieve. Galen called the bone the sieve-like bone because of the many small foramina that transmit the olfactory nerves to the nasal cavity. This unpaired bone is both complex and delicate and is the central bone of the nasal cavity. Wedged between the two orbits, the bone consists of a median vertical plate, a horizontal plate perforated by many small foramina, and bilateral pneumatic, labyrinthine regions. The labyrinthine regions form most of the medial walls of the orbit and the superior and middle nasal conchae. This bone consists of thin laminae of compact bone surrounding many small air sinuses, which communicate with the nasal cavity. The ethmoid bone articulates with thirteen bones, more articulations than any other cranial bone.

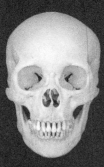

1 Cribriform plate
2 Cribriform foramina
3 Crista galli
4 Perpendicular plate
5 Ethmoidal air cells
6 Orbital plate
7 Superior nasal concha
8 Middle nasal concha
9 Ethmoidal bulla
10 Uncinate process
11 Ethmoidal infundibulum

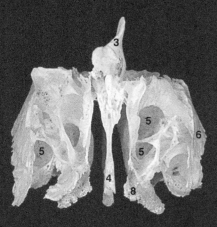

Ethmoid bone
Anterior view

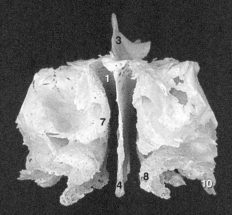

Ethmoid bone
Posterior view

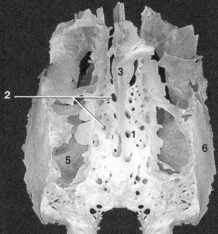

Ethmoid bone
Superior view, anterior at top

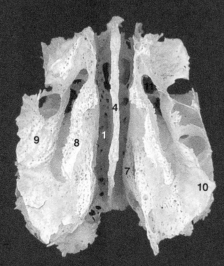

Ethmoid bone
Inferior view, anterior at top

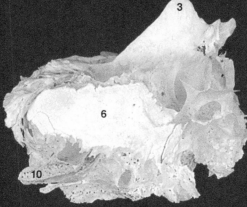

Ethmoid bone
Lateral view, anterior at right

Cranial Bones – Zygomatic

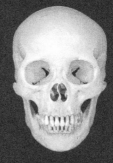

The zygomatic bone, originally named by Galen the os zygoma, comes from the Greek word zygon meaning yoke, after its resemblance to a yoke placed on oxen. This yoke-shaped bone has three distinct surfaces, five borders, and two processes. It is situated anterolateral on the face as the "cheekbone", and contributes to the lateral and inferior walls of the orbit. It consists of external and internal laminae of compact bone with an inner core of spongy bone. The zygomatic bone articulates with four bones.

1 Orbital surface
2 Temporal surface
3 Lateral surface
4 Temporal process
5 Frontal process
6 Zygomatico-orbital foramen
7 Zygomaticofacial foramen
8 Zygomaticotemporal foramen

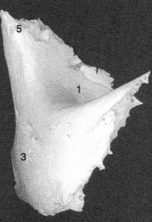

Right zygomatic bone
Anterior view

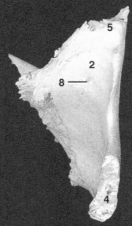

Right zygomatic bone
Posterior view

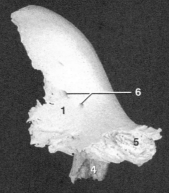

Right zygomatic bone
Superior view, anterior to top

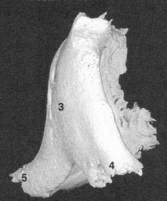

Right zygomatic bone
Inferior view, anterior to top

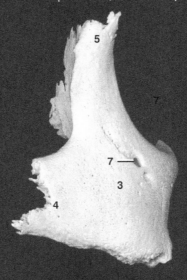

Right zygomatic bone
Lateral view, anterior to right

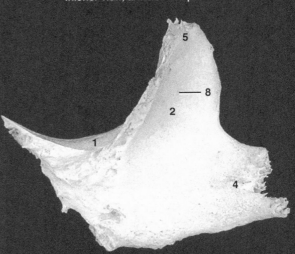

Right zygomatic bone
Medial view, anterior to left

Cranial Bones – Palatine

The palatine bone is a delicate and intricate bone that forms the shape of the letter L. It sits deep in the posterior facial region where it contributes to the roof of the mouth, floor of the orbit, floor and lateral walls of the nasal cavity, and to the pterygopalatine fossa. It has a strong horizontal plate with a delicate vertical lamina that projects superiorly. The palatine bone articulates with six bones.

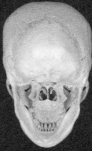

1 Perpendicular plate
2 Sphenopalatine notch
3 Greater palatine groove
4 Pyramidal process
5 Orbital process
6 Lesser palatine foramina
7 Posterior nasal spine
8 Conchal crest
9 Horizontal plate

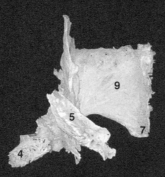

Left palatine bone
Superior view, anterior at top

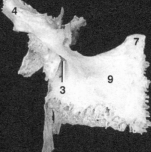

Left palatine bone
Inferior view, anterior at bottom

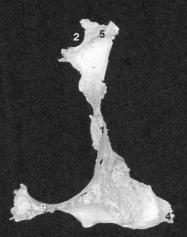

Left palatine bone
Anterior view, lateral at right

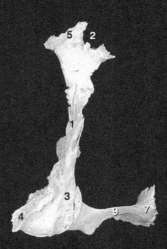

Left palatine bone
Posterior view, lateral at left

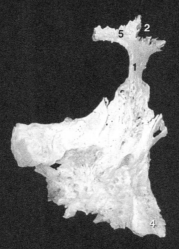

Left palatine bone
Lateral view, anterior at left

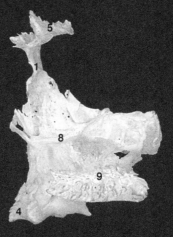

Left palatine bone
Medial view, anterior at right

Cranial Bones – Inferior Nasal Concha

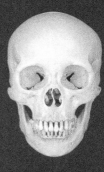

This is a small, delicate bone that projects from the lateral wall of the nasal cavity. It is scroll-like in appearance as it arches inferiorly and laterally from the nasal cavity's lateral wall. The medial surface of the bone is convex and furrowed by many longitudinal grooves that transport blood vessels beneath the thick nasal mucosa that covers this surface. The lateral surface of the bone is concave and forms most of the superior and medial boundary of the inferior nasal meatus. The inferior border of the bone has a rough, spongy appearance. Superiorly the bone forms an articular border with four bones.

1 Lacrimal process
2 Maxillary process
3 Ethmoidal process
4 Lateral surface
5 Medial surface

Left inferior nasal concha
Anterior view, lateral at left

Left inferior nasal concha
Posterior view, lateral at right

Left inferior nasal concha
Lateral view, anterior at right

Left inferior nasal concha
Medial view, anterior at left

Left inferior nasal concha
Superior view, anterior at right

Left inferior nasal concha
Inferior view, anterior at right

Cranial Bones – Lacrimal

The lacrimal bone derives its name from the Latin word meaning tear because the bone houses the "tear duct." This small, delicate, quadrate-shaped bone has a vertical axis that is slightly longer than its horizontal axis. It is extremely thin. When it is held up to a light source, the light easily penetrates the bone. The bone sits in the anterior part of the medial wall of the orbit. The orbital surface is smooth and flat in its posterior half where it contributes to the medial wall of the orbit. Anteriorly this surface has a longitudinal groove that ends posteriorly in a longitudinal crest that is hook-shaped inferiorly. This groove supports the nasolacrimal duct. Covered with mucous membrane, the slightly rough, medial surface of the bone contributes to the nasal cavity. The lacrimal bone articulates with four bones.

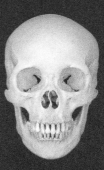

1 Posterior lacrimal crest
2 Lacrimal groove
3 Lacrimal hamulus

Left lacrimal bone
Anterior view, lateral at right

Left lacrimal bone
Posterior view, lateral at left

Left lacrimal bone
Lateral view, anterior at left

Left lacrimal bone
Medial view, anterior at right

Left lacrimal bone
Superior view, lateral at right

Left lacrimal bone
Inferior view, lateral at left

Cranial Bones – Auditory Ossicles

The auditory ossicles are the smallest bones of the human skeleton. These three small bones occupy the middle ear cavity, where they transmit and amplify the sound waves from the tympanic membrane to the inner ear. From lateral to medial the bones are the malleus, the incus, and the stapes, or in layman's terms the hammer, the anvil, and the stirrup, because of their striking resemblance to these structures.

1 Malleus
2 Incus
3 Stapes
4 Handle of malleus
5 Head of malleus
6 Neck of malleus
7 Lateral process
8 Anterior process
9 Body of incus
10 Long limb
11 Lenticular process
12 Short limb
13 Head of stapes
14 Anterior limb
15 Posterior limb
16 Footplate

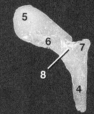
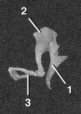

Left auditory ossicles
Anterior view, lateral at left

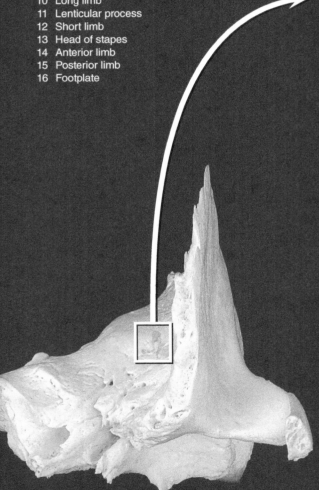

Auditory ossicles in situ within temporal bone
Anterior view, left temporal bone

Left malleus
Anterior view, lateral at left

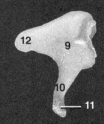

Left incus
Lateral view, anterior at left

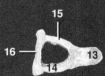

Left stapes
Superior view, lateral at left

Cranial Bones – Hyoid

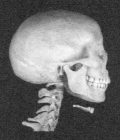

Suspended from the styloid processes of the temporal bones by the stylohyoid ligaments, the U-shaped hyoid bone occupies the ventrosuperior neck just inferior to the mandible. It serves as a skeletal attachment site for muscles associated with the tongue, larynx, and pharynx. It consists of five elements — a body and bilateral lesser and greater cornua. The body is the rectangular ventral element that sits in the transverse plane. Projecting posterolaterally from the body are the paired, long, slender greater cornua. At the junction of the greater cornua and the body are smaller superior projections, the lesser cornua.

1 Body
2 Lesser horn
3 Greater horn

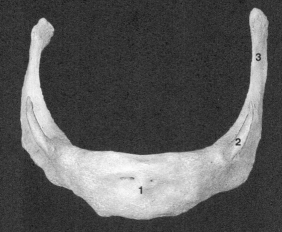

Hyoid bone
Anterior view

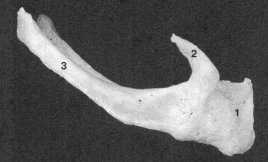

Hyoid bone
Lateral view, anterior at right

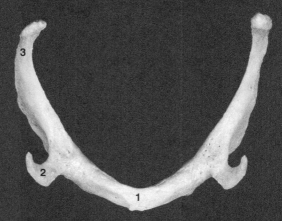

Hyoid bone
Superior view, anterior at bottom

Vertebral Column

The vertebral column consists of 26 bones that develop from a series of 33 identical embryonic body segments. Because they develop from similar repeating segments, each of the vertebrae has a similar structure. The bones of the vertebral column are grouped into seven cervical vertebrae, twelve thoracic vertebrae, five lumbar vertebrae, the sacrum consisting of five fused segments, and the coccyx comprised of three to five fused segments, most typically four. The column is the central axis of the body that supports the limbs and the cranium, protects the spinal cord, and provides attachment for muscles that move this flexible column of bones.

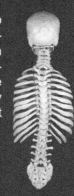

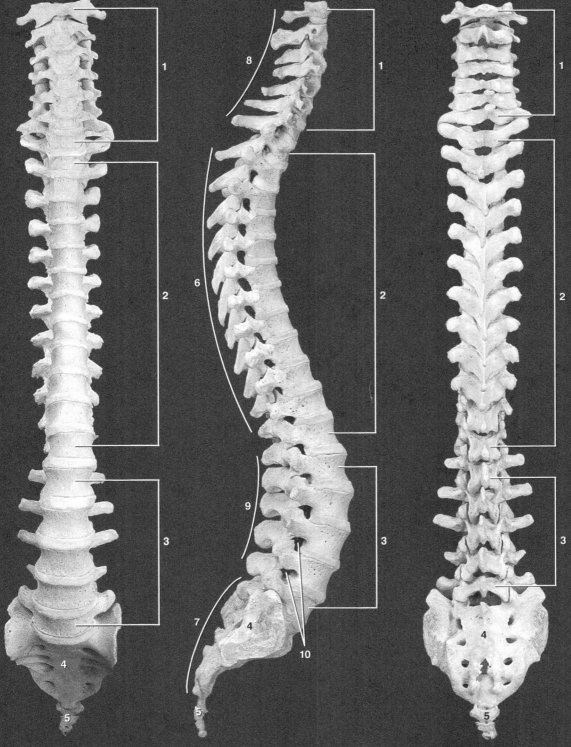

1 Cervical vertebrae
2 Thoracic vertebrae
3 Lumbar vertebrae
4 Sacrum
5 Coccyx
6 Thoracic kyphosis
7 Sacral kyphosis
8 Cervical lordosis
9 Lumbar lordosis
10 Intervertebral foramen

Vertebral column
Anterior view

Vertebral column
Lateral view, anterior at right

Vertebral column
Posterior view

Cervical Vertebrae

There are seven cervical vertebrae, which are the vertebrae with the greatest variation in shape. They form a delicate column of bones having a wide range of mobility at their joint surfaces. This is due to the fact that the first two cervical vertebrae, the atlas and axis, have forms that differ significantly from the remaining five vertebrae in the series. These differences arise as they become modified to provide the support and movement of the skull. The remaining cervical vertebrae show a lesser degree of mobility and have more uniform shapes. With few exceptions, the cervical vertebrae can be readily distinguished by the presence of a foramen in their transverse processes.

1	Vertebral body	12	Anterior tubercle of costal process
2	Pedicle	13	Posterior tubercle of costal process
3	Lamina	14	Lateral mass
4	Superior vertebral notch	15	Anterior arch
5	Inferior vertebral notch	16	Anterior tubercle of anterior arch
6	Vertebral foramen	17	Facet for dens
7	Spinous process	18	Posterior arch
8	Transverse process	19	Posterior tubercle of posterior arch
9	Superior articular process/facet	20	Groove for vertebral artery
10	Inferior articular process/facet	21	Dens
11	Transverse foramen	22	Anterior articular facet of dens

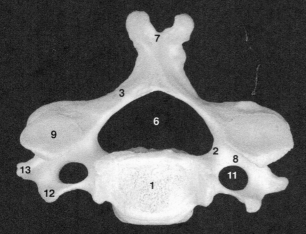

Typical cervical vertebra
Superior view, anterior at bottom

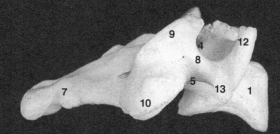

Typical cervical vertebra
Lateral view, anterior at right

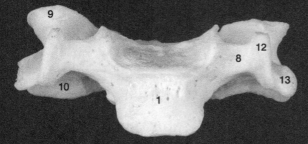

Typical cervical vertebra
Anterior view, superior at top

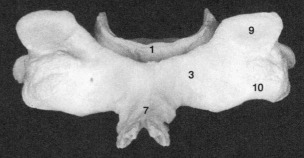

Typical cervical vertebra
Posterior view, superior at top

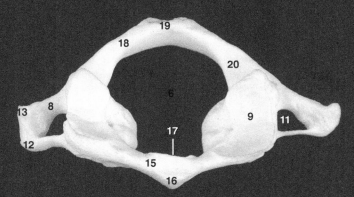

Atlas, 1st cervical vertebra
Superior view, anterior at bottom

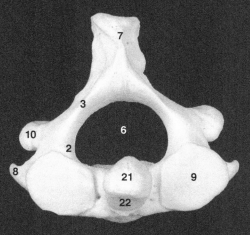

Axis, 2nd cervical vertebra
Superior view, anterior at bottom

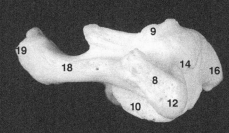

Atlas, 1st cervical vertebra
Lateral view, anterior at right

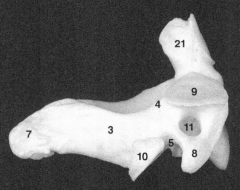

Axis, 2nd cervical vertebra
Lateral view, anterior at right

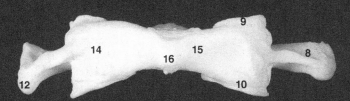

Atlas, 1st cervical vertebra
Anterior view, superior at top

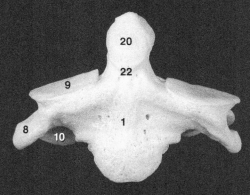

Axis, 2nd cervical vertebra
Anterior view, superior at top

Atlas, 1st cervical vertebra
Posterior view, superior at top

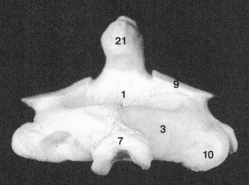

Axis, 2nd cervical vertebra
Posterior view, superior at top

Thoracic Vertebrae

The thoracic portion of the vertebral column, consisting of the twelve thoracic vertebrae, get progressively larger from the cranial end to the caudal end of the series. Except at its junction with the lumbar vertebrae, the thoracic region is the least mobile region of vertebral column. In addition to articulating with each other, the thoracic vertebrae also articulate with the ribs. Additionally, the laminae and spines of these vertebrae project inferiorly to overlap the next vertebra below. This suite of characters produces a strong imbricated column of bone that forms the impressive thoracic rib cage. Because of their association with the ribs, the thoracic vertebrae are readily identified by the costal articular facets, which are present on the bodies and transverse processes.

1 Vertebral body
2 Pedicle
3 Lamina
4 Superior vertebral notch
5 Inferior vertebral notch
6 Spinous process
7 Transverse process
8 Superior articular process/facet
9 Inferior articular process/facet
10 Superior costal facet
11 Inferior costal facet
12 Transverse costal facet

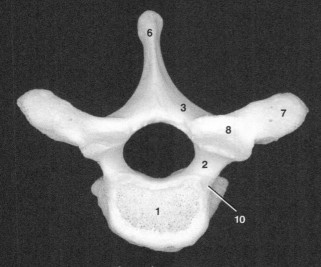

Thoracic vertebra
Superior view, anterior at bottom

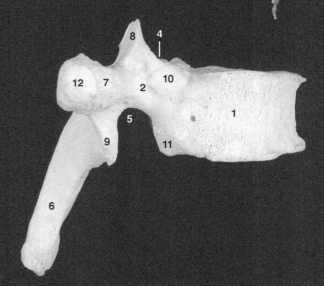

Thoracic vertebra
Lateral view, anterior at right

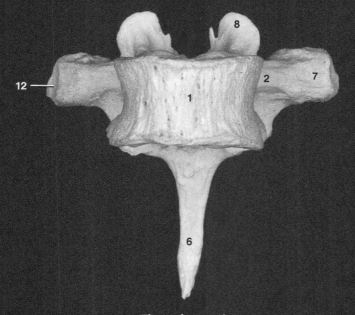

Thoracic vertebra
Anterior view, superior at top

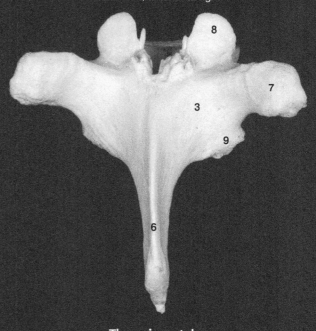

Thoracic vertebra
Posterior view, superior at top

Lumbar Vertebrae

There are five lumbar vertebrae that form the lumbar portion of the vertebral column. The mobile vertebrae of this region are the largest of the true or mobile vertebrae. Their large size and lack of transverse foramina and costal facets are their diagnostic features. They form a strong column of support at the base of the vertebral column. The articular processes of the lumbar vertebrae are robust and have their facets oriented in the sagittal plane to provide for the flexion and extension movements characteristic of the lumbar vertebral column. They have thick pedicles arising from the superior aspect of the vertebral body. The laminae are thick and short and project posteriorly to unite as thick, quadrilateral spinous processes. The vertebral bodies have a large elliptical shape when viewed from above.

1   Vertebral body
2   Pedicle
3   Lamina
4   Superior vertebral notch
5   Inferior vertebral notch
6   Spinous process
7   Transverse process (costal process)
8   Superior articular process/facet
9   Inferior articular process/facet
10 Accessory process (morphological transverse process)
11 Mammillary process

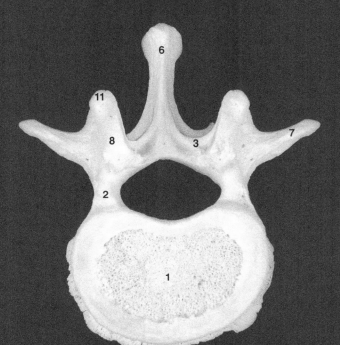

Lumbar vertebra
Superior view, anterior at bottom

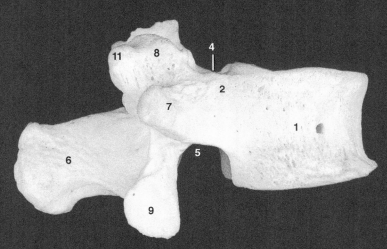

Lumbar vertebra
Lateral view, anterior at right

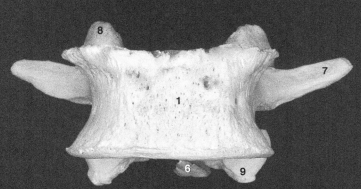

Lumbar vertebra
Anterior view, superior at top

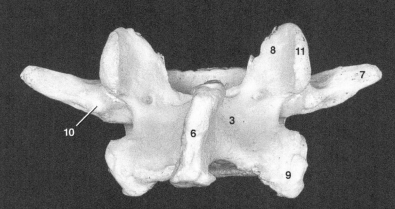

Lumbar vertebra
Posterior view, superior at top

Sacrum and Coccyx

The sacrum is a large triangular-shaped mass that forms from the fusion of five vertebrocostal segments. The base of the triangle is superior and tapers to a flattened apex inferiorly. It is concave anteriorly and convex posteriorly. The lateral margins of the triangle are widest superiorly where the bone articulates with the two ilia. Forming the large basal portion of the vertebral column, the bone wedges between the two os coxae to form the posterior element of the pelvic skeleton. Its ventral surface, smoother than the rough dorsal surface, forms the posterior wall of the pelvis. Within this triangular mass of bone is a hollow sacral canal. This canal opens through foramina onto the ventral and dorsal surfaces of the bone. It forms a large oval surface superiorly that articulates with the fifth lumbar vertebra and a smaller oval facet at its apex for articulation with the coccyx.

The coccyx is the terminal end of the vertebral column. It is a triangular bone that forms from the fusion of three to five vertebral segments, most commonly from four fused vertebrae. The superior surface of the first segment's body forms an oval articular surface with the inferior surface of the fifth sacral segment.

1 Promontory
2 Ala or wing
3 Superior articular process
4 Auricular surface
5 Sacral tuberosity
6 Pelvic surface
7 Transverse ridges
8 Anterior sacral foramina
9 Posterior sacral foramina
10 Median sacral crest
11 Intermediate sacral crest
12 Lateral sacral crest
13 Sacral cornu
14 Sacral canal
15 Sacral hiatus
16 Apex
17 Coccygeal cornu

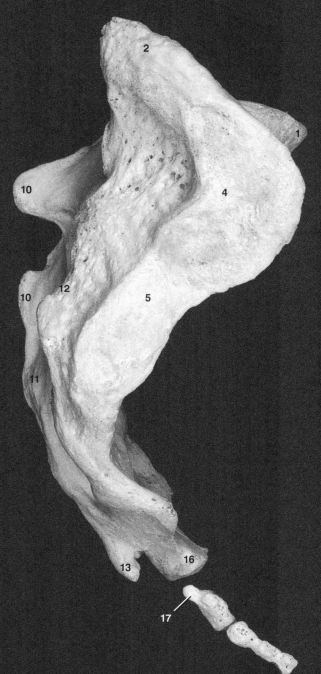

Sacrum and coccyx
Lateral view, anterior at right

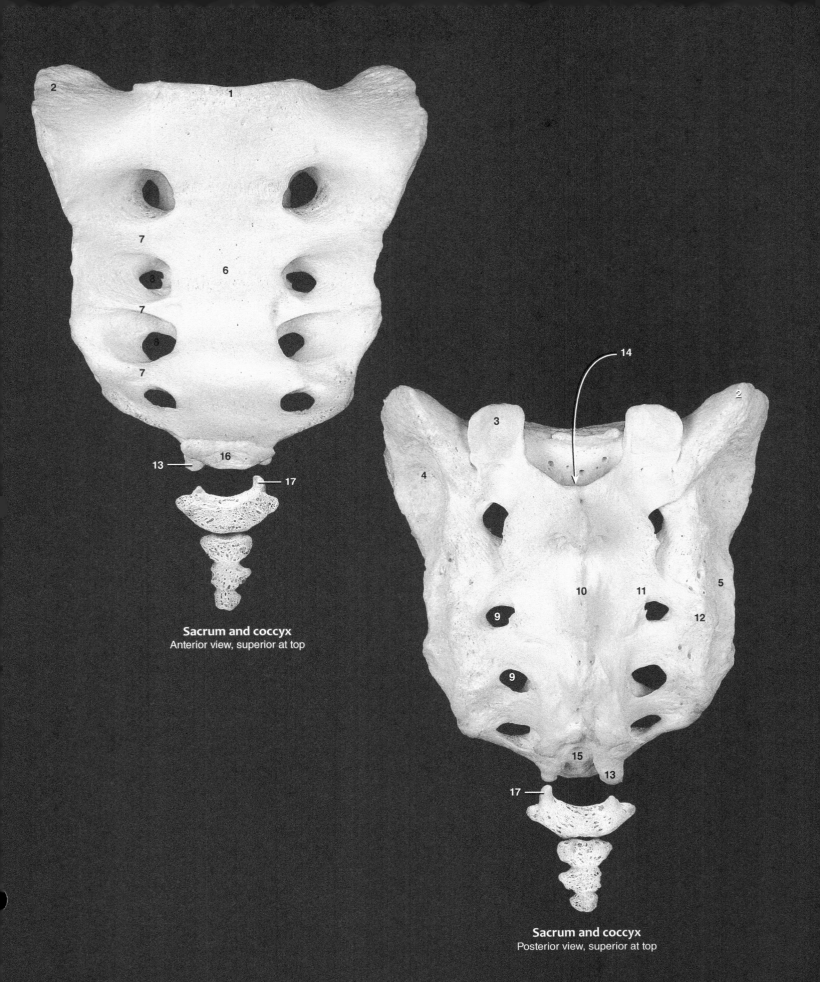

Sacrum and coccyx
Anterior view, superior at top

Sacrum and coccyx
Posterior view, superior at top

Ribs

There are twelve paired ribs, a pair for each of the twelve thoracic vertebrae. The ribs unite the thoracic vertebrae to the sternum via costal cartilages to form the thoracic skeleton, a flexible, bony wall that protects thoracic viscera and facilitates respiratory function. Although only the twelve thoracic ribs are named ribs, there are in reality ribs at every vertebral level. The cervical, lumbar, sacral, and coccygeal ribs fuse to their corresponding vertebrae to contribute to the formation of the transverse process. The ribs can be divided into two groups — true ribs and false ribs. The last two false ribs are called floating ribs. True ribs, ribs one through seven, are those that have their costal cartilages attached directly to the sternum. False ribs, ribs eight through twelve, have costal cartilages that do not attach directly to the sternum. The costal cartilage of each of the first three false ribs attaches to the cartilage of the rib superior to it. The last two false ribs do not attach to other ribs and are therefore called floating ribs.

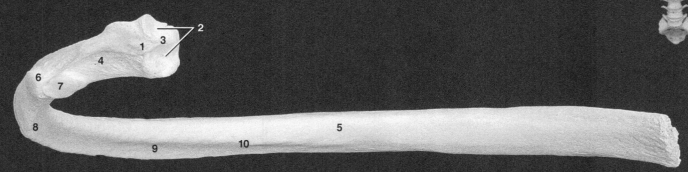

Left sixth rib
Posterior view, superior at top

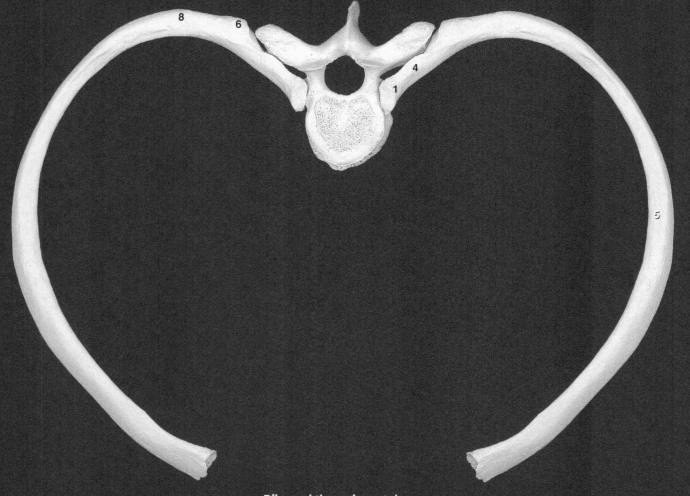

Ribs and thoracic vertebra
Superior view, posterior at top

1 Head
2 Articular facets of head
3 Crest of head
4 Neck
5 Body or shaft
6 Tubercle
7 Articular facet of tubercle
8 Angle
9 Costal groove
10 Crest of body
11 Scalene tubercle (first rib)
12 Tuberosity of serratus anterior (second rib)
13 Costal cartilage
14 True ribs [I-VII]
15 False ribs [VII-XII]
16 Floating ribs [XI-XII]

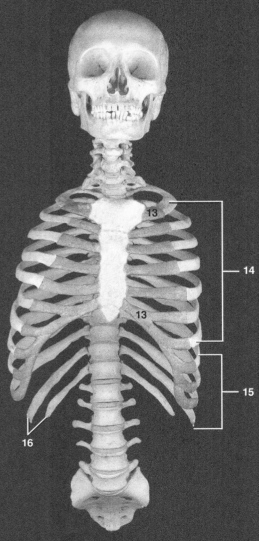

Rib cage
Anterior view

Left ribs 1 through 12
Superior view, first rib at top, posterior to right

Sternum

The sternum is the anterior bone of the thoracic wall. It forms from six segmental elements, or sternebrae, that fuse during development. The bone has the appearance of a sword with a wide handle called the manubrium, a tapering blade or body, and a sharp point-like apex named the xiphoid process. A distinct angle forms at the junction of the manubrium and the body. This angle is called the sternal angle. A horizontal plane extended posteriorly intersects the disc between the fourth and fifth thoracic vertebrae and marks the top of the heart in the thoracic cavity. The lateral margins of the bone are notched for reception of the costal cartilages and clavicles. Its anterior surface is slightly convex, while the posterior surface is weakly concave. The sternum articulates with sixteen bones, more articulations than any other bone in the body.

1 Manubrium
2 Clavicular notch
3 Jugular or suprasternal notch
4 Sternal angle

5 Body
6 Xiphoid process
7 Costal notches

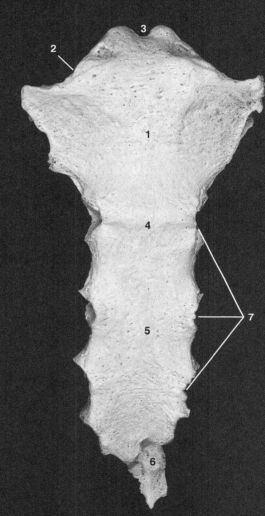

Sternum
Anterior view, superior at top

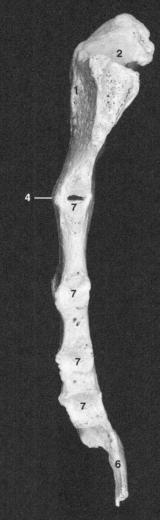

Sternum
Lateral view, anterior at left

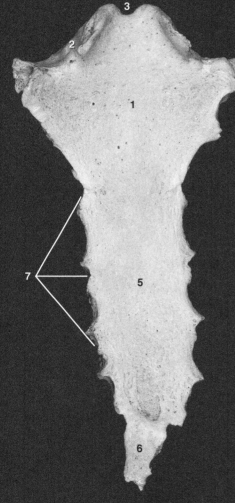

Sternum
Posterior view, superior at top

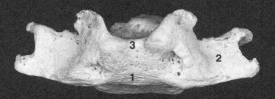

Sternum
Superior view, posterior at top

6 | Appendicular Skeleton

The appendicular portion of the skeleton forms the framework of the limbs. It includes the limb girdles, or fixed portion of the appendicular skeleton, and the series of bones that extend distally from the girdles into the limb proper, or free portion of the limb. The limb girdles, pectoral and pelvic, help anchor the limb to the axial skeleton. The free portion of each limb consists of a large proximal element, the humerus and femur, forming the skeleton of the arm and thigh, respectively. Next in sequence are the ulna and radius of the forearm, and the fibula and tibia of the leg. The distal-most regions of the limbs are the hand and foot consisting of the short carpal and tarsal bones, respectively, along with the metacarpals, metatarsals, and phalanges of the digits.

As the tetrapod (land) vertebrates evolved, a major difference emerged between the two limbs. The anterior, or upper limb, evolved as a steering device, while the posterior, or lower limb, became the locomotor limb. Accompanying these evolutionary modifications in limb function were important morphological differences. The powerful locomotor hind limb developed strong attachments to the axial skeleton. The strong iliosacral joint, with its accompanying ligaments, transfers the powerful forces generated by the posterior limb to the axial skeleton to propel the body forward. On the other hand, the anterior limb developed minimal, weak skeletal attachments between the girdle and axial skeleton while becoming a more mobile limb.

As you study the skeleton of the limbs in the photos that follow, note the similarities and differences that exist between the bones of the superior and inferior limb skeletons and think about the functional differences mentioned above.

Find more information about the appendicular skeleton in

REAL ANATOMY

Upper Limb

Each superior limb consists of 32 bones. The proximal end of the superior limb, the clavicle and scapula, form the pectoral or shoulder girdle. This girdle of bones provides a broad base of support that is primarily anchored to the axial skeleton by muscles rather than ligaments. The free part of the upper limb consists of the humerus, radius, ulna, and hand. The humerus forms the skeletal framework for the brachium. Distal to the brachium is the antebrachium containing the radius and ulna. The distal-most region of the superior limb is the hand consisting of a wrist region of eight carpal bones, the palm region consisting of five metacarpal bones, and the fourteen phalanges of the fingers and thumb.

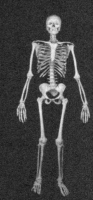

1 Scapula
2 Clavicle
3 Humerus
4 Ulna
5 Radius
6 Carpals
7 Metacarpals
8 Phalanges

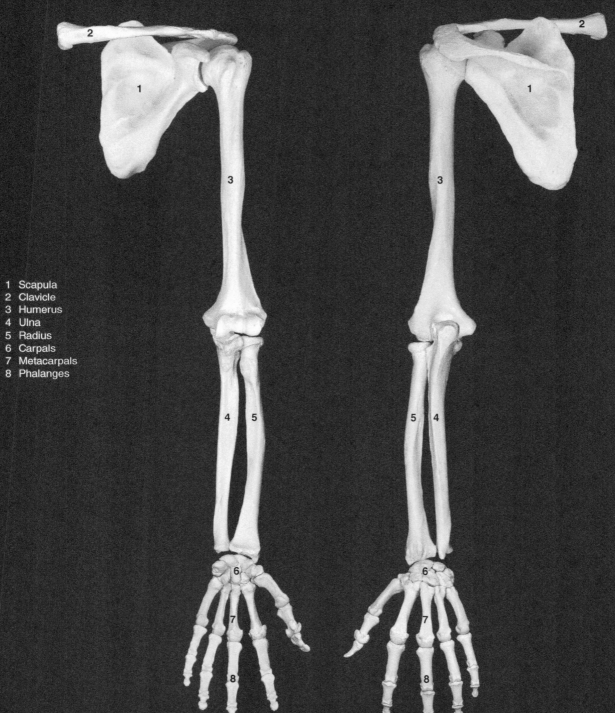

Left upper limb
Anterior view

Left upper limb
Posterior view

Pectoral Girdle

The pectoral, or shoulder, girdle consisting of the scapula and the clavicle forms the base of the upper limb skeleton. The rod-like clavicle forms a horizontal strut that links the scapula to the sternum of the axial skeleton. The large triangular scapula presents an extensive surface area for muscle attachment and a large lateral fossa that articulates with the humerus of the free part of the upper limb. Except for the weak joint formed between the clavicle and the sternum, the pectoral girdle is essentially unattached by ligaments or joints to the axial skeleton. This was paramount in the evolutionary role of this limb as a steering device and shock absorber during locomotion.

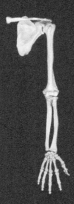

1 Scapula
2 Clavicle

Left pectoral girdle
Lateral view

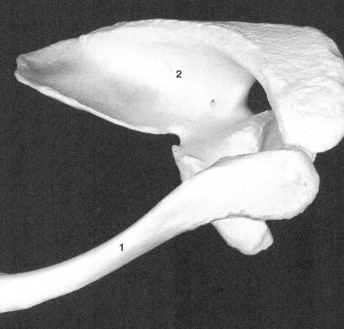

Left pectoral girdle
Superior view

Clavicle

The clavicle has an S-shaped appearance that can range from an almost straight, shallow S-curve shape to a deeper, more prominent S-curve shape. The curve at the medial or sternal end of the bone is concave posteriorly, while the curve at the lateral or acromial end is concave anteriorly. This is one of the more variable bones of the skeleton. It is typically smooth and straight in females and rougher and more curved in males. The bone forms the ventral strut of the pectoral girdle that props the shoulder joint away from the rib cage. It is subcutaneous and easily palpable throughout its length. This combination of features makes it susceptible to fracture from falls onto the limb. The clavicle articulates with the three bones — the scapula, sternum, and first rib.

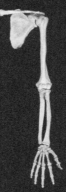

1 Sternal end
2 Sternal facet
3 Impression for costoclavicular ligament
4 Shaft or body
5 Subclavian groove
6 Acromial end
7 Acromial facet
8 Tuberosity for coracoclavicular ligament
9 Conoid tubercle
10 Trapezoid line

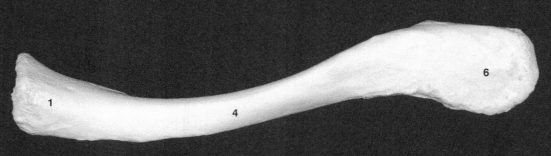

Left clavicle
Superior view, lateral to right

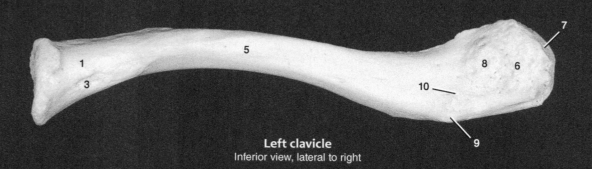

Left clavicle
Inferior view, lateral to right

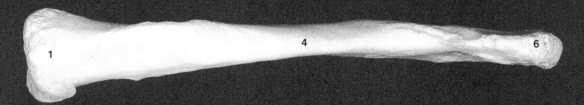

Left clavicle
Anterior view, lateral to right

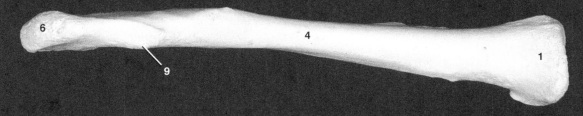

Left clavicle
Posterior view, lateral to left

Left clavicle
Lateral view, anterior to left

Left clavicle
Medial view, anterior to right

Scapula

The scapula is a flat, triangular bone with three prominent projections. The flattened triangular portion of the bone, the body, spans from the second to the seventh rib and consists of three borders (superior, lateral, and medial) and three angles (superior, inferior, and lateral) and is typically a very thin plate of bone. Its lateral angle is conspicuous as it forms the glenoid fossa, or shoulder socket that articulates with the head of the humerus. Its three prominent projections are the anterior projecting coracoid process, the posterior projecting ridge called the spine, and the flat laterally projecting acromion, which forms the lateral expansion of the spine. The scapula articulates with two bones — the clavicle and the humerus.

1 Subscapular fossa
2 Spine
3 Deltoid tubercle
4 Supraspinous fossa
5 Infraspinous fossa
6 Acromion
7 Clavicular facet
8 Acromial angle
9 Medial border
10 Lateral border
11 Superior border
12 Suprascapular notch
13 Inferior angle
14 Superior angle
15 Glenoid cavity
16 Supraglenoid tubercle
17 Infraglenoid tubercle
18 Neck
19 Coracoid process

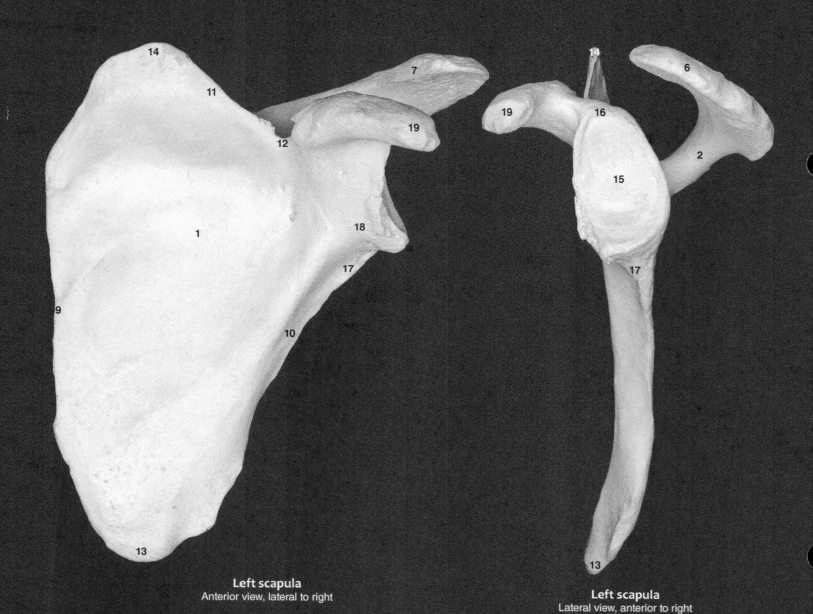

Left scapula
Anterior view, lateral to right

Left scapula
Lateral view, anterior to right

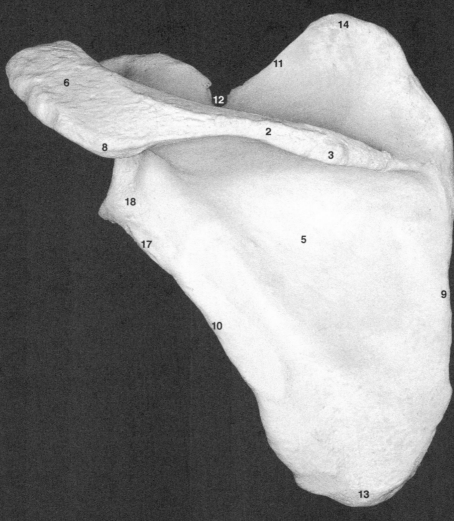

Left scapula
Posterior view, Lateral to left

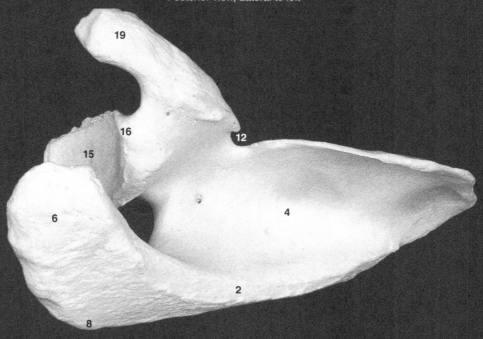

Left scapula
Superior view, lateral to left

Humerus

The humerus is the skeletal element of the brachium and it is the largest bone of the upper limb. It has a long cylindrical shaft that expands at the proximal and distal ends. The proximal end is rounded, while the distal end is flattened from anterior to posterior. The ends consist of a spongy core of bone covered with a thin lamina of compact bone. The shaft is a cylinder of thick compact bone surrounding a large medullary cavity. The humerus articulates with three bones — the scapula, ulna, and radius.

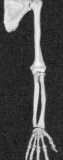

1 Head
2 Anatomical neck
3 Surgical neck
4 Greater tubercle
5 Lesser tubercle
6 Intertubercular sulcus or groove
7 Crest of greater tubercle
8 Crest of lesser tubercle
9 Shaft or body
10 Groove for radial nerve
11 Medial supracondylar ridge
12 Deltoid tuberosity
13 Capitulum
14 Trochlea
15 Olecranon fossa
16 Coronoid fossa
17 Radial fossa
18 Medial epicondyle
19 Groove for ulnar nerve
20 Lateral epicondyle

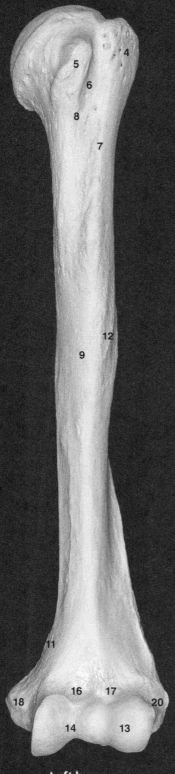

Left humerus
Anterior view, lateral to right

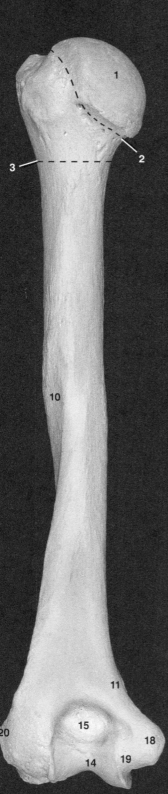

Left humerus
Posterior view, lateral to left

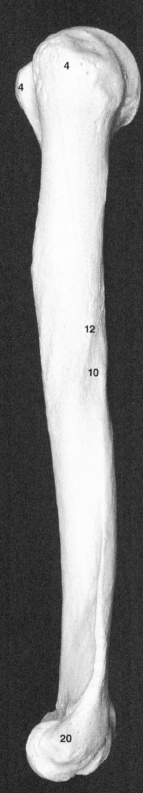

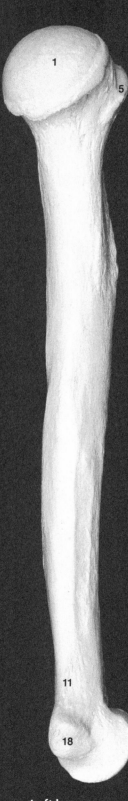

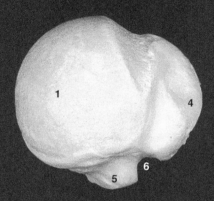

Left humerus
Superior view, lateral to left

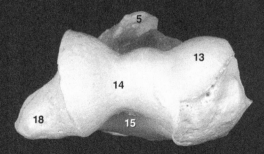

Left humerus
Inferior view, lateral to right

Left humerus
Lateral view, anterior to left

Left humerus
Medial view, anterior to right

Ulna

The ulna is the medial and longer bone of the antebrachium. It is thick and notched at its proximal end where it is a major contributor to the elbow joint. From the notched proximal end it tapers to a thin shaft that ends distally as a small rounded head. The ulna articulates with two bones— the humerus and the radius.

1 Olecranon
2 Coronoid process
3 Ulnar tuberosity
4 Radial notch
5 Trochlear notch
6 Shaft or body
7 Interosseous border
8 Anterior border
9 Posterior border
10 Supinator crest
11 Head
12 Articular circumference
13 Ulnar styloid process

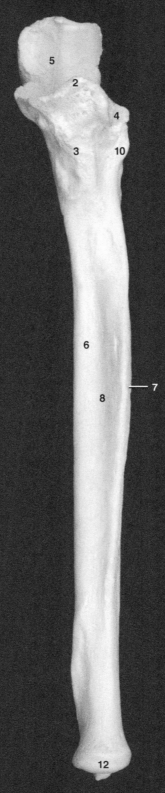

Left ulna
Anterior view, lateral to right

Left ulna
Posterior view, lateral to left

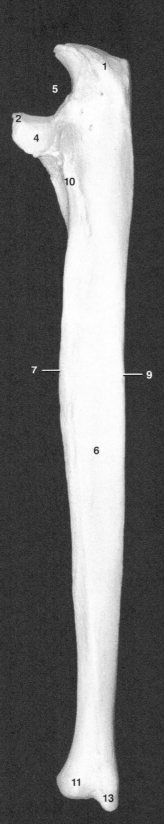

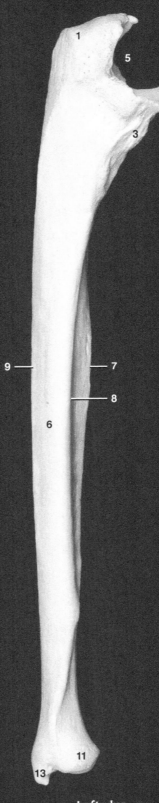

Left ulna
Superior view, lateral to left

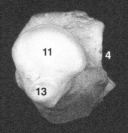

Left ulna
Inferior view, lateral to right

Left ulna
Lateral view, anterior to left

Left ulna
Medial view, anterior to right

Radius

The radius is the lateral, slender, rod-like bone of the antebrachium. The rod-like shaft expands at both ends. The proximal end forms a wheel-like head with a proximal concavity, while the distal end expands from medial to lateral to form the widest part of the bone. The distal end is concave anteriorly and convex and grooved posteriorly. The ridge-like borders of the shaft give it a triangular shape in cross section. The radius articulates with four bones — the humerus, ulna, scaphoid, and lunate.

1 Head
2 Articular facet
3 Articular circumference
4 Neck
5 Shaft or body
6 Radial tuberosity
7 Pronator tuberosity
8 Interosseous border
9 Anterior border
10 Posterior border
11 Radial styloid process
12 Suprastyloid crest
13 Dorsal tubercle
14 Groove for extensor muscle tendons
15 Ulnar notch
16 Carpal articular surface

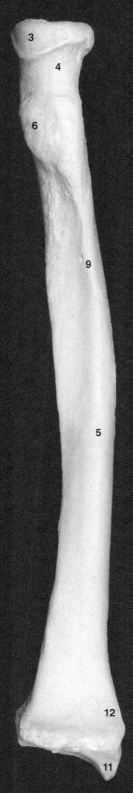

Left radius
Anterior view, lateral to right

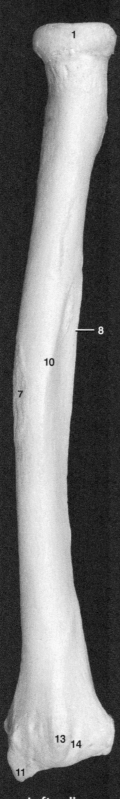

Left radius
Posterior view, lateral to left

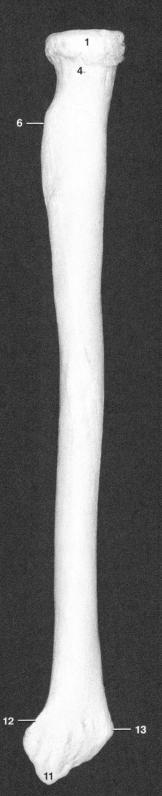

Left radius
Lateral view, anterior to left

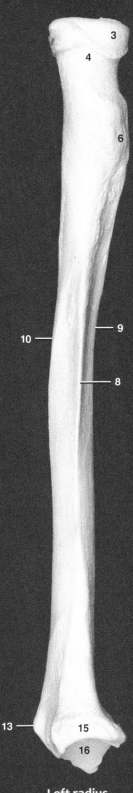

Left radius
Medial view, anterior to right

Left radius
Superior view, lateral to left

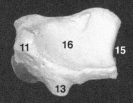

Left radius
Inferior view, lateral to right

Hand Skeleton

The hand is a composite structure consisting of 27 bones. The proximal end of the hand is the carpus or wrist. The carpal bones are eight in number and are arranged in two rows of four, a distal row and a proximal row. Distal to the carpus are the five digital rays. Each digit, called a finger of which there are four, consists of a metatarsal bone and three phalanges. The remaining digit, the thumb or pollex, has a metatarsal bone and only two phalanges. The photos of the hands below and on the opposing page are positioned as if you were looking at your own hand.

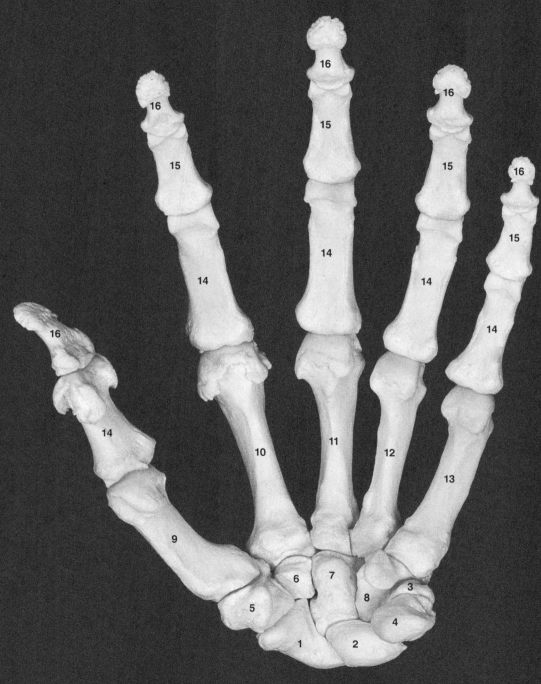

Left hand
Anterior view, lateral to left

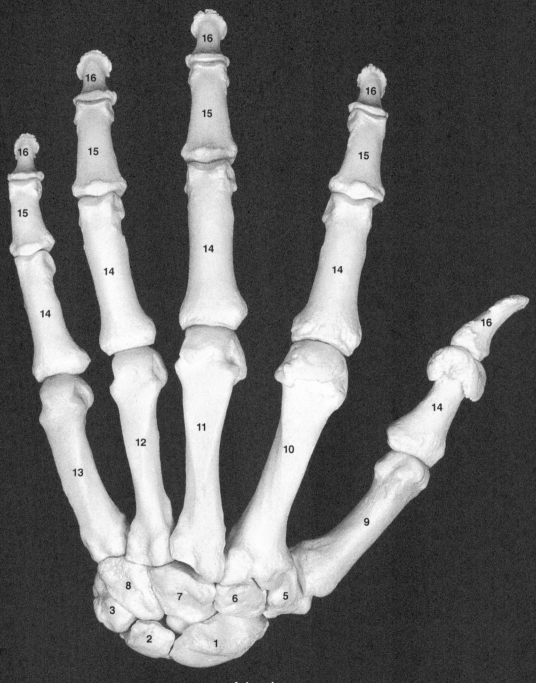

1 Scaphoid	9 Metacarpal I
2 Lunate	10 Metacarpal II
3 Triquetrum	11 Metacarpal III
4 Pisiform	12 Metacarpal IV
5 Trapezium	13 Metacarpal V
6 Trapezoid	14 Proximal phalanx
7 Capitate	15 Middle phalanx
8 Hamate	16 Distal phalanx

Left hand
Posterior view, lateral to right

Carpal Bones

The eight carpal bones form the proximal end of the hand skeleton. The main features of this complex little series of bones are the numerous articular surfaces they form with one another and with the metacarpal and antebrachial bones. The carpal bones form two rows of four bones each. The two largest bones of the proximal row, the scaphoid and the lunate, articulate with the distal end of the radius. The row of distal bones form the skeletal foundation for the fingers and articulate with the metacarpal bones of the fingers and thumb. The anterior surface of the carpal bones forms the floor of the carpal tunnel that supports the major digital flexor tendons that enter the hand.

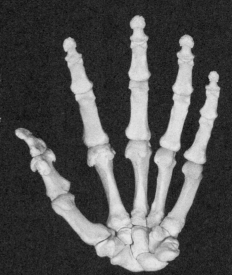

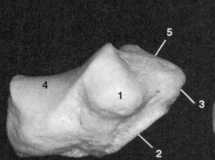

Left trapezium
Anterior view, lateral to left

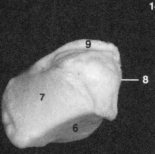

Left trapezoid
Anterior view, lateral to left

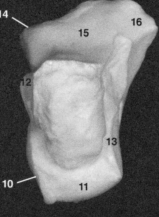

Left capitate
Anterior view, lateral to left

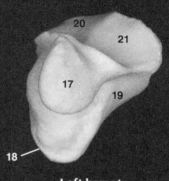

Left hamate
Anterior view, lateral to left

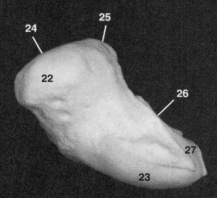

Left scaphoid
Anterior view, lateral to left

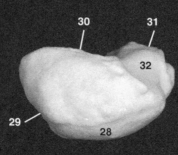

Left lunate
Anterior view, lateral to left

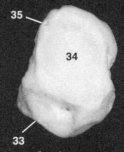

Left triquetrum
Anterior view, lateral to left

Left pisiform
Anterior view, lateral to left

Trapezium
1 Tubercle of trapezium
2 Articular surface with scaphoid
3 Articular surface with trapezoid
4 Articular surface with first metacarpal
5 Articular surface with second metacarpal

Trapezoid
6 Articular surface with scaphoid
7 Articular surface with trapezium
8 Articular surface with capitate
9 Articular surface with first metacarpal

Capitate
10 Articular surface with scaphoid
11 Articular surface with lunate
12 Articular surface with trapezoid
13 Articular surface with hamate
14 Articular surface with second metacarpal
15 Articular surface with third metacarpal
16 Articular surface with fourth metacarpal

Hamate
17 Hook of hamate or hamulus
18 Articular surface with lunate
19 Articular surface with triquetrum
20 Articular surface with fourth metacarpal
21 Articular surface with fifth metacarpal

Scaphoid
22 Scaphoid tubercle
23 Articular surface with radius
24 Articular surface with trapezium
25 Articular surface with trapezoid
26 Articular surface with capitate
27 Articular surface with lunate

Lunate
28 Articular surface with radius
29 Articular surface with scaphoid
30 Articular surface with capitate
31 Articular surface with hamate
32 Articular surface with triquetrum

Triquetrum
33 Articular surface with lunate
34 Articular surface with pisiform
35 Articular surface with hamate

Pisiform
36 Articular surface with triquetrum

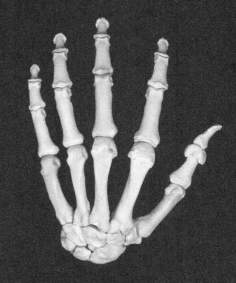

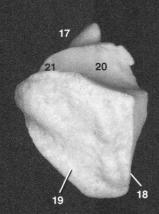

Left hamate
Posterior view, lateral to right

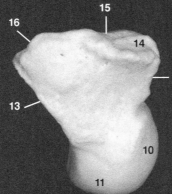

Left capitate
Posterior view, lateral to right

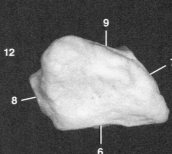

Left trapezoid
Posterior view, lateral to right

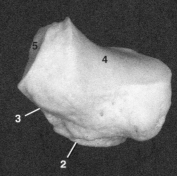

Left trapezium
Posterior view, lateral to right

Left pisiform
Posterior view, lateral to right

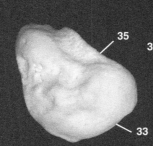

Left triquetrum
Posterior view, lateral to right

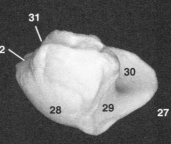

Left lunate
Posterior view, lateral to right

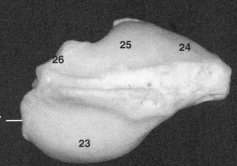

Left scaphoid
Posterior view, lateral to right

Metacarpals and Phalanges

The five digital rays of the hand consist of a series of four bones, except in the thumb where there are only three bones, that decrease in length from proximal to distal. Forming the skeleton of the palmar region of the hand are the stout metacarpal bones. Note their saddle-like bases and rounded heads. The anterior-posterior flattened phalanges project into the proper portion of the fingers and thumb from the metacarpal bones.

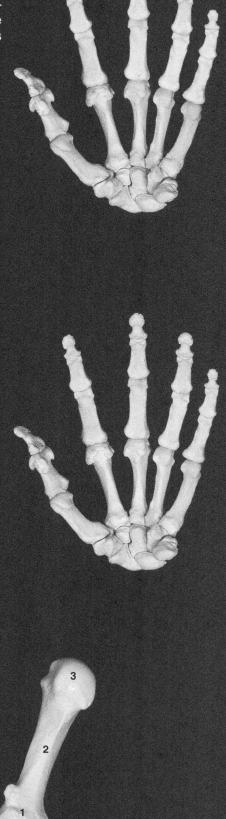

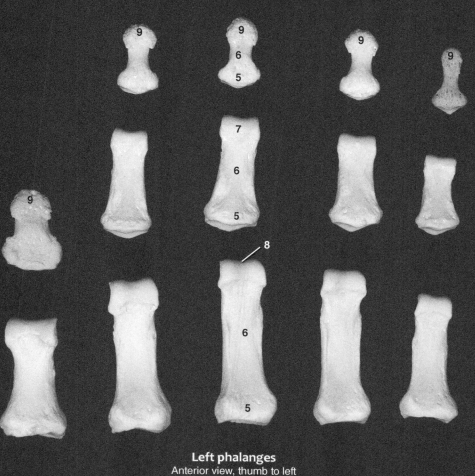

Left phalanges
Anterior view, thumb to left

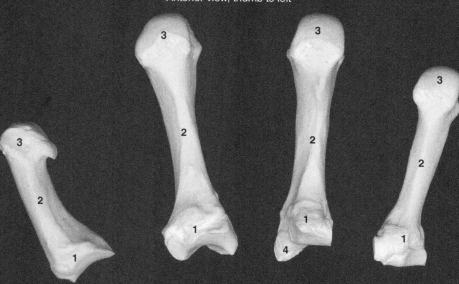

Left metacarpal bones, numbered I to V from lateral to medial
Anterior view, thumb to left

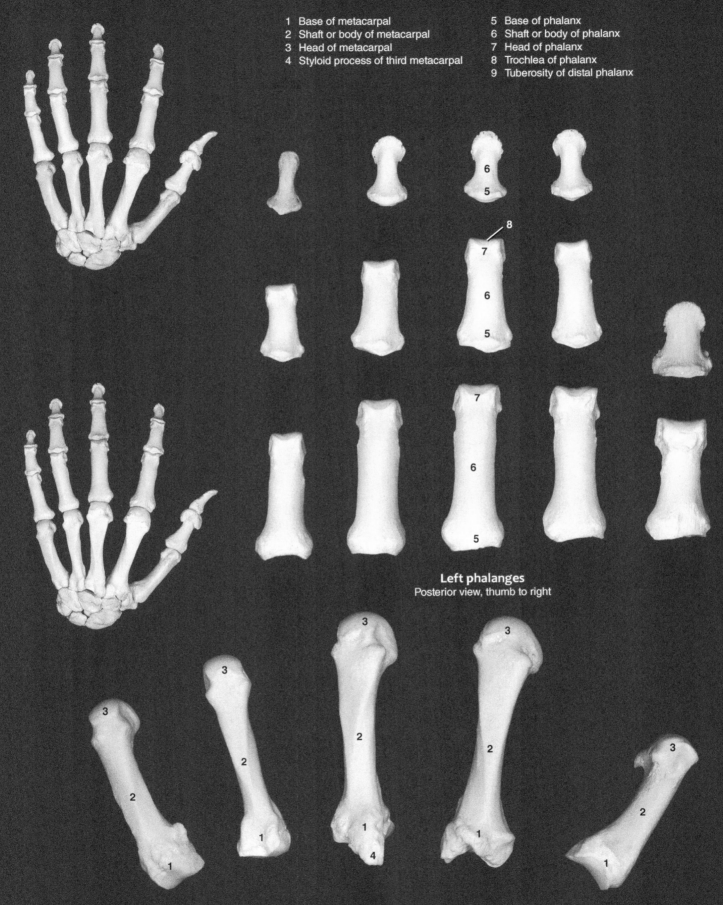

1 Base of metacarpal
2 Shaft or body of metacarpal
3 Head of metacarpal
4 Styloid process of third metacarpal

5 Base of phalanx
6 Shaft or body of phalanx
7 Head of phalanx
8 Trochlea of phalanx
9 Tuberosity of distal phalanx

Left phalanges
Posterior view, thumb to right

Left metacarpal bones, numbered I to V from lateral to medial
Posterior view, thumb to right

Pelvis - Female

The characteristic features of the female pelvis are related to the role of the female pelvis in childbirth. While there are numerous diagnostic features that help distinguish a female pelvis, some of the most obvious are those that increase the diameter of the pelvic outlet. For example, note the wider pubic angle (1) and greater sciatic notch (2) of the female pelvis.

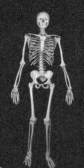

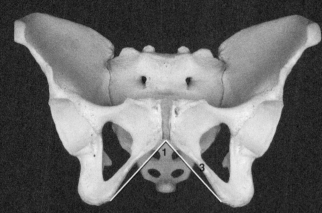

Female pelvis
Anterior view, superior to top

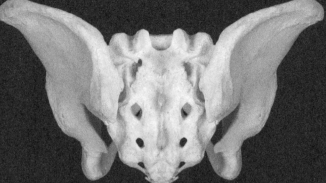

Female pelvis
Posterior view, superior to top

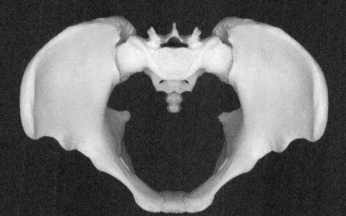

Female pelvis
Superior view, anterior to bottom

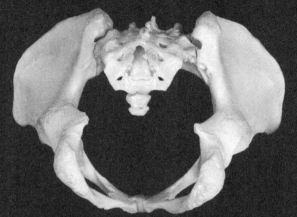

Female pelvis
Inferior view, anterior to bottom

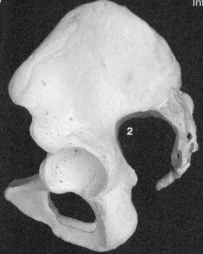

Female pelvis
Lateral view, anterior to left

Pelvis - Male

The male pelvis tends to have a more narrow profile than the pelvis of the female. Compare the diameter of the outlet, the angle of the pubic arch, and the width of the greater sciatic notch with those of the female pelvis. Also, note the stout, thick ischiopubic ramus (3) of the male compared to the slender ischiopubic ramus of the female pelvis.

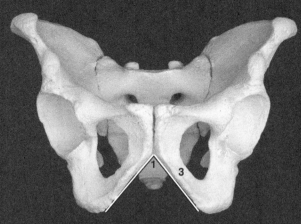

Male pelvis
Anterior view, superior to top

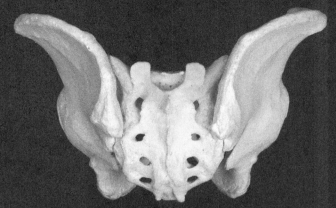

Male pelvis
Posterior view, superior to top

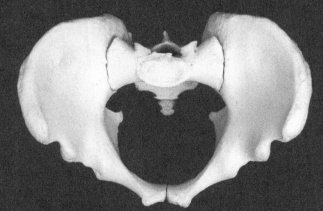

Male pelvis
Superior view, anterior to bottom

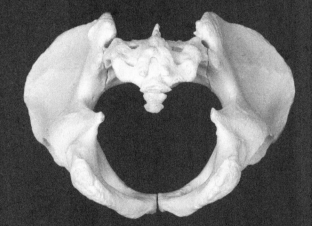

Male pelvis
Inferior view, anterior to bottom

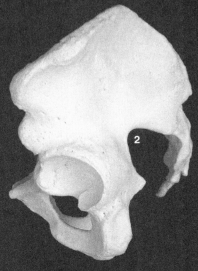

Male pelvis
Lateral view, anterior to left

Inferior Limb

Each inferior appendage consists of 31 bones. The broad base of the inferior limb is the pelvic girdle. This girdle is the strong fusion of three bones, the ilium, ischium, and pubis, to form the os coxae or hip bone. The os coxae is firmly anchored to the sacrum via strong ligaments and a synovial joint. Distal to the girdle is the free part of the lower limb. The bony framework of the thigh is the femur with the sesamoid patella at its distal end. Distal to the femur, the tibia and fibula form the skeleton of the crus or leg. The distal-most region of the inferior limb is the foot consisting of seven tarsal bones, five metatarsal bones, and fourteen phalanges.

1 Os coxae or hip bone
2 Femur
3 Patella
4 Tibia
5 Fibula
6 Tarsal bones
7 Metatarsal bones
8 Phalanges

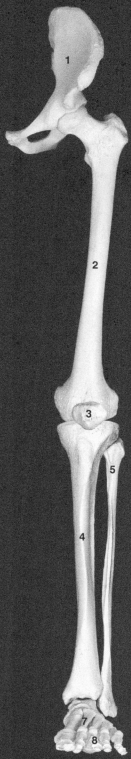

Left lower limb
Anterior view, lateral to right

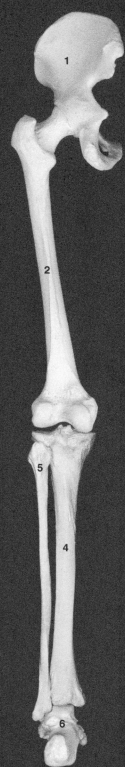

Left lower limb
Posterior view, lateral to left

Os Coxae

Each os coxae forms from three separate bony elements that fuse during development at their site of union within the acetabulum. The three bony elements are the ilium, ischium, and pubis. This strong girdle of bone unites the inferior limb to the axial skeleton and transfers the forces of locomotion from the inferior limb to the vertebral column. Each os coxae articulates with three bones — the femur, sacrum, and opposite os coxae. The photo on this page depicts the three bones of the os coxae — the ilium (green), the ischium (blue), and the pubis (red). Landmarks that are shared by the bones are depicted on this image. The following two pages show all the landmarks of the individual bones of the os coxae.

1 Acetabulum
2 Acetabular notch
3 Lunate surface
4 Ischiopubic ramus
5 Obturator foramen
6 Greater sciatic notch

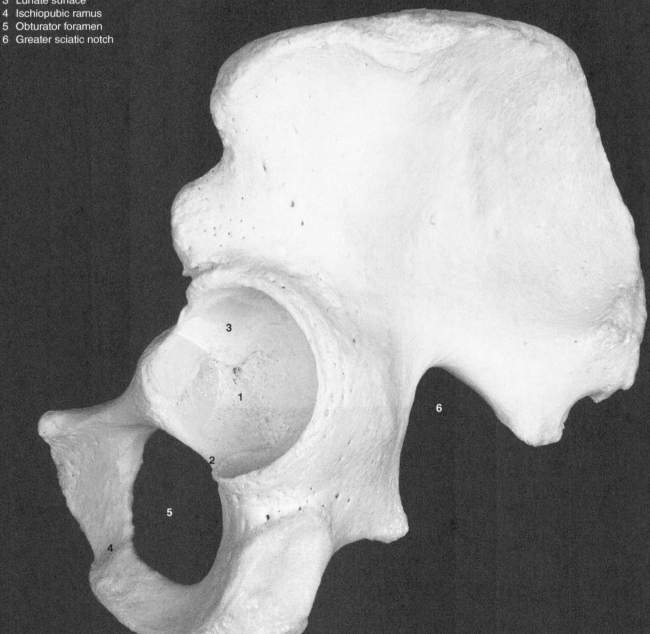

Left os coxae showing individual bones
Lateral view, anterior to left

Os Coxae

Ilium
1. Body of ilium
2. Supra-acetabular groove
3. Ala or wing
4. Arcuate line
5. Iliac crest
6. Outer lip of crest
7. Intermediate zone of crest
8. Inner lip of crest
9. Tuberculum of crest
10. Anterior superior iliac spine
11. Anterior inferior iliac spine
12. Posterior superior iliac spine
13. Posterior inferior iliac spine
14. Iliac fossa
15. Anterior gluteal line
16. Posterior gluteal line
17. Inferior gluteal line
18. Auricular surface
19. Iliac tuberosity

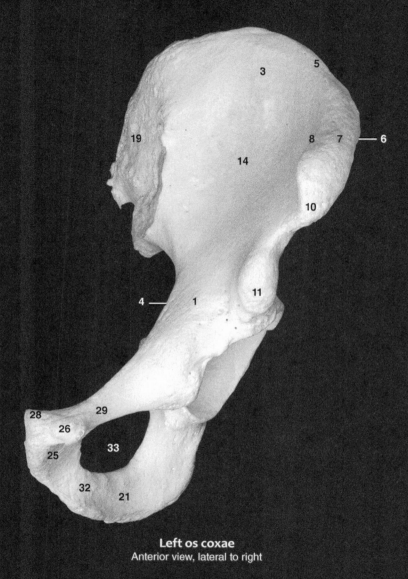

Left os coxae
Anterior view, lateral to right

Left os coxae
Posterior view, lateral to right

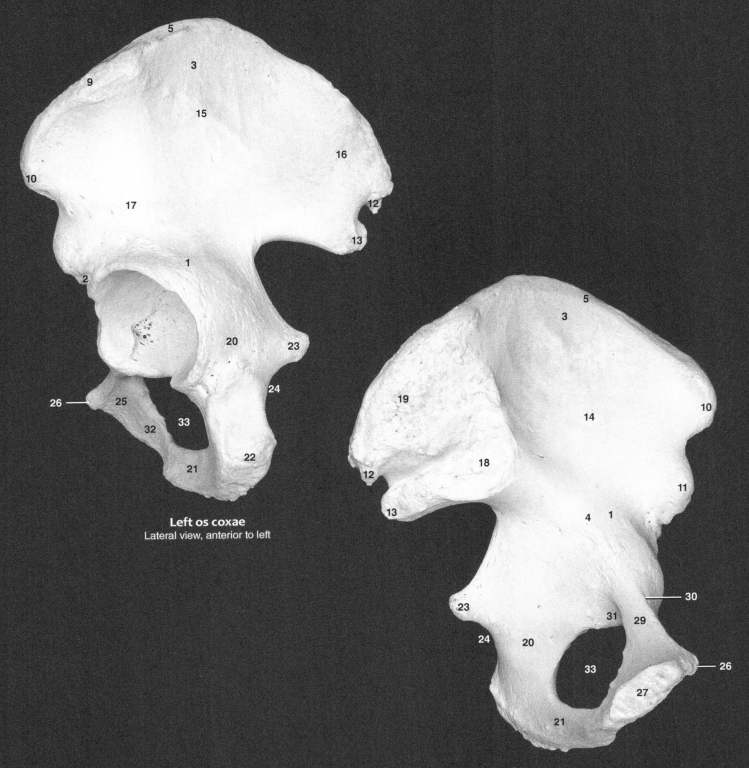

Ischium
20 Body of ischium
21 Ischial ramus
22 Ischial tuberosity
23 Ischial spine
24 Lesser sciatic notch

Pubis
25 Body of pubis
26 Pubic tubercle
27 Symphysial surface
28 Pubic crest
29 Superior pubic ramus

30 Pecten pubis or pectineal line
31 Obturator groove
32 Inferior pubic ramus
33 Obturator foramen

Left os coxae
Lateral view, anterior to left

Left os coxae
Medial view, anterior to right

107

Femur

The femur is the longest bone of the body. The strong shaft forms a long cylindrical tube with a slight forward bow. The strong wall of the shaft is thickest near the narrow center of the bone where the medullary cavity is also the most spacious. As the shaft becomes progressively wider toward each end, the compact wall of bone becomes thinner and the medullary cavity accumulates spongy bone. The proximal end consists of a short cantilevered neck capped by a smooth, round articular head. Projections of bone, the trochanters, form at the base of the cantilevered neck. The distal end consists of two large, knuckle-like processes separated by an intermediate groove. The femur articulates with three bones: the os coxae, patella, and tibia.

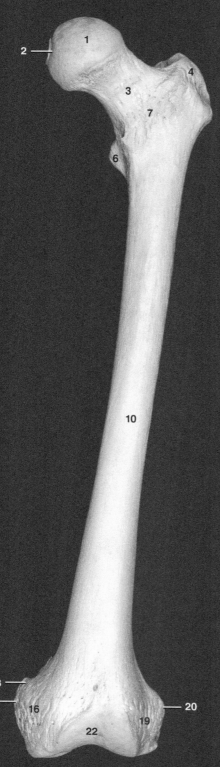

Left femur
Anterior view, lateral to rigjt

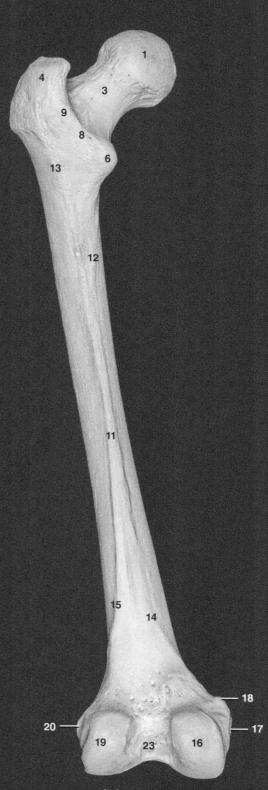

Left femur
Posterior view, lateral to left

1 Head
2 Fovea for ligament of head
3 Neck
4 Greater trochanter
5 Trochanteric fossa
6 Lesser trochanter
7 Intertrochanteric line
8 Intertrochanteric crest
9 Quadrate tubercle
10 Shaft or body
11 Linea apsera
12 Pectineal or spiral line
13 Gluteal tuberosity
14 Medial supracondylar line
15 Lateral supracondylar line
16 Medial condyle
17 Medial epicondyle
18 Adductor tubercle
19 Lateral condyle
20 Lateral epicondyle
21 Groove for popliteus
22 Patellar surface
23 Intercondylar fossa

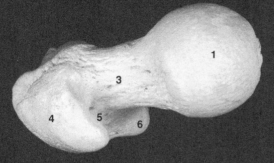

Left femur
Superior view, lateral to left

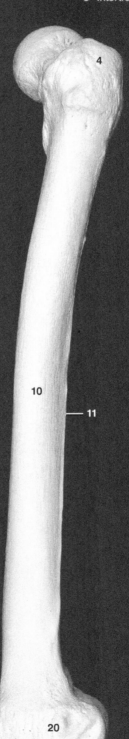

Left femur
Lateral view, anterior to left

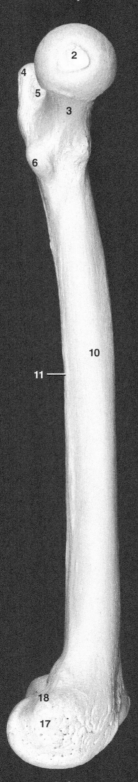

Left femur
Medial view, anterior to right

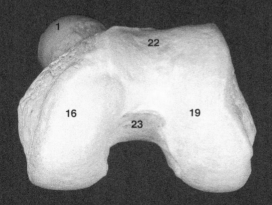

Left femur
Inferior view, lateral to right

109

Tibia

The tibia is the large, medial bone of the leg skeleton. It is the second longest bone of the body, only exceeded in length by the femur. Its strong shaft, consisting of thick walls of compact bone, is triangular in cross-section. The shaft expands proximally into a fluted extremity of spongy bone with a flat plateau-like superior surface largely covered with articular cartilage. The smaller distal end is more knob-like with a pronounced medial projection, the malleolus. The shaft has a strong anterior crest with sloping surfaces to either side. The bone is easily palpable throughout its length. The tibia articulates with three bones — the femur, fibula, and talus.

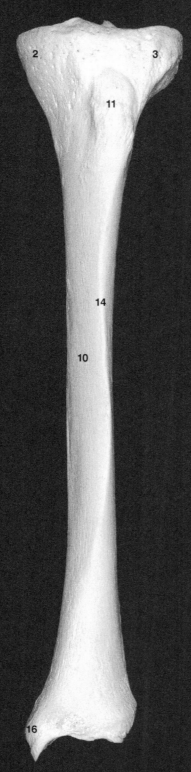

Left tibia
Anterior view, lateral to right

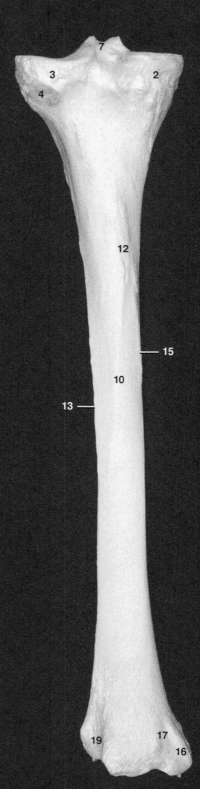

Left tibia
Posterior view, lateral to left

1 Superior articular surface
2 Medial condyle
3 Lateral condyle
4 Fibular articular facet
5 Anterior intercondylar area
6 Posterior intercondylar area
7 Intercondylar eminence
8 Medial intercondylar tubercle
9 Lateral intercondylar tubercle
10 Shaft or body
11 Tibial tuberosity
12 Soleal line
13 Interosseous border
14 Anterior border
15 Posterior border
16 Medial malleolus
17 Malleolar groove
18 Malleolar articular facet
19 Fibular notch
20 Inferior articular surface

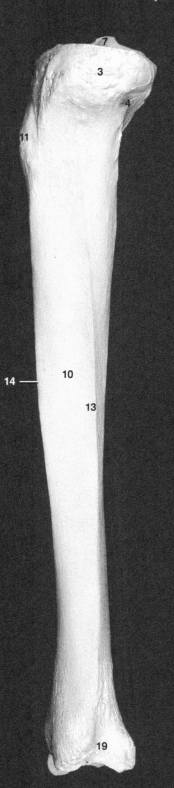

Left tibia
Lateral view, anterior to left

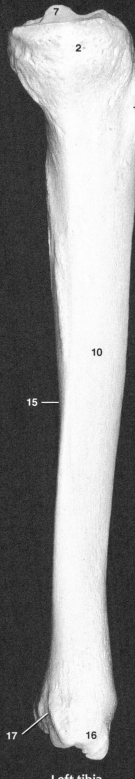

Left tibia
Medial view, anterior to right

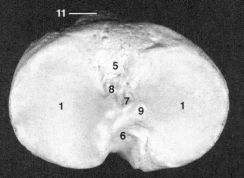

Left tibia
Superior view, lateral to left

Left tibia
Close-up of lateral view

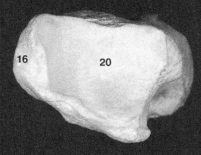

Left tibia
Inferior view, lateral to right

Fibula

The fibula is the lateral bone of the leg skeleton. It is a slender, splint-like bone that is slightly expanded at both ends. It plays no role in the weight-bearing function of the lower limb, but serves as a significant site of muscle attachment. It is not easily palpable except at its proximal and distal ends, the shaft being totally surrounded with muscle. The fibula articulates with two bones — the tibia and talus.

1 Head
2 Articular facet for tibia
3 Apex of head
4 Neck
5 Shaft or body
6 Interosseous border
7 Anterior border
8 Posterior border
9 Lateral malleolus
10 Articular facet for talus
11 Malleolar fossa
12 Malleolar groove

Left fibula
Anterior view, lateral to right

Left fibula
Posterior view, lateral to left

Left fibula
Lateral view, anterior to left

Left fibula
Medial view, anterior to right

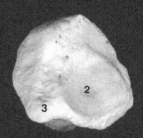

Left fibula
Superior view, lateral to left

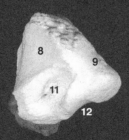

Left fibula
Inferior view, lateral to right

Foot Skeleton

Like the hand, the foot is a composite structure comprised of 26 bones, not counting the small sesamoid bones that are found in certain tendons. The proximal end of the foot is the tarsus or ankle. There are seven tarsal bones that show a greater range in size and shape than their carpal counterparts in the hand. Distal to the tarsals are the five digital rays. The four lateral digits consist of a metatarsal bone and three phalanges. The large medial digit, the hallux or great toe, has a metatarsal bone and only two phalanges. Two prominent sesamoid bones (bones that form in tendons) are present on the plantar surface at the head end of the first metatarsal.

1 Talus	7 Cuboid	12 Metatarsal V
2 Calcaneus	8 Metatarsal I	13 Proximal phalanx
3 Navicular	9 Metatarsal II	14 Middle phalanx
4 Medial cuneiform	10 Metatarsal III	15 Distal phalanx
5 Intermediate cuneiform	11 Metatarsal IV	16 Sesamoid bones
6 Lateral cuneiform		

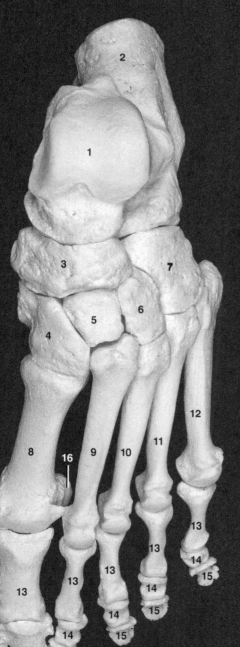

Left foot
Dorsal view, lateral to right

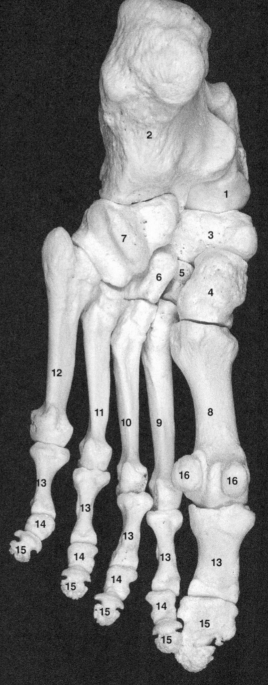

Left foot
Plantar view, lateral to left

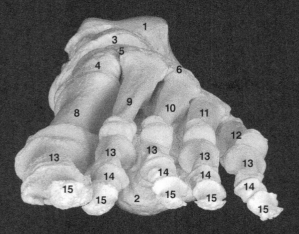

Left foot
Anterior view, lateral to right

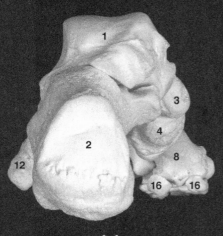

Left foot
Posterior view, lateral to left

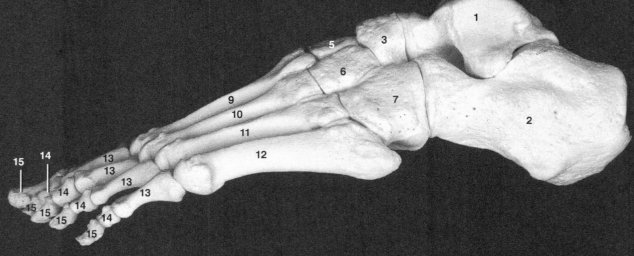

Left foot
Lateral view, anterior to left

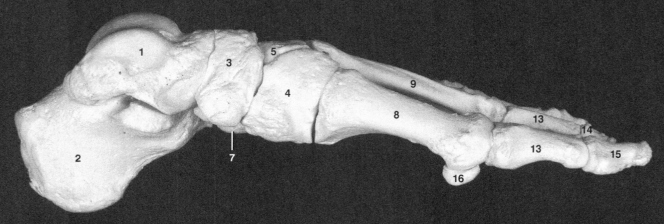

Left foot
Medial view, anterior to right

Tarsal Bones - Talus

The next four pages depict the tarsal bones. Like the carpals, this is a complex series of bones that form numerous articulations with one another. All the tarsal bones were photographed at the same scale so you can see their relative sizes. The talus is the second largest and most proximal of the tarsal bones. It forms the ankle joint with the distal end of the leg skeleton. It consists of a cuboid body, a distally directed neck capped by a convex, oval head, a proximolateral facet for the fibular malleolus, and a proximal trochlea for the tibia. It articulates with four bones — the tibia, fibula, calcaneus, and navicular.

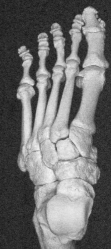

1 Head
2 Navicular articular surface
3 Anterior facet for calcaneus
4 Neck
5 Middle facet for calcaneus
6 Sulcus tali
7 Body
8 Trochlea of talus
9 Lateral malleolar facet
10 Lateral process
11 Medial malleolar facet
12 Posterior process
13 Groove for flexor hallucis longus
14 Lateral tubercle
15 Medial tubercle
16 Posterior calcaneal articular facet

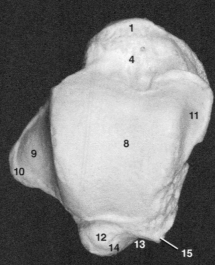

Left talus
Superior view, lateral to left

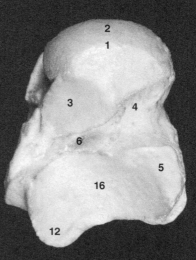

Left talus
Inferior view, lateral to right

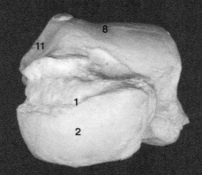

Left talus
Anterior view, lateral to left

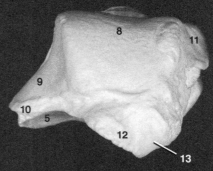

Left talus
Posterior view, lateral to right

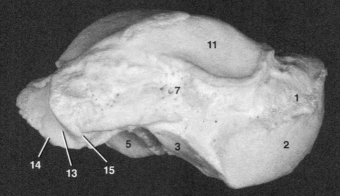

Left talus
Medial view, anterior to right

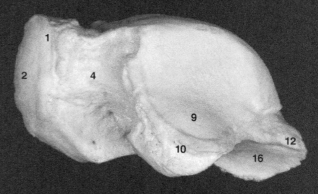

Left talus
Lateral view, anterior to left

Tarsal Bones - Calcaneus

The calcaneus is the largest bone of the foot and its long axis parallels the long axis of the foot. Its distal end forms a series of articular surfaces with neighboring bones. Its posterior or proximal end is box-like and forms a roughened calcaneal tubercle at the posterior surface. The calcaneus articulates with two bones — the talus and the cuboid.

1 Calcaneal tuberosity
2 Calcaneal tubercle
3 Sustentaculum tali
4 Groove for flexor hallucis longus
5 Calcaneal sulcus
6 Tarsal sinus
7 Anterior talar articular surface
8 Middle talar articular surface
9 Posterior talar articular surface
10 Groove for fibularis longus
11 Fibular trochlea
12 Articular surface for cuboid

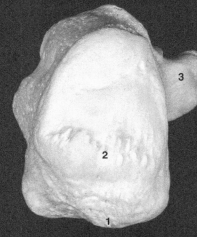

Left calcaneus
Posterior view, lateral to left

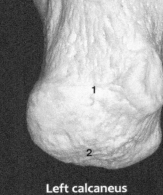

Left calcaneus
Inferior view, lateral to right

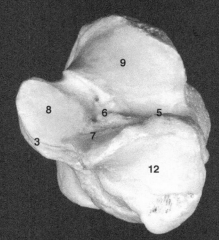

Left calcaneus
Anterior view, lateral to right

Left calcaneus
Superior view, lateral to left

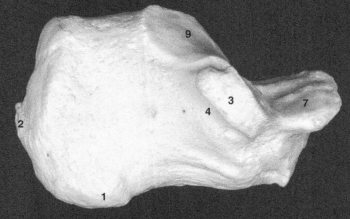

Left calcaneus
Medial view, anterior to right

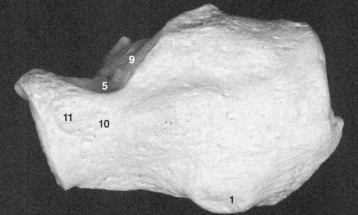

Left calcaneus
Lateral view, anterior to left

Tarsal Bones - Cuboid and Navicular

The cuboid bone, like its name suggests, has a cube shape when viewed from above, but has ridges and grooves on its plantar surface. It is the lateral bone in the distal series of tarsal bones and articulates with the fourth and fifth metatarsals. With a good imagination one can visualize the hull of a ship when observing the navicular bone. This ship-shaped bone is an intermediate bone between the talus and the three cuneiforms on the medial aspect of the foot.

Cuboid
1 Groove for fibularis longus
2 Cuboid tuberosity
3 Calcaneal process
4 Articular surface for calcaneus
5 Articular surface for navicular
6 Articular surface for lateral cuneiform
7 Articular surface for fourth metatarsal
8 Articular surface for fifth metatarsal

Navicular
9 Tuberosity
10 Articular surface for talus
11 Articular surface for cuboid
12 Articular surface for medial cuneiform
13 Articular surface for intermediate cuneiform
14 Articular surface for lateral cuneiform

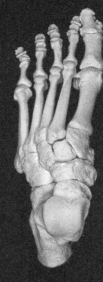

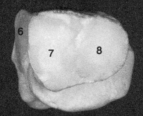

Left cuboid
Superior view, lateral to left

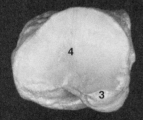

Left cuboid
Inferior view, lateral to right

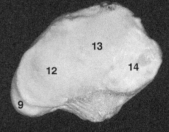

Left navicular
Superior view, lateral to left

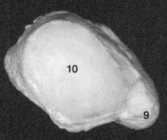

Left navicular
Inferior view, lateral to right

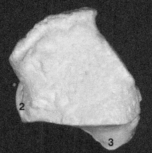

Left cuboid
Anterior view, lateral to right

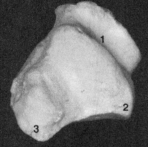

Left cuboid
Posterior view, lateral to left

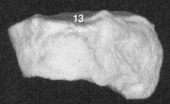

Left navicular
Anterior view, lateral to right

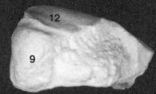

Left navicular
Posterior view, lateral to left

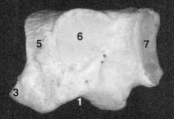

Left cuboid
Medial view, anterior to right

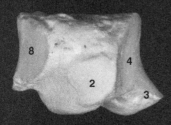

Left cuboid
Lateral view, anterior to left

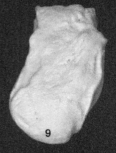

Left navicular
Medial view, anterior to right

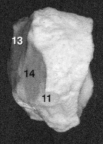

Left navicular
Lateral view, anterior to left

Tarsal Bones - Cuneiforms

The wedge-shaped cuneiforms are the distal tarsal bones on the medial aspect of the ankle. They articulate with the three medial metatarsal bones. Their wedge shapes contribute to the formation of the transverse arch of the foot.

Lateral cuneiform
1 Articular surface for cuboid
2 Articular surface for navicular
3 Articular surface for middle cuneiform
4 Articular surface for second metatarsal
5 Articular surface for third metatarsal
6 Articular surface for fourth metatarsal

Middle cuneiform
7 Articular surface for navicular
8 Articular surface for medial cuneiform
9 Articular surface for lateral cuneiform
10 Articular surface for second metatarsal

Medial cuneiform
11 Articular surface for navicular
12 Articular surface for middle cuneiform
13 Articular surface for second metatarsal
14 Articular surface for first metatarsal

Left lateral cuneiform
Superior view, lateral to left

Left middle cuneiform
Superior view, lateral to left

Left lateral cuneiform
Inferior view, lateral to right

Left middle cuneiform
Inferior view, lateral to right

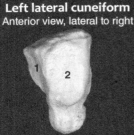

Left lateral cuneiform
Anterior view, lateral to right

Left middle cuneiform
Anterior view, lateral to right

Left medial cuneiform
Superior view, lateral to left

Left medial cuneiform
Inferior view, lateral to right

Left lateral cuneiform
Posterior view, lateral to left

Left middle cuneiform
Posterior view, lateral to left

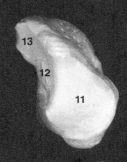

Left medial cuneiform
Anterior view, lateral to right

Left medial cuneiform
Posterior view, lateral to left

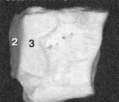

Left lateral cuneiform
Medial view, anterior to right

Left middle cuneiform
Medial view, anterior to right

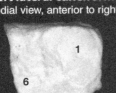

Left lateral cuneiform
Lateral view, anterior to left

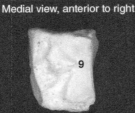

Left middle cuneiform
Lateral view, anterior to left

Left medial cuneiform
Medial view, anterior to right

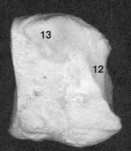

Left medial cuneiform
Lateral view, anterior to left

Metatarsal Bones

The five metatarsal bones form the central portion of the foot skeleton. The three central metatarsals most closely resemble one another, while the first and fifth metatarsal bones are the most distinct. The first metatarsal is short and thick compared to its counterparts, while the distinguishing feature of the fifth metatarsal bone is the projecting tuberosity at its proximal end.

1 Base
2 Shaft or body
3 Head
4 Tuberosity of first metatarsal
5 Tuberosity of fifth metatarsal

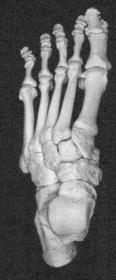

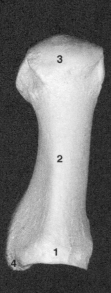

Left metatarsal bones, numbered I to V from medial to lateral
Dorsal view, lateral to left

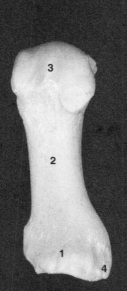

Left metatarsal bones, numbered I to V from medial to lateral
Plantar view, lateral to right

Phalanges

Similar in number to the phalanges of the hand, the phalanges of the foot are much smaller than those of the hand, with the exception of the large first toe. The proximal phalanges have broad bases that form the widest part of the bone. From the base a narrow shaft projects to a rounded head with a trochlear articular surface. The middle and distal phalanges are short bones that can be easily distinguished by their distal ends. The middle phalanges have a trochlear articular surface on their distal head, while the distal phalanges have a broad tuberosity at their distal ends.

1 Base
2 Shaft or body
3 Head
4 Trochlea
5 Tuberosity of distal phalanx

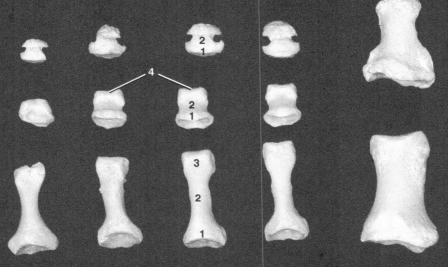

Left phalanges
Dorsal view, lateral to left

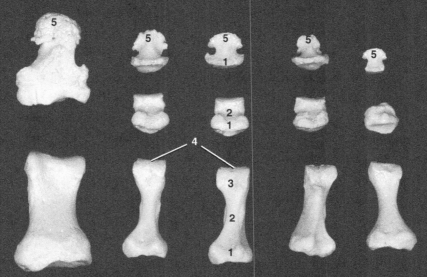

Left phalanges
Plantar view, lateral to right

Patella

The patella is the largest sesamoid bone of the body. A sesamoid bone is a bone that forms within a tendon. The patella occupies the posterior half of the quadriceps tendon just anterior to the knee joint. It is a disc-like bone with a curved superior margin and a triangular inferior border. The posterior surface of the bone is smooth and articulates with the femur, while the anterior surface of the bone is rough by its attachment to the quadriceps tendon.

1 Base
2 Apex
3 Articular surface
4 Anterior surface

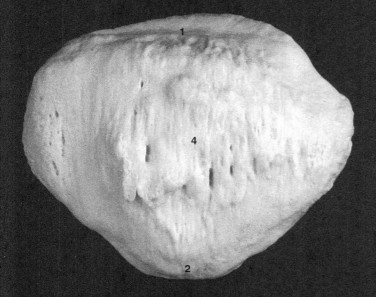

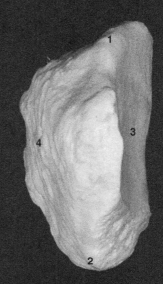

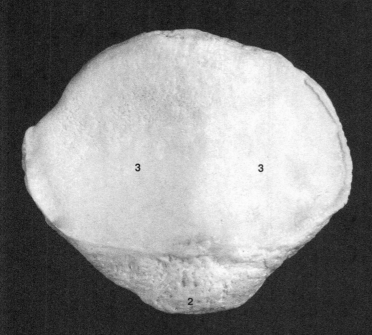

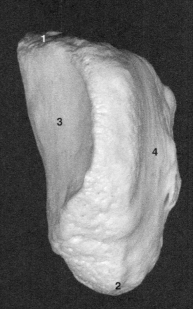

7 | Articular System

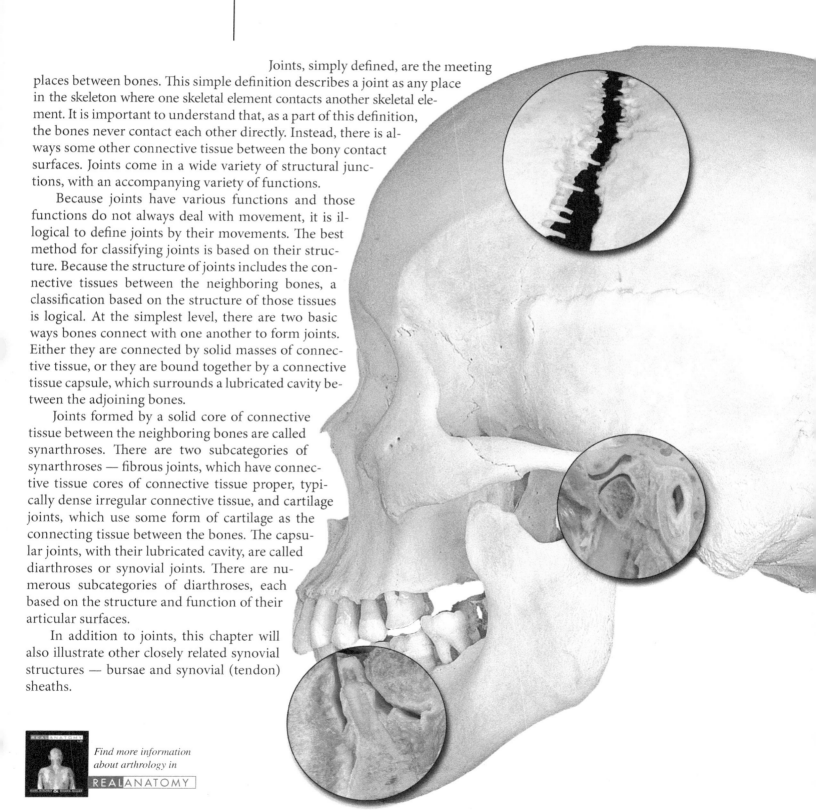

Joints, simply defined, are the meeting places between bones. This simple definition describes a joint as any place in the skeleton where one skeletal element contacts another skeletal element. It is important to understand that, as a part of this definition, the bones never contact each other directly. Instead, there is always some other connective tissue between the bony contact surfaces. Joints come in a wide variety of structural junctions, with an accompanying variety of functions.

Because joints have various functions and those functions do not always deal with movement, it is illogical to define joints by their movements. The best method for classifying joints is based on their structure. Because the structure of joints includes the connective tissues between the neighboring bones, a classification based on the structure of those tissues is logical. At the simplest level, there are two basic ways bones connect with one another to form joints. Either they are connected by solid masses of connective tissue, or they are bound together by a connective tissue capsule, which surrounds a lubricated cavity between the adjoining bones.

Joints formed by a solid core of connective tissue between the neighboring bones are called synarthroses. There are two subcategories of synarthroses — fibrous joints, which have connective tissue cores of connective tissue proper, typically dense irregular connective tissue, and cartilage joints, which use some form of cartilage as the connecting tissue between the bones. The capsular joints, with their lubricated cavity, are called diarthroses or synovial joints. There are numerous subcategories of diarthroses, each based on the structure and function of their articular surfaces.

In addition to joints, this chapter will also illustrate other closely related synovial structures — bursae and synovial (tendon) sheaths.

Find more information about arthrology in

REAL ANATOMY

Synarthrosis - Fibrous Joints

Fibrous joints are synarthrotic joints that bind bone to bone with collagenous connective tissue. The amount of connective tissue binding the neighboring bones can vary considerably. Examples of fibrous joints are depicted on this and the facing page. Gomphoses and sutures (the four different suture types are shown on the opposite page) have a very thin membrane of collagenous connective tissue anchoring neighboring bony structures to one another. On the other hand, the syndesmoses between the tibia and fibula — both the interosseous membrane and the tibiofibular ligaments at the distal end — have considerably more binding connective tissue. There is also an example of another syndesmosis, the interspinous ligament, in the next section.

1 Periodontal membrane
2 Crown of tooth
3 Root of tooth
4 Gingiva
5 Mandible
6 Tibia
7 Fibula
8 Interosseous membrane
9 Anterior tibiofibular ligament of tibiofibular syndesmosis
10 Patellar ligament (cut)

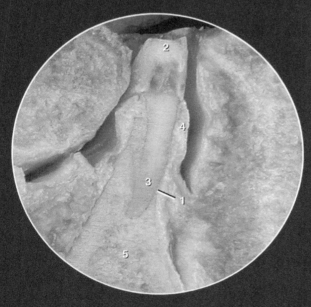

Dento-alveolar syndesmosis or gomphosis
Sagittal section of tooth in mandible

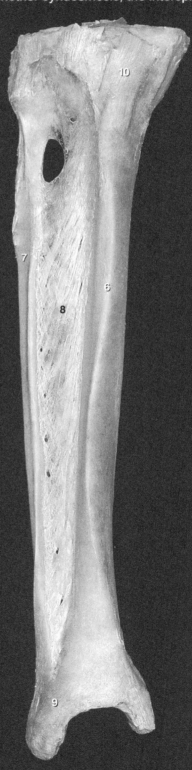

Crural skeleton – tibia and fibula
Anterior view

Squamous-type suture
Squamous or temporoparietal suture

Serrate-type suture
Coronal or frontoparietal suture

Denticulate-type suture
Lamboidal or parieto-occipital suture

Plane-type suture
Internasal suture

125

Synarthrosis - Cartilaginous Joints

Like the fibrous joints, the cartilaginous joints join neighboring skeletal elements with a solid mass of connective tissue, but the uniting tissue is some type of cartilage instead of collagenous connective tissue proper. The three types of cartilaginous joints are: 1) synchondroses, 2) symphyses, and 3) epiphysial cartilages or primary cartilaginous joints. The photos on these facing pages depict the different categories of cartilaginous joints. A few syndesmoses from the fibrous joint category are also evident.

1 Intervertebral disc (symphysis)
2 Nucleus pulposus of intervertebral disc
3 Anulus fibrosus of intervertebral disc
4 Pubic symphysis
5 Manubriosternal synchondrosis
6 Spheno-occipital synchondrosis
7 Epiphysial cartilage or primary cartilaginous joint
8 Sternocostal (synchondrosis)
9 Sternocostal (typically synovial but can be symphysial)
10 Interchondral (synovial)
11 Interchondral (synchondrosis)
12 Costochondral (synchondrosis)
13 Interspinous ligament (vertebral syndesmosis)
14 Nuchal ligament (vertebral syndesmosis)
15 Anterior longitudinal ligament (vertebral syndesmosis)
16 Posterior longitudinal ligament (vertebral syndesmosis)
17 Body of vertebra
18 Spinous process of vertebra
19 Lamina of vertebra
20 Psoas major muscle
21 Aorta
22 Inferior vena cava

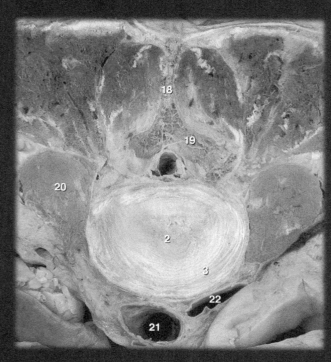

Transverse section of lumbar intervertebral disc
Inferior view

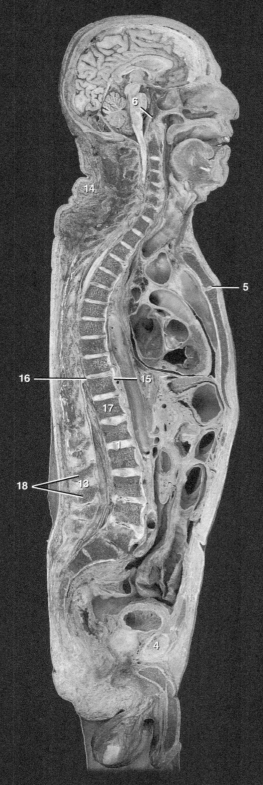

Sagittal section of head and trunk
Medial view

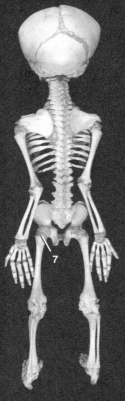

Fetal skeleton
Posterior view

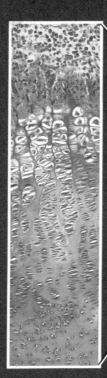

Epiphysial cartilage
200x

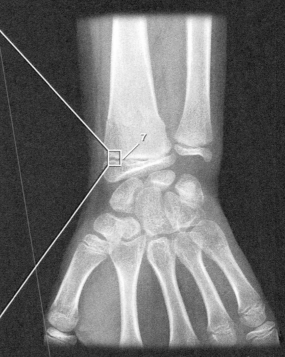

Radiograph of juvenile wrist region
Anterior view

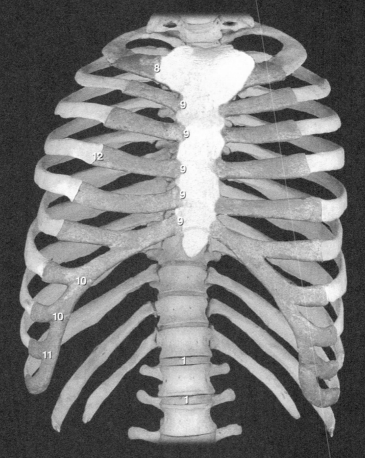

Joints of the thoracic cage
Anterior view

Diarthroses or Synovial Joints

Diarthroses differ from synarthroses in one major way: instead of connecting neighboring bones by a solid mass of connectve tissue, the bony connection consists of a double-layered connective tissue capsule that surrounds a lubricated cavity between the bones. Within the capsule the ends of neighboring bony surfaces are covered by a smooth layer of hyaline cartilage. As a result of this design there is typically a much greater range of motion present in synovial joints, and they form the joints of the skeleton that are responsible for the major movements of the body. The outer layer of the capsule, the fibrous membrane, is continuous with the periosteum on the adjoining bones, while the inner layer of the capsule, the synovial membrane, attaches from the border of the articular cartilage on one bone to the border of the articular cartilage on the other bone. Additionally, the synovial membrane secretes synovial fluid, a lubricant that reduces friction between the mobile cartilage-covered articular surfaces of the bones. The section through a finger joint below and the dissections of the knee joint on the opposite page illustrate the basic features of a synovial joint. The pages that follow depict the major synovial joints of the skeleton. One other key feature among synovial joints that is responsible for their varied range of motion is the shape of the adjoining bone surfaces. It is this feature that anatomists use to describe the different types of synovial joints.

1 Middle phalanx of index finger
2 Proximal phalanx of index finger
3 Fibrous membrane of joint capsule
4 Synovial membrane of joint capsule
5 Articular cartilage
6 Joint cavity
7 Collateral ligament
8 Quadriceps tendon
9 Patellar ligament
10 Suprapatellar bursa
11 Synovial fold
12 Meniscus
13 Periosteum
14 Junction of periosteum (removed) with fibrous membrane
15 Junction of synovial membrane (removed) with articular cartilage
16 Femur with periosteum removed
17 Tibia with periosteum removed
18 Fibula with periosteum removed
19 Patella within quadriceps tendon

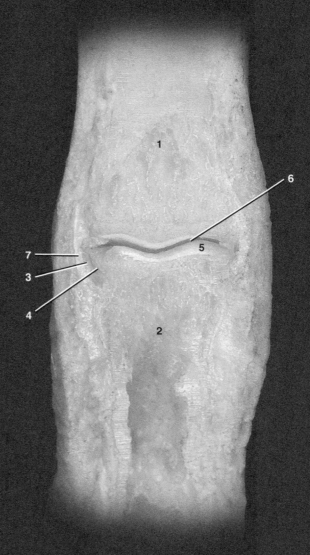

Proximal interphalangeal joint showing design of synovial joint
Frontal section, anterior view

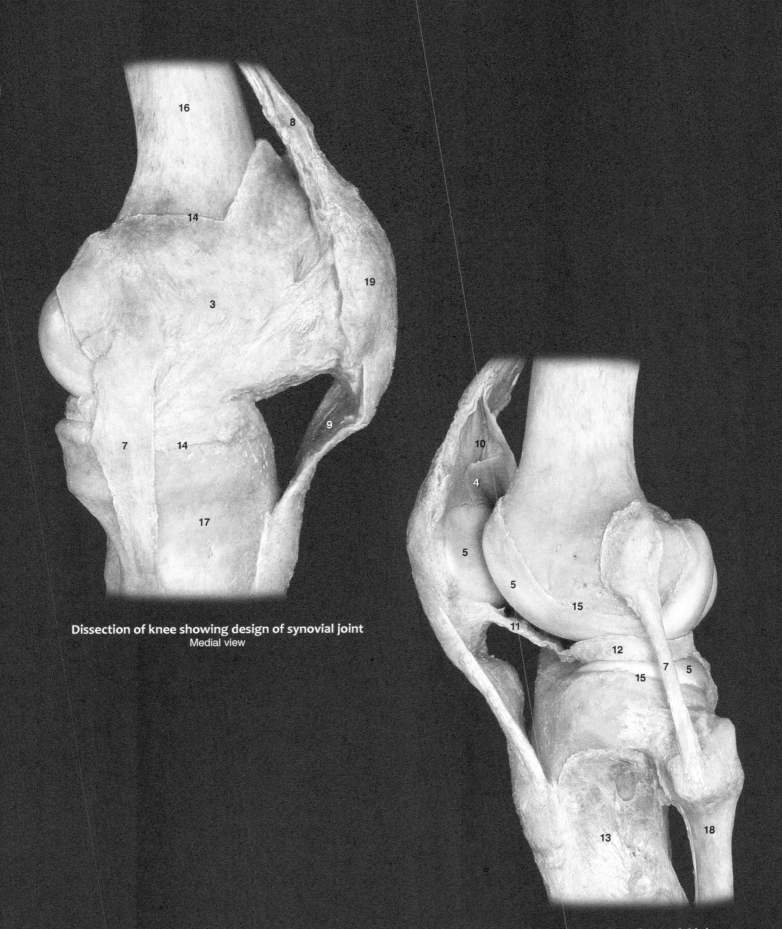

Dissection of knee showing design of synovial joint
Medial view

Dissection of knee showing design of synovial joint
Lateral view

Types of Synovial Joints

There are seven types of synovial joints in the body. Each of the different synovial joints has the basic structural features common to all synovial joints but is further classified based on the shape of and motion that occurs at the articular surfaces of the joint. The different types of synovial joint are depicted below and on the opposite page. Note the shapes of the reciprocal surfaces as you study these photos.

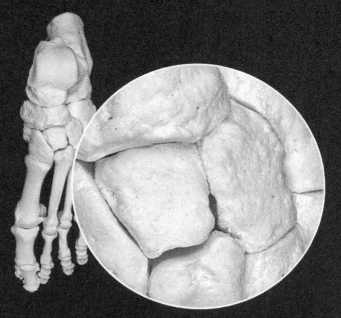

Plane joint examples
Intertarsal joints

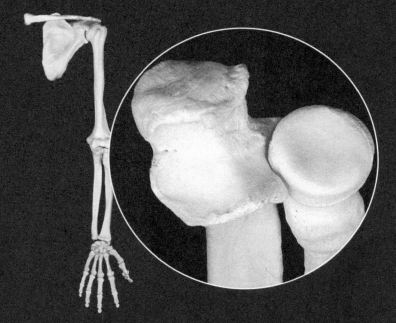

Pivot joint examples
Proximal radio-ulnar joint of elbow

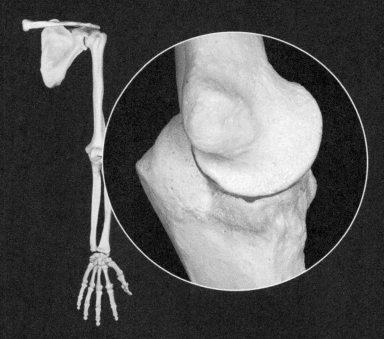

Hinge joint example
Humero-ulnar joint of elbow

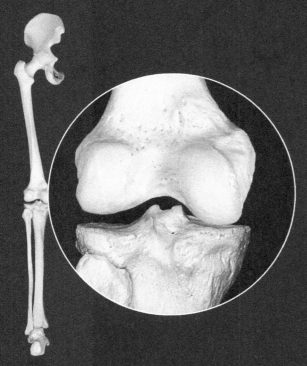

Bicondylar joint example
Knee joint

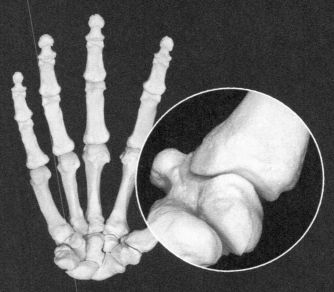

Saddle joint example
Metacarpal-carpal joint of thumb

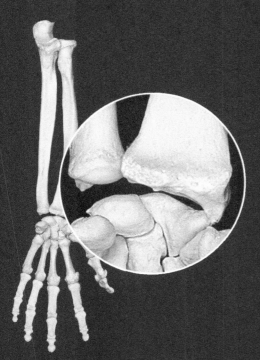

Condylar joint example
Wrist joint

Ball and socket joint example
Shoulder joint

Temporomandibular Joint

The complex temporomandibular joint differs from other synovial joints by having an articular disc that usually separates the joint into two separate synovial capsules, one above and one below the disc. The articular surfaces have a covering of dense fibrocartilage rather than the typical hyaline cartilage of most synovial joints. With its associated ligaments this joint structure accounts for the complex series of movements that are essential during the activities of eating and speech. Each temporomandibular joint is a condylar joint and both joints together form a bicondylar joint. The fibrous membrane of the articular capsule spans from temporal bone to mandible only on the lateral side. Anteriorly, medially, and posteriorly the fibers attach from mandible and temporal bone to the articular disc. Extrinsic ligaments that help stabilize the joint are the lateral temporomandibular ligament, sphenomandibular ligament, and stylomandibular ligament.

1 Mandibular condyle
2 Mandibular ramus
3 Articular tubercle of temporal bone
4 Mastoid process of temporal bone
5 Mastoid air cells
6 Superior compartment of articular cavity
7 Inferior compartment of articular cavity

8   Articular disc
9   Joint (articular) capsule
10 Masseter muscle
11 Parotid gland
12 Brain
13 External acoustic meatus
14 Sigmoid venous sinus

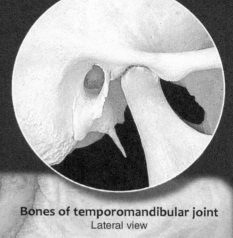

Bones of temporomandibular joint
Lateral view

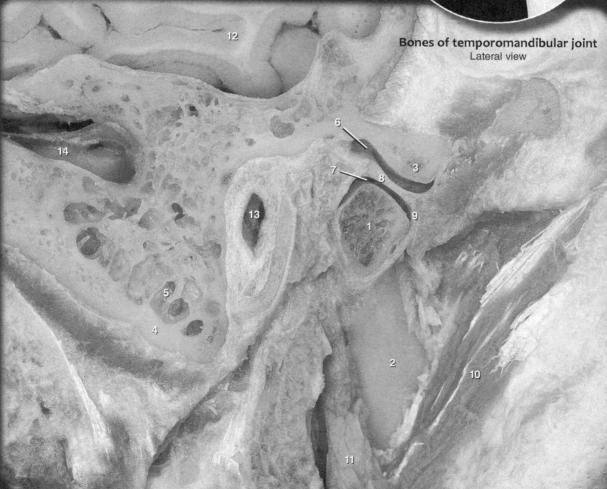

Section of right temporomandibular joint
Lateral view of sagittal section

Glenohumeral Joint

The glenohumeral or shoulder joint is a ball and socket joint and is the most mobile joint in the body. The tremendous range of motion at this joint is the result of few external ligaments that present little limitation to movement, and shallow, ovoid articular surfaces that make movements in all planes of space possible. In fact, surrounding muscles and tendons play a more significant role in joint support than do the joint structures. The capsular ligament is extremely lax, providing limited support to the joint. Blending with the capsule are the tendons of four muscles. Together the capsule and tendons form the rotator cuff, which is the major support structure of the joint.

1 Articular cartilage
2 Synovial membrane
3 Fibrous membrane
4 Glenoid labrum
5 Acromioclavicular ligament
6 Clavicle
7 Humerus
8 Glenoid of scapula
9 Acromion of scapula
10 Supraspinatus muscle
11 Subscapularis muscle
12 Deltoid muscle
13 Tendon of long head of biceps brachii
14 Skin
15 Subcutaneous layer

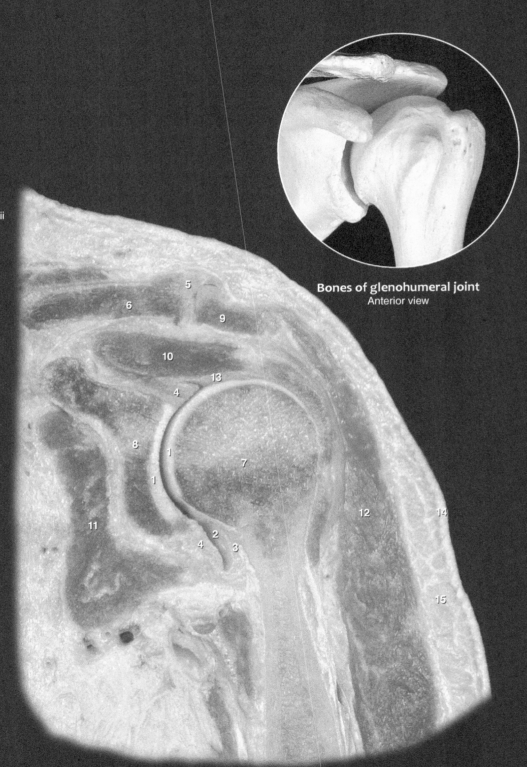

Bones of glenohumeral joint
Anterior view

Section of left glenohumeral joint
Anterior view of frontal section

Elbow Joint

The elbow joint is a complex joint comprised of multiple articular surfaces within one articular capsule. The elbow joint can be subdivided into three distinct articular interfaces — the humero-ulnar joint (hinge), the humeroradial joint (combined hinge and pivot), and the proximal radioulnar joint (pivot). Two distinct pairs of movements occur as a result of the articulations within the elbow joint — the hinged movements of flexion and extension, and the rotational movements of pronation and supination. Unlike the shoulder joint, the joints fo the elbow have strong extrinsic ligaments that limit movemnts and stabilize the articulating bones. The fibrous capsule is thin anteriorly and posteriorly, allowing for free range of motion during flexion and extension. On either side the capsule is reinforced by strong extrinsic ligaments, the ulnar collateral and radial collateral ligaments. Wrapping from the back of the ulna at the base of the olecranon to the front of the ulna at the lateral surface of the coronoid process is the semicircular anular ligament. With the radial notch of the ulna this ligament forms a fibro-osseous ring for the pivoting action of the radial head.

1　Articular cartilage
2　Joint (articular) capsule
3　Articular (synovial) cavity
4　Capitulum of humerus
5　Olecranon of ulna
6　Head of radius
7　Anular ligament
8　Biceps brachii muscle
9　Brachialis muscle
10　Triceps brachii muscle
11　Brachioradialis muscle

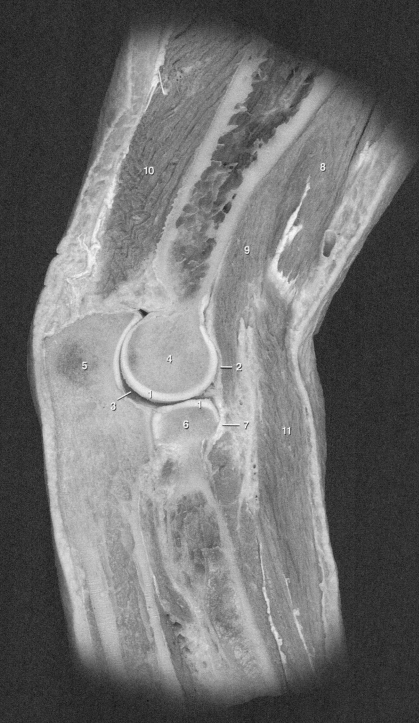

Section of pronated left elbow joint
Medial view of sagittal section

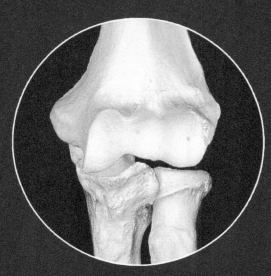

Bones of elbow joint
Anterior view

Hip Joint

Like the shoulder joint the hip joint, also a ball and socket joint, allows for great freedom of motion, although the range of motion is not quite as great as that of the shoulder. This comparative decrease in mobility results from the deep hip socket with its extended labrum, which almost completely engulfs the head of the femur. In addition, thick extrinsic ligaments tightly surround the joint to form a strong, reinforced capsule. The three major ligaments of the hip joint, the iliofemoral, pubofemoral, and ischiofemoral, form a sheath around the fibrous capsule. The iliofemoral ligament is argued to be the strongest ligament in the human body. Often called the Y-shaped ligament it passes superior and anterior to the joint, running from the anterior inferior iliac spine to the intertrochanteric line. With the thinner pubofemoral and ischiofemoral ligaments it spirals around the joint to stabilize this powerful joint. In additon to these large ligaments, a triangular flat band, the ligament of the head of the femur, extends from the fovea of the femoral head to the margins of the acetabular fossa. This ligament is also important because it functions as a pathway for blood vessels that supply the bone tissue in the head of the femur.

1 Ligament of head of femur
2 Joint (articular) capsule
3 Articular cartilage of acetabulum
4 Articular cartilage of femur
5 Articular (synovial) cavity
6 Acetabular labrum
7 Fovea capitis of femur
8 Head of femur
9 Greater trochanter of femur
10 Os coxae
11 Psoas major muscle
12 Iliacus muscle
13 Adductor muscles
14 Vastus lateralis muscle
15 Gluteus medius muscle
16 Gluteus minimis muscle
17 Obturator internus muscle
18 Obturator externus muscle
19 Skin
20 Subcutaneous layer
21 External iliac artery
22 Intestine

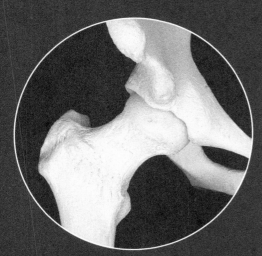

Bones of hip joint
Anterior view

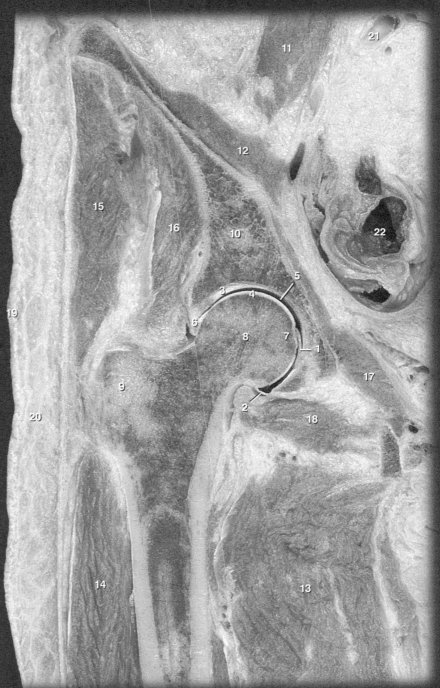

Section of right hip joint
Anterior view of frontal section

135

Knee Joint

The knee joint is a combined bicondylar and saddle joint. The relationships between the femur and the tibia provide no interlocking joint mechanisms or stability between the neighboring bones, and from this perspective the knee joint is completely unstable. The strength of the knee joint is dependent on strong ligaments and surrounding muscles. Although its primary motions are of a hinge nature, it is a complex joint with subtle rotational and sliding movements also. The major stabilizers of the joint are four strong ligaments. Two collateral ligaments support the joint on either side, while two cruciate ligaments criss-cross through the middle of the joint. The tibial or medial collateral ligament is a strong, flat band that stretches from the femoral epicondyle to the tibial condyle. Posteriorly it firmly attaches to the joint capsule and the medial meniscus, while anteriorly bursae separate it from these structures. The fibular or lateral collateral ligament is a strong cord that runs from the lateral femoral

1 Articular (synovial) cavity
2 Articular cartilage
3 Medial meniscus
4 Suprapatellar bursa
5 Prepatellar bursa
6 Infrapatellar bursa
7 Infrapatellar fat pad
8 Fibrous membrane of joint capsule
9 Synovial membrane of joint capsule
10 Lateral meniscus
11 Fibular collateral ligament
12 Tibial collateral ligament
13 Anterior cruciate ligament
14 Posterior cruciate ligament
15 Oblique popliteal ligament
16 Patellar ligament
17 Quadriceps tendon
18 Femur
19 Tibia
20 Fibula
21 Patella
22 Periosteum
23 Semimembranosus muscle
24 Gastrocnemius muscle
25 Soleus muscle
26 Popliteal fat

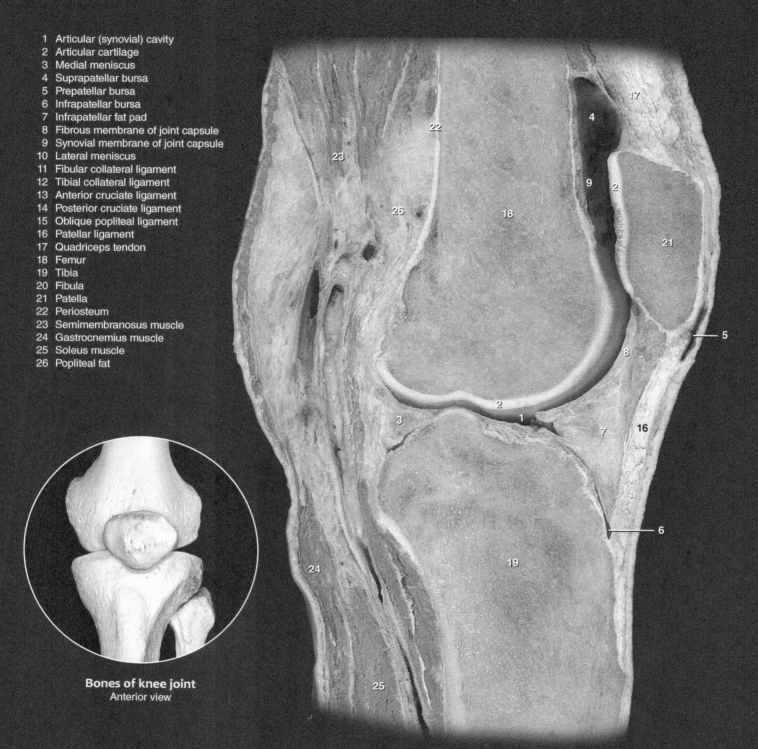

Bones of knee joint
Anterior view

Section of right knee joint
Lateral view of sagittal section

epicondyle to the head of the fibula. Unlike the tibial collateral ligament it does not attach to the lateral meniscus or joint capsule. The cruciate ligaments stabilize the knee from excessive anterior-posterior and rotational movements. The anterior cruciate ligament ascends posterolaterally from the medial aspect of the intercondylar area to the medial aspect of the lateral condyle of the femur. The shorter posterior cruciate ligament ascends from the posterior intercondylar area to the medial femoral condyle. Both cruciates have fibers that blend with the lateral meniscus. In additon to these ligamentous structures, two semilunar menisci project into the capsule between the femoral condyles and the articular plateaus of the tibia. The large, extensive articular capsule connects the femur, patella, and tibia.

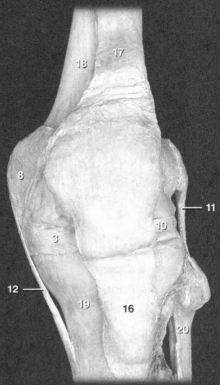

Dissection of left knee joint
Anterior view

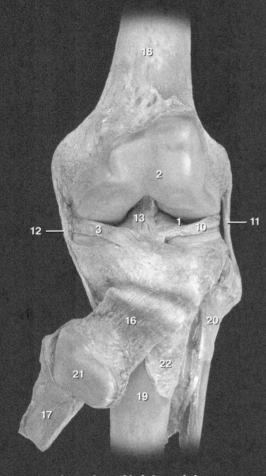

Dissection of left knee joint
Anterior view

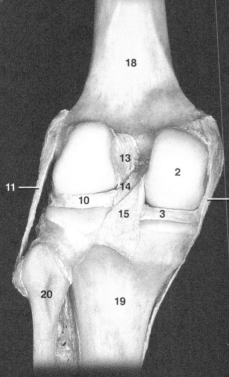

Dissection of left knee joint
Posterior view

Synovial Bursae and Sheaths

A synovial bursa is a small sac-like structure interposed between structures that generate significant amounts of friction. Bursae have a similar design to the articular capsule of a synovial joint. These small bags have an outer fibrous membrane of dense irregular collagenous connective tissue and an inner lining of synovial membrane. The synovial membrane produces a small amount of synovia as a lubricant inside the sac. The fibrous membrane binds to surrounding tissues, allowing the juxtaposed walls of synovial membrane to rub together in a frictionless manner. Many bursae arise as outgrowths of synovial joint cavities. In some cases these pinch off from the joint forming sacs that are independent from the joint, while other bursal sacs retain their connections with the joint cavity. A synovial sheath is a modified bursa that wraps around a tendon to protect it from friction on all sides. In the tight confines of the wrist, ankle, and digits, tendons often pass beneath fibrous bands called retinacula. The retinaculum is a connective tissue band that crosses over the tendons and keeps them from being displaced upward when the muscle shortens and bends the joints. Because the retinaculum and bone create a fibro-osseous tunnel around the tendon, considerable friction can occur on all surfaces of the tendon at these locations. As the tendon moves through the tunnel, the juxtaposed synovial membranes smoothly glide over each other with minimal friction.

1 Suprapatellar bursa
2 Prepatellar bursa
3 Infrapatellar bursa
4 Synovial (tendon) sheath
5 Retinaculum
6 Flexor digitorum superficialis tendon
7 Flexor digitorum profundus tendon
8 Lumbrical muscles
9 Flexor digiti minimi brevis muscle
10 Abductor digiti minimi muscle

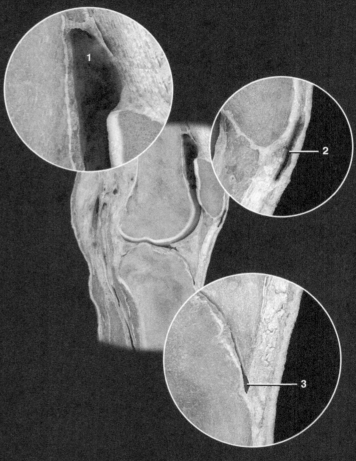

Synovial bursae around the knee joint
Medial view of sagittal section

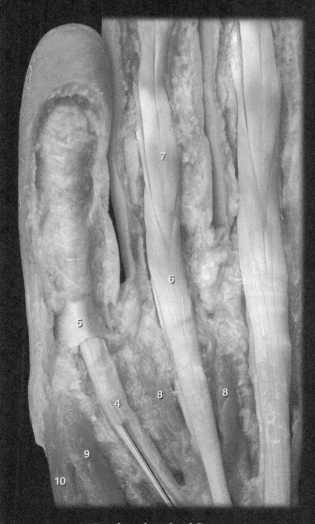

Tendon sheath of fingers
Anterior view, pin inserted into tendon sheath

138

8 | Muscular System

Bodies are designed to move! We move when we walk, jog, or run, activities that transport our bodies from one location to another. In addition to moving from location to location we also move in other ways. For example, think about grasping something with your hands and placing it in your mouth, or protecting yourself by kicking at something with your lower limb. How about throwing something? All of these activities are forms of movement that occur without moving from one location to another, yet they are movements nonetheless. Like moving about, these other types of movements are not only essential for survival, but define the broad spectrum for the majority of human movement. Reflect for a moment on the wide variety of movements that you make without moving from place to place. For example, think about the variety of intricate movements required to eat a meal, movements such as grasping, manipulating, cutting, chewing, and swallowing. Another example is getting dressed for the day. From the simple movements of pulling on clothing to the intricate movements of buttoning shirts and tying shoelaces, getting dressed involves a wide variety of movements. And here is something else to ponder — how about all the movements involved in communication? Think of the wide array of movements that you produce as you communicate with others — whether the communication involves writing a note on a piece of paper, typing a letter on the keyboard of a computer, signaling pleasure and happiness with a smile, or using your voice to talk to a friend on the telephone.

We could go on and on discussing the wide variety of movement and its importance, but the bottom line is all movement results from the combined activity of individual muscles. The most detailed movements you make can be broken down into the simple actions of individual muscles moving the bones of the skeleton at the joints. This chapter introduces the muscular system. On the pages that follow you will see the structural design of a typical muscle and whole body views of the muscles of the body. Our approach to the skeletal muscles of the body is based on their embryonic origins. The four chapters that follow this chapter cover each of the developmental groups of muscles — muscles of the head, muscles of the trunk, muscles of the upper limb, and muscles of the lower limb. The logic of this approach will be further discussed as we introduce each chapter.

Find more information about the muscular system in

REALANATOMY

Anatomy of a Muscle

While there is a wide variety to the shape, size, and architecture of the skeletal muscles of the body, most muscles share a common basic design — a tendon of origin, a muscle body or belly, and a tendon of insertion. The tendons, projecting from the muscle belly, are a continuation of the connective tissue surrounding the muscle cells within the belly of the muscle. As the connective tissue projects beyond the muscle cells, it condenses to become the tendons, which merge and blend with the periosteum to attach the muscle to bone.

1   Muscle belly or body
2   Tendon of origin
3   Tendon of insertion
4   Collagen fiber
5   Muscle cell or fiber
6   Nucleus
7   Biceps brachii muscle
8   Brachialis muscle
9   Triceps brachii muscle
10   Epimysium
11   Perimysium
12   Endomysium
13   Blood vessels in perimysium
14   Nerve in perimysium
15   Fascia
16   Sucutaneous layer
17   Skin
18   Periosteum

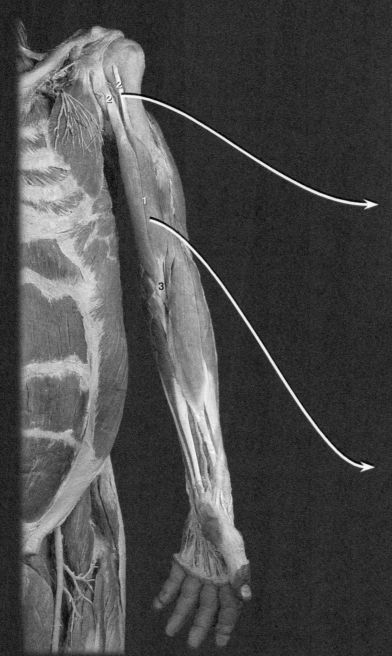

Dissection of brachium highlighting biceps brachii as example of muscle anatomy
Anterior view

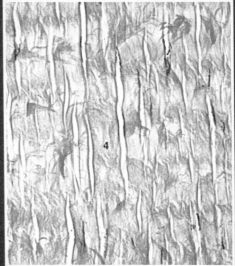

Dense regular connective tissue of tendon
200x

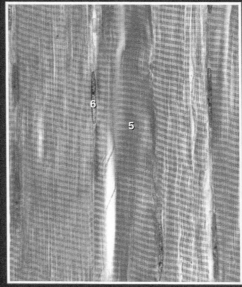

Skeletal muscle tissue of muscle belly
400x

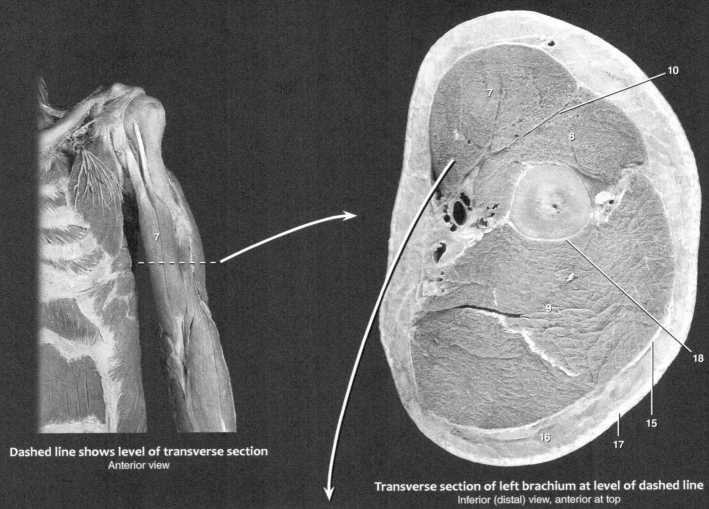

Dashed line shows level of transverse section
Anterior view

Transverse section of left brachium at level of dashed line
Inferior (distal) view, anterior at top

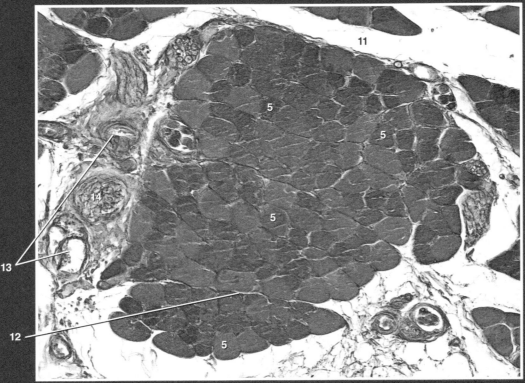

Photomicrograph of muscle fasciculus
Transverse section, 100x

Skeletal Muscles

In the dissections below, the integument and fascia were removed to reveal the superficial skeletal muscles. Some of the larger muscles are identified here. More detailed muscle labeling will occur in the next four chapters.

1 Platysma
2 Pectoralis major
3 Deltoid
4 Rectus abdominis
5 External oblique
6 Biceps brachii

7 Triceps brachii
8 Trapezius
9 Brachioradialis
10 Latissimus dorsi
11 Gluteus maximus
12 Biceps femoris

13 Sartorius
14 Vastus medialis
15 Rectus femoris
16 Adductor magnus
17 Tibialis anterior
18 Gastrocnemius

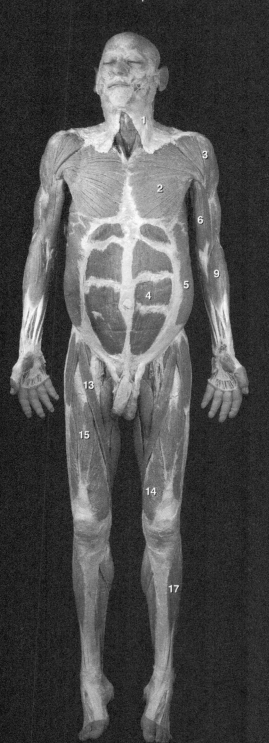

Skeletal muscles of the body
Anterior view

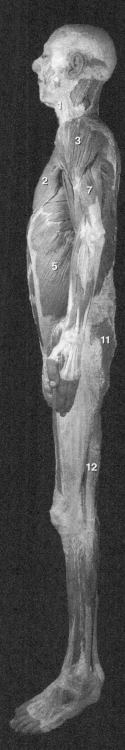

Skeletal muscles of the body
Lateral view

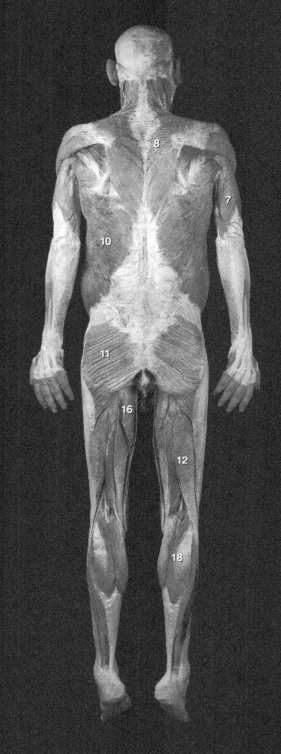

Skeletal muscles of the body
Posterior view

9 | Head Muscles

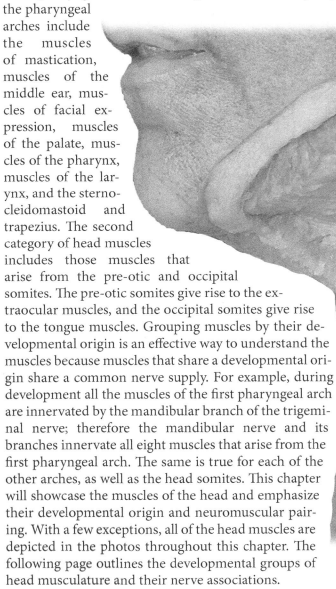

Head muscles, like the platysma and risorius seen in the photo on this page, arise from two sources during embryonic development. One source is the pharyngeal arches, which give rise to the majority of the head muscles. Muscles of the pharyngeal arches include the muscles of mastication, muscles of the middle ear, muscles of facial expression, muscles of the palate, muscles of the pharynx, muscles of the larynx, and the sternocleidomastoid and trapezius. The second category of head muscles includes those muscles that arise from the pre-otic and occipital somites. The pre-otic somites give rise to the extraocular muscles, and the occipital somites give rise to the tongue muscles. Grouping muscles by their developmental origin is an effective way to understand the muscles because muscles that share a developmental origin share a common nerve supply. For example, during development all the muscles of the first pharyngeal arch are innervated by the mandibular branch of the trigeminal nerve; therefore the mandibular nerve and its branches innervate all eight muscles that arise from the first pharyngeal arch. The same is true for each of the other arches, as well as the head somites. This chapter will showcase the muscles of the head and emphasize their developmental origin and neuromuscular pairing. With a few exceptions, all of the head muscles are depicted in the photos throughout this chapter. The following page outlines the developmental groups of head musculature and their nerve associations.

Find more information about the muscles of the head in
REAL ANATOMY

Head Muscles

This chapter presents numerous dissections of the head and neck that depict the muscles of the head. We define the head muscles as all muscles that arise from the pharyngeal (branchial) arches or the head somites (pre-otic and occipital). All of these muscles arise from the paraxial mesoderm of the embryonic head. Unlike many anatomy sources that mix these muscles into multiple groups, with no logic to their innervation, we choose to present them based on their embryonic origins. Taking this approach makes it very easy to learn the innervation patterns of the head muscles because each developmental group is associated with a distinct cranial nerve or set of cranial nerves (see groups below). Accompanying each labeled dissection photograph on the pages that follow are small reference photos that clearly depict each of the developmental muscle groups of the head. Since some of the head muscles migrate into the neck, we also depict the somitic muscles of the neck in the reference photos, to help distinguish them from the true head muscles. The somitic muscles of the neck will be the subject of the next chapter. For example, the first photo (see opposite page) labels numerous head muscles. The reference photos clearly reveal that the labeled muscles are primarily from two sources — the first pharyngeal arch and the second pharyngeal arch (accounting for the majority of the muscles). The third reference photo shows that some muscles are from neck somites.

Muscles of the First Pharyngeal Arch
(Nerve supply - mandibular branch of the trigeminal nerve CN V)
    Temporalis
    Masseter
    Medial pterygoid
    Lateral pterygoid
    Anterior digastricus
    Mylohyoid
    *Tensor tympani
    Tensor veli palatini

Muscles of the Second Pharyngeal Arch
(Nerve supply - facial nerve CN VII)
    Occipitofrontalis
    Temporoparietalis
    Transversus nuchae
    Procerus
    Nasalis
    *Depressor septi nasi
    Orbicularis oculi
    Corrugator supercilii
    Depressor supercilii
    Auricularis anterior
    Auricularis superior
    Auriculalris posterior
    Intrinsic auricular muscles
      Helicis major muscle
      Helicis minor muscle
      Tragicus muscle
      *Pyramidal muscle of auricle
      Antitragicus muscle
      *Transverse muscle of auricle
      *Oblique muscle of auricle
    Orbicularis oris
    Depressor anguli oris
    Transversus menti
    Risorius
    Zygomaticus major
    Zygomaticus minor
    Levator labii superioris
    Levator labii superioris alaeque nasi
    Depressor labii inferioris
    Levator anguli oris
    Buccinator
    Mentalis
    *Stapedius
    Stylohyoid
    Posterior digastricus
    Platysma

Muscle of the Third Pharyngeal Arch
(Nerve supply - glossopharyngeal nerve CN IX)
    Stylopharyngeus

Muscles of the Fourth Pharyngeal Arch
(Nerve supply - vagus nerve CN X)
    Levator veli palatini
    Palatoglossus
    Palatopharyngeus
    Musculus uvulae
    Superior pharyngeal constrictor
    Middle pharyngeal constrictor
    Inferior pharyngeal constrictor
    Cricothyroid
    Salpingopharyngeus

Muscles of the Sixth Pharyngeal Arch
(Nerve supply - vagus nerve CN X)
    Posterior crico-arytenoid
    Lateral crico-arytenoid
    Vocalis
    Thyro-arytenoid
    Oblique arytenoid
    Transverse arytenoid

Muscles of the Posterior Pharyngeal Arch
(Nerve supply - accessory nerve CN XI)
    Sternocleidomastoid
    Trapezius

Muscles of the Pre-otic Somites
(Nerve supply - oculomotor CN III, trochlear CN IV, and abducens CVI)
    Superior rectus
    Inferior rectus
    *Medial rectus
    Lateral rectus
    Superior oblique
    Inferior oblique
    Levator palpebrae superioris

Muscles of the Occipital Somites
(Nerve supply - hypoglossal nerve CN XII)
    Genioglossus
    Hyoglossus
    Styloglossus
    Superior longitudinal muscle
    Inferior longitudinal muscle
    Transverse muscle
    Vertical muscle

All the muscles listed above are depicted in photos in this chapter except those marked with an asterisk.

1 Masseter
2 Anterior belly of digastricus (cut)
3 Mylohyoid
4 Frontal belly of occipitofrontalis
5 Temporoparietalis
6 Procerus
7 Nasalis
8 Orbicularis oculi
9 Corrugator supercilii
10 Depressor supercilii
11 Auricularis anterior
12 Auricularis superior
13 Orbicularis oris
14 Depressor anguli oris
15 Transversus menti
16 Zygomaticus major
17 Zygomaticus minor
18 Levator labii superioris
19 Levator labii superioris alaeque nasi
20 Depressor labii inferioris
21 Levator anguli oris
22 Buccinator
23 Mentalis
24 Posterior digastricus
25 Epicranial aponeurosis
26 Temporal fascia
27 Parotid gland (cut)

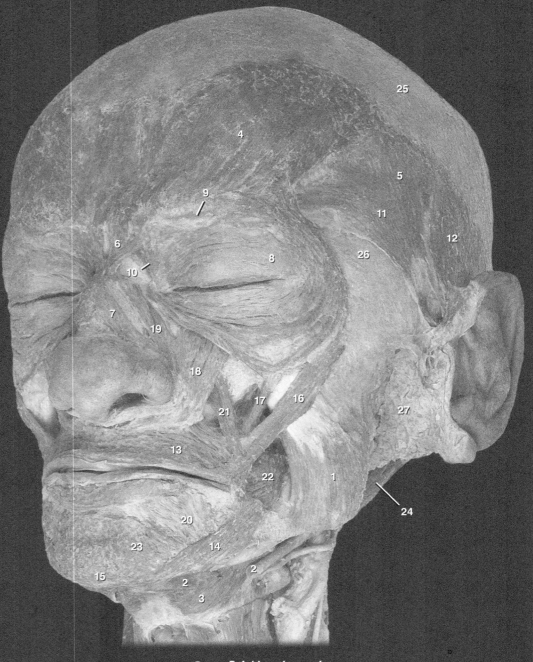

Superficial head muscles
Anterolateral view

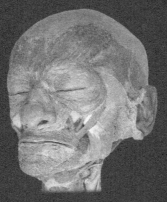

First arch muscles

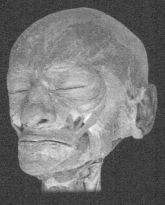

Second arch muscles

Somitic muscles
of neck

Head Muscles

The dissections depicted on this page and the facing page represent two stages in a dissection of the head. Below is a superficial dissection with the integument and some fascia removed. On the opposing page some superficial muscles were removed. Most of the head muscle groups are represented. Note also the somitic muscles of the neck that are visible.

1 Temporalis	7 Orbicularis oculi	13 Levator anguli oris
2 Masseter	8 Procerus	14 Orbicularis oris
3 Mylohyoid	9 Levator labii superioris alaeque nasi	15 Buccinator
4 Anterior belly of digastricus	10 Nasalis	16 Depressor anguli oris
5 Frontal belly of occipitofrontalis	11 Levator labii superioris	17 Depressor labii inferioris
6 Temporoparietalis	12 Zygomaticus major	18 Mentalis

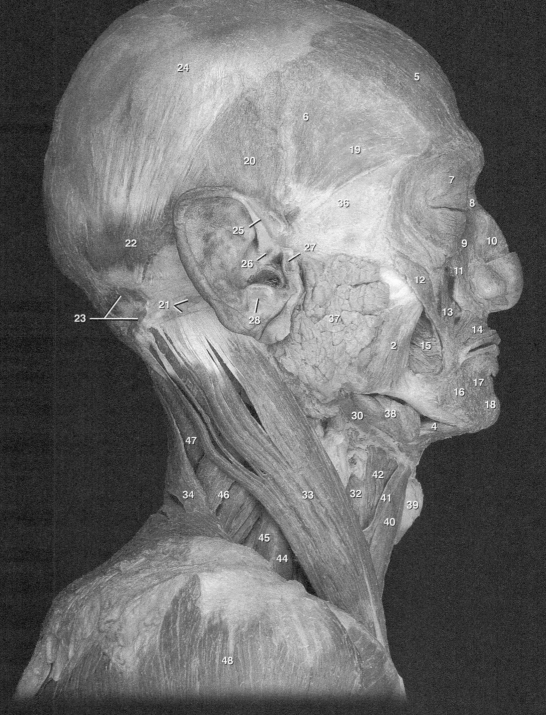

Head muscles, superficial dissection
Lateral view

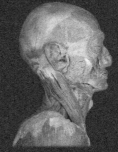

First arch muscles

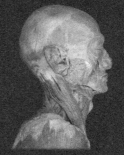

Second arch muscles

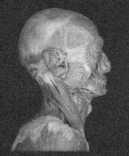

Fourth arch muscles

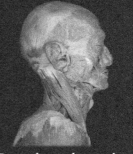

Posterior arch muscles

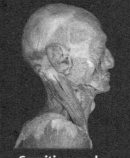

Somitic muscles
of neck

19 Auricularis anterior
20 Auricularis superior
21 Auricularis posterior
22 Occipital belly of occipitofrontalis
23 Transversus nuchae
24 Epicranial aponeurosis
25 Helicis major
26 Helicis minor
27 Tragicus
28 Antitragicus

29 Posterior belly of digastricus
30 Stylohyoid
31 Middle pharyngeal constrictor
32 Inferior pharyngeal constrictor
33 Sternocleidomastoid
34 Trapezius
35 Styloglossus
36 Temporal fascia
37 Parotid gland
38 Submandibular gland

39 Thyroid cartilage
40 Sternohyoid
41 Omohyoid
42 Thyrohyoid
43 Longus colli
44 Middle scalene
45 Posterior scalene
46 Levator scapulae
47 Splenius capitis
48 Deltoid

Head muscles, masticatory muscles exposed
Lateral view

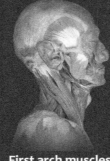

First arch muscles

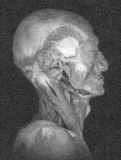

Second arch muscles

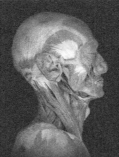

Fourth arch muscles

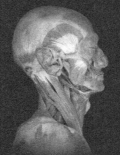

Posterior arch muscles

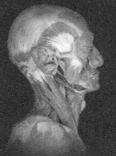

Somitic muscles
of head and neck

Head Muscles

The lateral head dissections below and opposite are deeper dissections that expose the deep masticatory muscles (below) and the extraocular muscles (opposite).

1 Temporalis
2 Masseter
3 Medial pterygoid
4 Lateral pterygoid
5 Anterior belly of digastricus
6 Mylohyoid
7 Frontal belly of occipitofrontalis
8 Occipital belly of occipitofrontalis
9 Transversus nuchae
10 Procerus
11 Nasalis
12 Orbicularis oculi
13 Auricularis anterior (cut)
14 Auricularis superior (cut)
15 Auricularis posterior

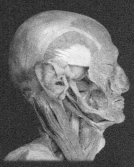

First arch muscles

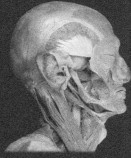

Second arch muscles

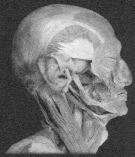

Fourth arch muscles

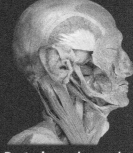

Posterior arch muscles

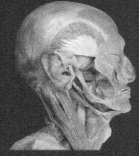

Somitic muscles
of head and neck

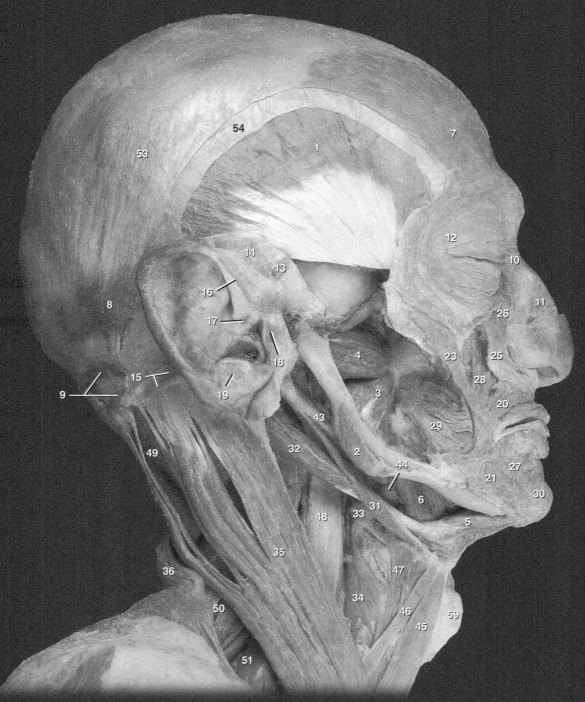

Head muscles, deep masticatory muscles exposed
Lateral view, portion of mandible removed

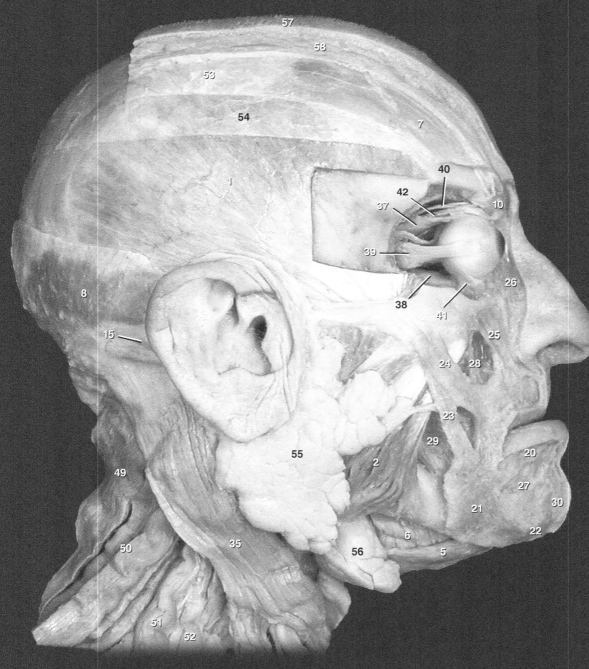

16 Helicis major	27 Depressor labii inferioris	38 Inferior rectus	49 Splenius capitis
17 Helicis minor	28 Levator anguli oris	39 Lateral rectus	50 Levator scapulae
18 Tragicus	29 Buccinator	40 Supra-orbital nerve	51 Posterior scalene
19 Antitragicus	30 Mentalis	41 Inferior oblique	52 Middle scalene
20 Orbicularis oris	31 Stylohyoid	42 Levator palpebrae superioris	53 Epicranial aponeurosis
21 Depressor anguli oris	32 Posterior belly of digastricus	43 Styloglossus	54 Temporal fascia (cut)
22 Transversus menti	33 Middle pharyngeal constrictor	44 Hyoglossus	55 Parotid gland
23 Zygomaticus major	34 Inferior pharyngeal constrictor	45 Sternohyoid	56 Submandibular gland
24 Zygomaticus minor	35 Sternocleidomastoid	46 Omohyoid	57 Skin
25 Levator labii superioris	36 Trapezius	47 Thyrohyoid	58 Subcutaneous layer
26 Levator labii superioris alaeque nasi	37 Superior rectus	48 Longus colli	59 Thyroid cartilage

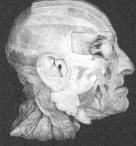

First arch muscles

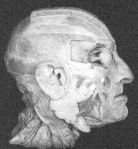

Second arch muscles

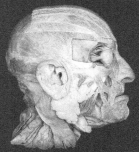

Posterior arch muscles

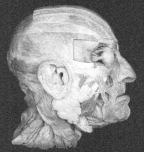

Somitic muscles
of head and neck

Head muscles, extraocular muscels exposed
Lateral view, lateral wall of orbit removed

Head Muscles

The dissections on this and the opposing page are deep dissections of the head and neck that expose many of the muscles of the palate, pharynx, and tongue. The palatal and pharyngeal muscles, along with the muscles of the larynx, are the deepest of the head muscles. These groups arise from the third, fourth, and sixth arches and form the muscular walls to the upper regions of the embryonic gut tube. All of the "true" tongue muscles (the palatoglossus is included by many with the tongue muscles, but it is a muscle of the palate from fourth arch origin) arise from the occipital somites and are innervated by the cranial nerve XII, the hypoglossal nerve. The hypoglossal nerve is the lowest of the ventral motor nerves arising from the brainstem and is developmentally paired with the occipital somites.

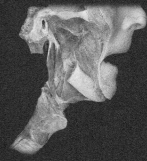

First arch muscles

Second arch muscles

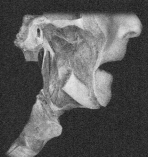

Third arch muscles

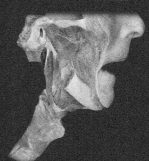

Fourth arch muscles

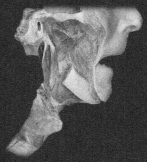

Somitic muscles
of head

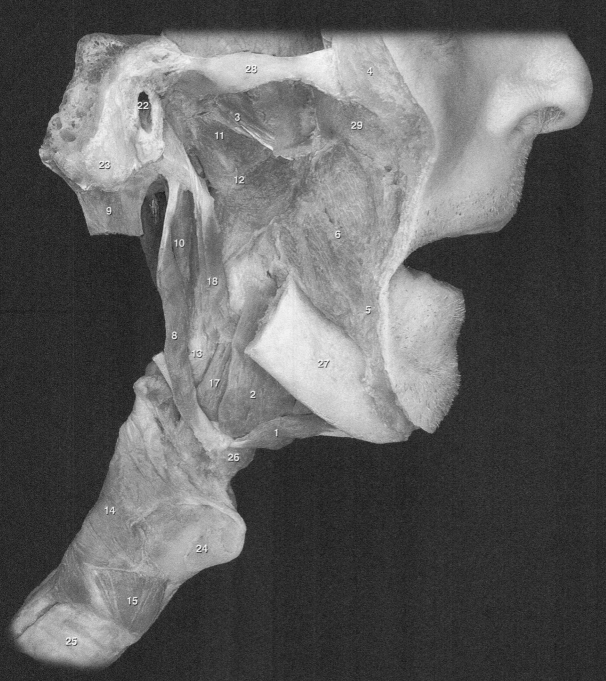

Head muscles, palatal and pharyngeal muscles exposed
Lateral view, mandibular ramus removed

1 Anterior belly of digastricus
2 Mylohyoid
3 Tensor veli palatini
4 Orbicularis oculi
5 Orbicularis oris
6 Buccinator
7 Mentalis
8 Stylohyoid
9 Posterior belly of digastricus (cut)
10 Stylopharyngeus
11 Levator veli palatini
12 Superior pharyngeal constrictor
13 Middle pharyngeal constrictor
14 Inferior pharyngeal constrictor
15 Cricothyroid
16 Genioglossus
17 Hyoglossus
18 Styloglossus
19 Inferior longitudinal muscle
20 Geniohyoid
21 Mucosa of tongue
22 External acoustic meatus
23 Mastoid process
24 Thyroid cartilage
25 Trachea
26 Hyoid bone
27 Mandible (cut)
28 Zygomatic arch
29 Maxilla

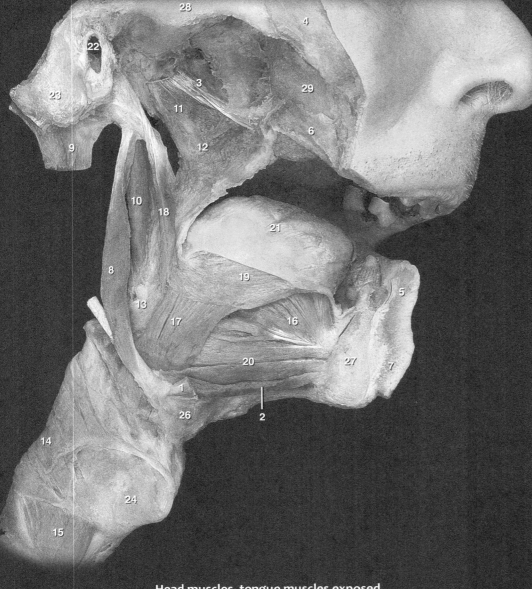

Head muscles, tongue muscles exposed
Lateral view, right half of mandible removed

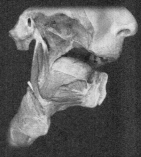

First arch muscles

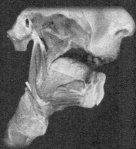

Second arch muscles

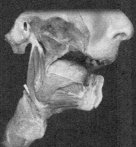

Third arch muscles

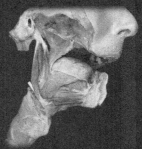

Fourth arch muscles

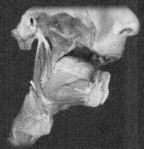

Somitic muscles
of head

Head Muscles

The dissections on this and the opposing page are deep dissections of the head and neck that expose the palate and muscular wall of the pharynx and larynx (muscles that arise from the third, fourth, and sixth pharyngeal arches). These are the deepest muscles of the head and neck, and they form the muscular walls of the upper end of the embryonic gut tube. The dissection below depicts the posterior wall of the pharynx. On the opposing page the pharyngeal wall has been sectioned to reveal the inside of the palate and larynx from behind.

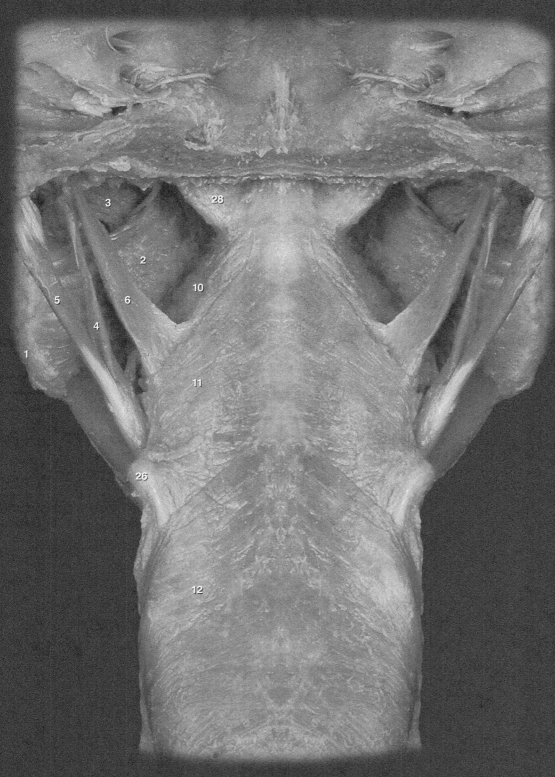

Head muscles, posterior wall of pharynx exposed
Posterior view, cervical vertebrae and occipital bone removed

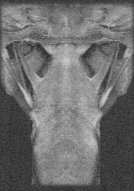

First arch muscles

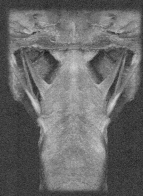

Second arch muscles

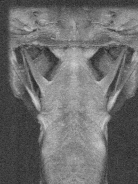

Third arch muscles

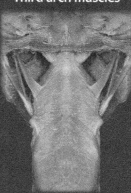

Fourth arch muscles

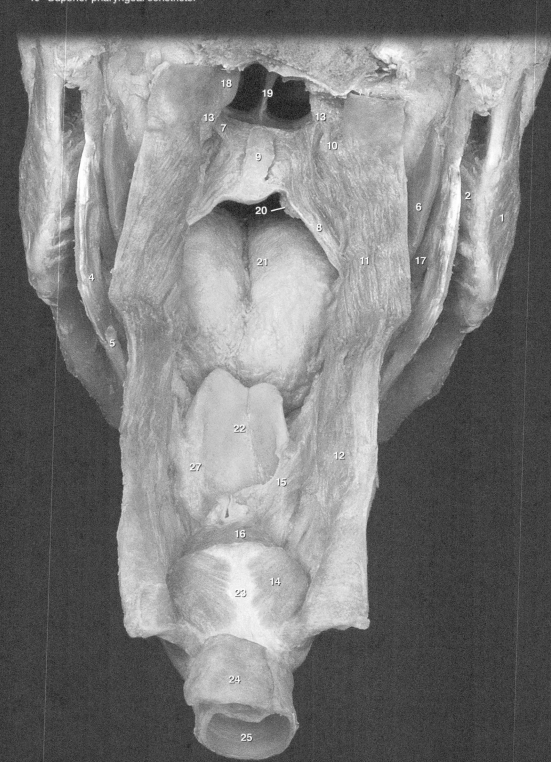

1 Masseter
2 Medial pterygoid
3 Lateral pterygoid
4 Stylohyoid
5 Posterior belly of digastricus
6 Stylopharyngeus
7 Levator veli palatini
8 Palatopharyngeus
9 Musculus uvulae
10 Superior pharyngeal constrictor
11 Middle pharyngeal constrictor
12 Inferior pharyngeal constrictor
13 Salpingopharyngeus
14 Posterior crico-arytenoid
15 Oblique arytenoid
16 Transverse arytenoid
17 Styloglossus
18 Pharyngotympanic tube
19 Bony nasal septum
20 Palatine tonsil
21 Tongue
22 Epiglottis
23 Cricoid cartilage
24 Esophagus
25 Trachea
26 Greater cornu of hyoid bone
27 Aryepiglottic fold
28 Pharyngobasilar fascia

Head muscles, posterior wall of pharynx cut and reflected
Posterior view, cervical vertebrae and occipital bone removed

First arch muscles

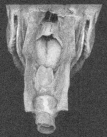

Second arch muscles

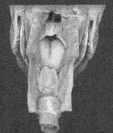

Third arch muscles

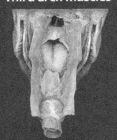

Fourth arch muscles

Sixth arch muscles

Somitic muscles
of head

153

Head Muscles

Sectional anatomy broadens perspective and showcases anatomical relationships in ways that are not possible to achieve by dissection alone. The frontal and parasagittal sections on these pages depict and clarify the relationships of many of the head muscles and show the relationships these muscles have with other structures of the head.

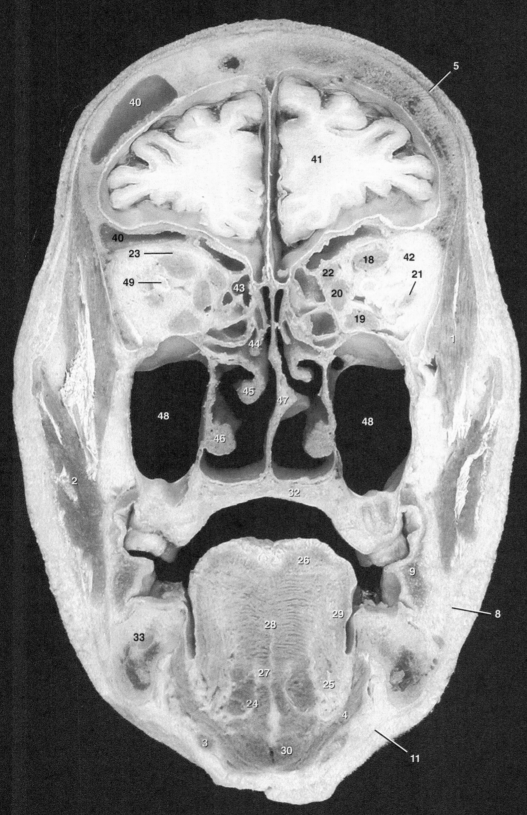

Head muscles, frontal section through orbits, nasal cavity, and oral cavity
Posterior view

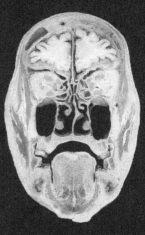

First arch muscles

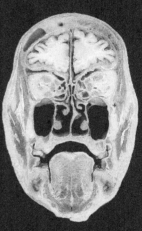

Second arch muscles

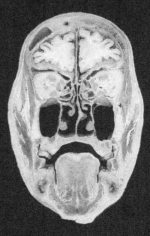

Somitic muscles
of head

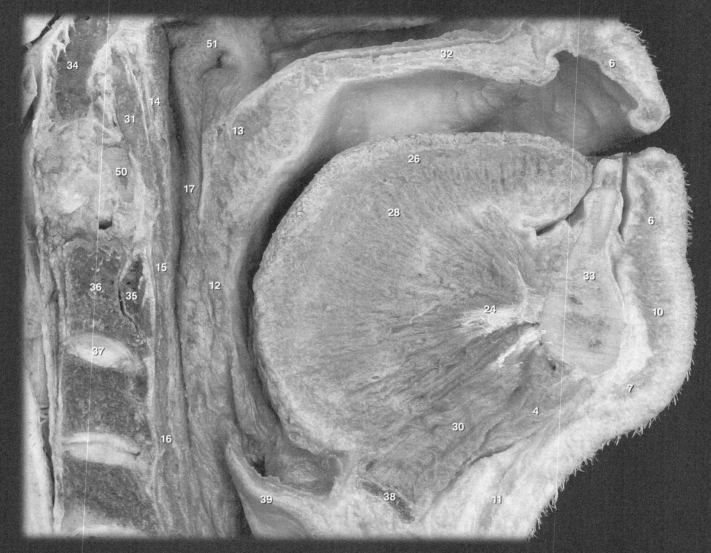

1 Temporalis	14 Superior pharyngeal constrictor	27 Inferior longitudinal muscle	40 Frontal sinus
2 Masseter	15 Middle pharyngeal constrictor	28 Transversus muscle	41 Frontal lobe of cerebrum
3 Anterior digastricus	16 Inferior pharyngeal constrictor	29 Vertical muscle	42 Periorbital fat
4 Mylohyoid	17 Salpingopharyngeus	30 Geniohyoid	43 Ethmoidal air cells
5 Frontal belly of occipitofrontalis	18 Superior rectus	31 Longus capitis	44 Superior nasal conchae
6 Orbicularis oris	19 Inferior rectus	32 Hard palate	45 Middle nasal conchae
7 Transversus menti	20 Medial rectus	33 Mandible	46 Inferior nasal conchae
8 Risorius	21 Lateral rectus	34 Occipital bone	47 Bony nasal septum
9 Buccinator	22 Superior oblique	35 Atlas	48 Maxillary sinus
10 Mentalis	23 Levator palpebrae superioris	36 Axis	49 Optic nerve
11 Platysma	24 Genioglossus	37 Intervertebral disc	50 Occipital condyle
12 Palatopharyngeus	25 Hyoglossus	38 Hyoid bone	51 Torus tubarius of
13 Musculus uvulae	26 Superior longitudinal muscle	39 Epiglottis	pharyngotympanic tube

Head muscles, parasagittal section through oral cavity and pharynx
Posterior view, section is 1.2 cm lateral to the midline

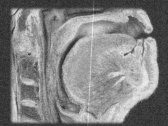

First arch muscles

Second arch muscles

Fourth arch muscles

Somitic muscles
of head and neck

Head Muscles

The dissection on this page exposes the deepest of the head muscles, those of the sixth pharyngeal arch. This group, found within the wall of the larynx, is the small series of muscles that are responsible for sound production. Contractions of these muscles vary the tension on the vocal folds and adjust the size of the rima glottidis. A cut anterior portion of the cricothyroid is also visible; however this muscle is actually the anterior continuation of the inferior pharyngeal constrictor and develops from the fourth pharyngeal arch.

1 Posterior crico-arytenoid
2 Lateral crico-arytenoid
3 Thyro-arytenoid
4 Thyro-epiglottic part of thyro-arytenoid
5 Oblique arytenoid
6 Ary-epiglottic part of oblique arytenoid
7 Transverse arytenoid
8 Cricothyroid (cut)
9 Hyoid bone
10 Epiglottis
11 Thyroid cartilage (cut)
12 Cricoid cartilage
13 Trachea
14 Thyrohyoid membrane

Dissection of the larynx, right lamina and horns removed
Posterolateral view

Fourth arch muscle

Sixth arch muscles

10 | Trunk Muscles

The trunk, which is defined by the span of the vertebral column, includes the neck (span of the cervical vertebrae), the thorax (span of the thoracic vertebrae), the abdomen (span of the lumbar vertebrae), and the pelvis (span of the sacral vertebrae). The muscles of the trunk are the most primitive muscles in the vertebrate body. This series of muscles arises as epithelial migrations from the myotomes of the embryonic somites and forms a distinct muscle pattern throughout the length of the trunk. The trunk muscle pattern has two distinct subdivisions, the epaxial muscles and the hypaxial muscles, which are separated by a transverse intermuscular septum. The epaxial muscles, situated posterior to the vertebral axis, are the extensor muscles of the vertebral column that develop from the epimere of the myotomes. The dorsal rami of the spinal nerves innervate these muscles. The hypaxial muscles, positioned primarily anterior and lateral to the vertebral axis, develop from the hypomere of the myotomes and are supplied by the ventral rami of the spinal nerves.

The epaxial muscles form a number of muscle layers that anatomists typically describe as a series of groups. From superficial to deep the groups are the spinotransversales muscles, the erector spinae muscles, the transversospinales muscles, and the deepest groups (most of which are intersegmental) consisting of the interspinales, intertransversarii, and suboccipital muscles.

The hypaxial muscles form a distinct pattern throughout the trunk wall. This pattern consists of a subvertebral musculature (positioned on the anterior and lateral aspect of the vertebral bodies), a four-layered lateral wall of muscles situated on the lateral aspect of the trunk wall, and a ventral strap of musculature on the anterior trunk wall.

The photos in this chapter clearly depict the trunk muscles and the patterns outlined above.

Find more information about the muscles of the trunk in

REAL ANATOMY

Epaxial Muscles

The epaxial muscles, or vertebral extensors, develop on the dorsal side of the vertebral column and skull. These muscles arise from the myotomal epimere of all the trunk somites and span the entire length of the vertebral column to the posterior aspect of the occipital bone. They comprise the intrinsic muscles of the vertebral column, which are often referred to as the "true back muscles." The vertebral extensors form four distinct muscle groups. These groups are, from superficial to deep, the spinotransversales (splenius muscles), the erector spinae, the transversospinales (three layers — the semispinalis, multifidus, and rotatores layers), and the intersegmental muscles. However, each of the four groups does not extend the entire length of the vertebral column, and in some regions not all four layers are represented. All epaxial muscles receive a nerve supply from the dorsal (posterior) rami of the spinal nerves.

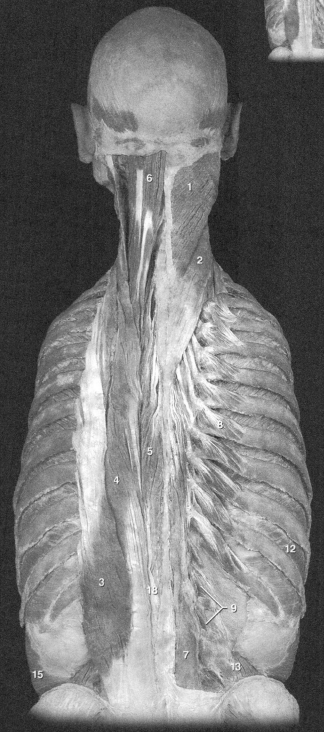

Epaxial Muscle Layers
    Spinotransversales — Splenius layer
    Erector spinae layer
    Transversospinalis — Semispinalis layer
    Transversospinalis — Multifidus layer
    Transversospinalis — Rotatores layer
    Deep intersegmental layer

Vertical muscle subdivisions within muscle layers
    Capitis Muscles
      Splenius capitis
      Erector spinae capitis
        Longissimus capitis
        Spinalis capitis
      Transversospinales capitis
        Semispinalis capitis
      Suboccipitales
        Rectus capitis posterior major
        Rectus capitis posterior minor
        Obliquus capitis superior
        Obliquus capitis inferior

    Cervical Muscles
      Splenius cervicis
      Erector spinae cervicis
        Iliocostalis cervicis
        Longissimus cervicis
        Spinalis cervicis
      Transversospinales cervicis
        Semispinalis cervicis
        Multifidus cervicis
        Rotatores cervicis
      Interspinales cervicis
      Intertransversarii posteriores cervicis medialis

    Thoracic Muscles
      Erector spinae thoracis
        Iliocostalis thoracis
        Longissimus thoracis
        Spinalis thoracis
      Transversospinales thoracis
        Semispinalis thoracis
        Multifidus thoracis
        Rotatores thoracis
      Interspinales thoracis
      Intertransversarii thoracis
      Levatores costarum

    Lumbar Muscles
      Erector spinae lumborum
        Iliocostalis lumborum
      Transversospinales lumborum
        Multifidus lumborum
        Rotatores lumborum
      Interspinales lumborum
      Intertransversarii lumborum medialis

Dissection of epaxial musculature
Posterior view

Spinotransversales Muscles

The spinotransversales muscles are the superficial-most epaxial muscles and are only present in the superior half of the vertebral column. This group is comprised of two named muscles — the splenus capitis and splenius cervicis. They span from the midthoracic region to the base of the occipital bone. As their name suggests, the fibers attach to the spinous processes of the vertebrae and course laterally to attach to the vertebral transverse processes. These flat bands of muscle are primary extensors of the upper vertebral column and head.

Splenius Musculature
1 Splenius capitis muscle
2 Splenius cervicis muscle

Other Muscles and Structures
3 Iliocostalis muscle
4 Longissimus muscle
5 Spinalis muscle
6 Semispinalis muscle
7 Multifidus muscle
8 Levatores costarum muscle
9 Intertransversarii muscle
10 Posterior scalene muscle
11 External intercostal muscle
12 Internal intercostal muscle
13 Quadratus lumborum muscle
14 External oblique muscle
15 Transversus abdominis muscle
16 Gluteus maximus muscle
17 Fascia of gluteus medius muscle
18 Supraspinous ligament
19 Nuchal ligament

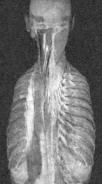

Dissection of splenius and erector spinae muscles
Posterior view

159

Erector Spinae Muscles

The erector spinae muscles comprise the second layer of epaxial muscles. Unlike the splenius muscles, the erector spinae muscle group spans the entire length of the vertebral column. The erector spinae is divided into three parts, which from medial to lateral are the spinalis muscle, the longissimus muscle, and the iliocostalis muscle. This strong group of epaxial muscles consists of muscle fibers that course vertically and somewhat laterally as they span multiple vertebral levels. They function as primary extensors of the vertebral column.

Erector Spinae and Semispinalis Musculature
1 Iliocostalis lumborum muscle - lumbar part
2 Iliocostalis lumborum muscle - thoracic part
3 Iliocostalis cervicis muscle
4 Longissimus thoracis muscle
5 Longissimus cervicis muscle
6 Longissimus capitis muscle
7 Spinalis thoracis muscle
8 Spinalis cervicis muscle
9 Spinalis capitis muscle
10 Semispinalis thoracis muscle
11 Semispinalis cervicis muscle
12 Semispinalis capitis muscle

Other Muscles and Structures
13 Multifidus muscle
14 Levatores costarum muscle
15 External intercostal muscle
16 Internal intercostal muscle
17 Middle scalene muscle
18 Nuchal ligament
19 Trapezius muscle
20 Rhomboideus major muscle
21 Latissimus dorsi muscle
22 Infraspinatus muscle
23 Teres major muscle
24 Deltoid muscle
25 Triceps muscle

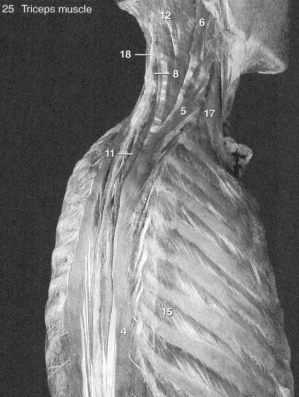

Dissection of erector spinae muscles
Posterolateral view

Dissection of erector spinae muscles
Posterior view

Transversospinales Muscles

The transversospinales muscles form the third layer of epaxial muscles. This deeper layer of muscles has shorter muscle fibers, on average, than its two superficial counterparts, and the fibers angle from lateral (transverse processes) to medial (spinous processes) as they course from sacrum to cranium. Within this group there are three muscles — the semispinalis, multifidus, and the rotatores muscles. The more superficial semispinalis muscle is depicted on this page.

Dissection of semispinalis muscles
Posterior view

Dissection of semispinalis muscles
Lateral view

Dissection of semispinalis layer on left and limb muscles on right
Posterolateral view

Transversospinales Muscles

The multifidus layer of the transversospinales musculature is highlighted on this page, and the deeper rotatores are evident on the opposite page along with the deeper intersegmental muscles. The multifidus muscles span three to five vertebral levels in their span from the sacrum to the second cervical vertebra, while the deepest member, the rotatores, typically span only one to two vertebrae. The transversospinales muscles assist their more superficial counterparts with extension of the vertebral column and play important roles in the maintenance of posture.

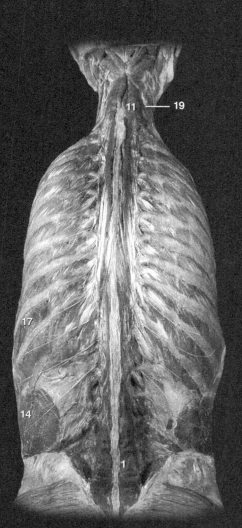

Dissection of multifidus muscles
Posterior view

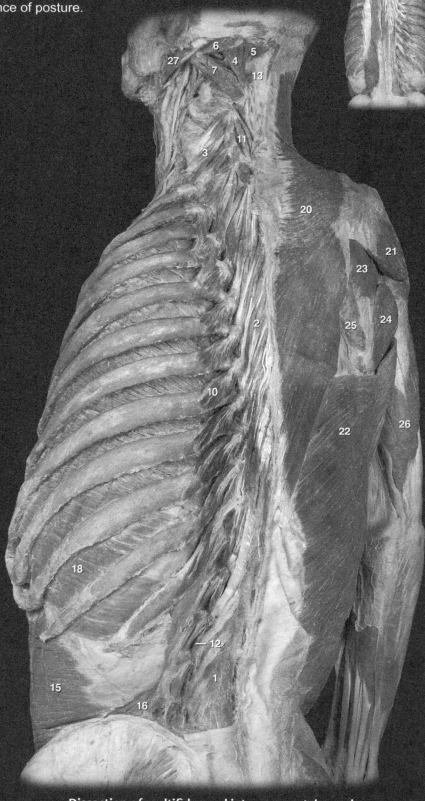

Dissection of multifidus and intersegmental muscles
Posterolateral view

Intersegmental Muscles

The small intersegmental muscles — the interspinales muscles, intertrans-versarii muscles, levatores costarum, and subocciptal muscles — in general span a single intervertebral joint. The interspinales and intertransversarii muscles contribute little to any significant vertebral movements. They contain large numbers of sensory neurons within their muscultendinous fasciculi. These spindle-like sensory receptors in the muscles monitor muscle tension. These small muscles, with their poor mechanical advantage, probably function as receptors that monitor the regional movements of the vertebral column and supply feedback that influences the action of the larger surrounding muscles. Associated deep in the junction of the cranium and vertebral column are the four suboccipital muscles. The suboccipital muscles are homologous to the other deep muscles at more inferior vertebral levels, but are developmentally modified and enlarged to function with their specialized vertebral counterparts — the axis, atlas, and occipital bone.

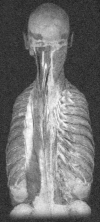

Multifidus, Rotatores, and Intersegemental Muscles
1   Multifidus lumborum muscle
2   Multifidus thoracis muscle
3   Multifidus cervicis muscle
4   Rectus capitis posterior major muscle
5   Rectus capitis posterior minor muscle
6   Obliquus capitis superior muscle
7   Obliquus capitis inferior muscle
8   Rotatores cervicis muscle
9   Rotatores thoracis muscle
10   Levatores costarum muscle

Other Muscles and Structures
11   Semispinalis cervicic muscle
12   Medial lumbar intertransversarii muscle
13   Nuchal ligament
14   External oblique muscle
15   Transversus abdominis muscle
16   Quadratus lumborum muscle
17   External intercostal muscle
18   Internal intercostal muscle
19   Middle scalene muscle
20   Trapezius muscle
21   Deltoid muscle
22   Latissimus dorsi muscle
23   Infraspinatus muscle
24   Teres major muscle
25   Rhomboideus major muscle
26   Triceps muscle
27   Posterior digastricus msucle
28   Auricularis posterior muscle
29   Transversus nuchae muscle

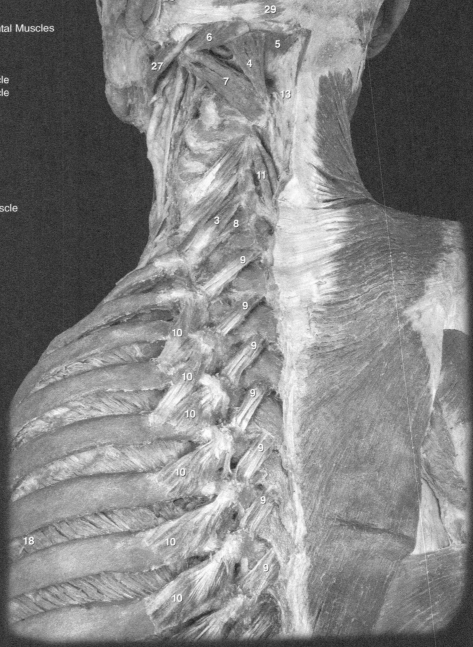

Dissection of upper deep intersegmental muscles on left
Posterior view

163

Intersegmental Muscles

The intertransversarii muscles are a mixed group that are technically misnamed. The epaxial intertransverse muscles (present at cervical, thoracic, and lumbar levels) are the "true intertransverse" muscles. They attach to the transverse elements of the vertebral arch. The hypaxial intertransverse muscles should be named intercostal muscles. They are only present in the cervical and lumbar regions and attach to the costal processes (ribs) of the cervical and lumbar vertebrae, which are unfortunately named transverse processes even though they are not homologous with the thoracic transverse processes. These cervical and lumbar transverse processes are homologous with the thoracic ribs. There are no thoracic hypaxial intertransverse muscles because they are already present as the intercostal muscles and in this region they are properly named.

Rotatores and Intersegemental Muscles
1   Rotatores thoracis muscle
2   Rotatores lumborum muscle
3   Levatores costarum muscle
4   Interspinales thoracis muscle
5   Interspinales lumborum muscle
6   Thoracic intertransversarii muscle
7   Medial lumbar intertransversarii muscle

Other Muscles and Structures
8   Intertransversarii laterales
     lumborum muscle - dorsal part
9   Intertransversarii laterales
     lumborum muscle - ventral part
10 Internal intercostal muscle
11 Quadratus lumborum muscle
12 Iliocostalis muscle (cut)
13 Multifidus muscle (cut)
14 Trapezius muscle
15 Latissimus dorsi muscle
16 Rib 12
17 Iliac crest
18 Thoracolumbar fascia
19 Supraspinous ligament

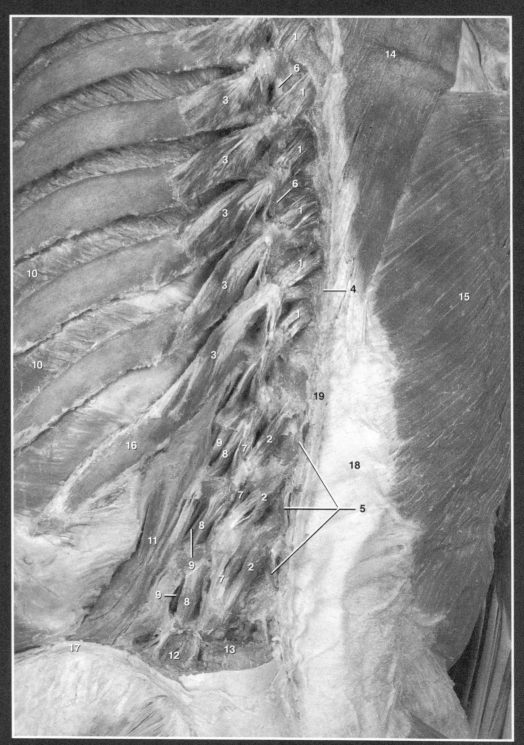

Dissection of lower deep intersegmental muscles on left
Posterolateral view

Hypaxial Muscles

The hypaxial muscles develop from the hypomere of each somite's myotome and form the lateral and ventral muscle wall of the trunk. As the hypomeres migrate to form the ventrolateral muscle wall of the trunk, a repeating segmental pattern emerges. This common muscle pattern is present in the anterior and lateral muscles of the neck, the thorax, the abdomen, and in a modified form in the wall and floor of the pelvis. Each hypomere contributes six basic muscles, per side, to the trunk wall. The six muscles are a ventral muscle, a series of four superficial to deep lateral muscles, and a subvertebral muscle. This simple, eloquent design runs the entire length of the trunk. Understanding and recognizing this pattern of design not only clarifies trunk wall anatomy, but also helps simplify the task of learning the myriad of hypaxial trunk muscles. These hypaxial trunk muscles are the flexors and rotators of the vertebral column. They also support the internal viscera of the abdomen and thorax and play important roles in respiration, vocalization, urination, and defecation. The ventral (anterior) ramus of each spinal nerve supplies all of the hypaxial muscles. The hypaxial muscle pattern and the muscles that form the pattern are summarized below. On the next two pages the pattern is clearly demonstrated.

Hypaxial Muscle Pattern
Ventral musculature
Four-layered lateral musculature
    Supracostal or outermost muscle layer
    External muscle layer
    Middle muscle layer
    Internal muscle layer
Subvertebral musculature

Cervical Hypaxial Muscles
Ventral musculature
    Geniohyoid muscle
    Thyrohyoid muscle
    Superior omohyoid muscle
    Inferior omohyoid muscle
    Sternothyroid muscle
    Sternohyoid muscle
Four-layered lateral musculature
    Supracostal layer
        Levator scapulae muscle
    External layer
        Posterior scalene muscle
    Middle layer
        Middle scalene muscle
        Lateral posterior cervical intertransversarii muscle
    Internal layer
        Anterior scalene muscle
        Anterior cervical intertransversarii muscle
Subvertebral musculature
    Longus capitis muscle
    Longus colli muscle

Thoracic Hypaxial Muscles
Ventral musculature
    Sternalis muscle (present in about 10% of people)
Four-layered lateral musculature
    Supracostal layer
        Serratus posterior superior muscle
        Serratus posterior inferior muscle
        Rhomboideus major muscle (annexed by the limb)
        Rhomboideus minor muscle (annexed by the limb)
        Serratus anterior muscle (annexed by the limb)
    External layer
        External intercostal muscle
    Middle layer
        Internal intercostal muscle
    Internal layer
        Innermost intercostal muscle
        Subcostal muscle
        Transversus thoracis muscle
        Diaphragm
Subvertebral musculature
    Longus capitis muscle

Lumbar Hypaxial Muscles
Ventral musculature
    Rectus abdominis muscle
    Pyramidalis muscle
Four-layered lateral musculature
    Supracostal layer
        External oblique muscle - superficial lamina
    External layer
        External oblique muscle - deep lamina
    Middle layer
        Internal oblique muscle
        Cremaster muscle
        Intertransversarii laterales lumborum muscle - dorsal part
    Internal layer
        Transversus abdominis muscle
        Quadratus lumborum muscle
        Intertransversarii laterales lumborum muscle - ventral part
Subvertebral musculature
    Psoas major muscle (annexed by the limb)
    Psoas minor muscle

Pelvis/Perineal Hypaxial Muscles
Ventral musculature
    Not present as it terminates on the pubic crest
Four-layered lateral musculature
    Supracostal layer
        Not present
    External layer
        Obturator externus muscle (annexed by the limb)
        Bulbospongiosus muscle
        Ischiocavernosus muscle
        Superficial transverse perinei muscle
        Superficial external anal sphincter
    Middle layer
        Obturator internus muscle (annexed by the limb)
        Deep transverse perinei - male
        Compressor urethrae - female
        Sphincter urethrovaginalis -female
        External urethral sphincter
        Deep external anal sphincter
    Internal layer
        Levator ani muscle
        Ischiococcygeus muscle
Subvertebral musculature
    Not present as psoas is annexed by the limb

165

Hypaxial Muscle Pattern

The dissection photos on this and the facing page clearly depict the pattern of design that arises from the hypomere migration in the trunk wall. Note that both the ventral and subvertebral muscles are reduced in the thorax because the sturdy thoracic cage leads to a lack of mobility in the thoracic vertebral column. Also, note that the lateral supracostal muscles of the neck and thorax are annexed by the pectoral girdle to support the unattached upper limb. The clear relationship of the serratus anterior and its abdominal homologue – the superficial lamina of the external oblique muscle – is also evident, as well as the continuity of the deep lamina of the external oblique and its homologue, the external intercostal muscle. Finally, note how the subvertebral psoas major is annexed away from the sacrum and onto the lower limb.

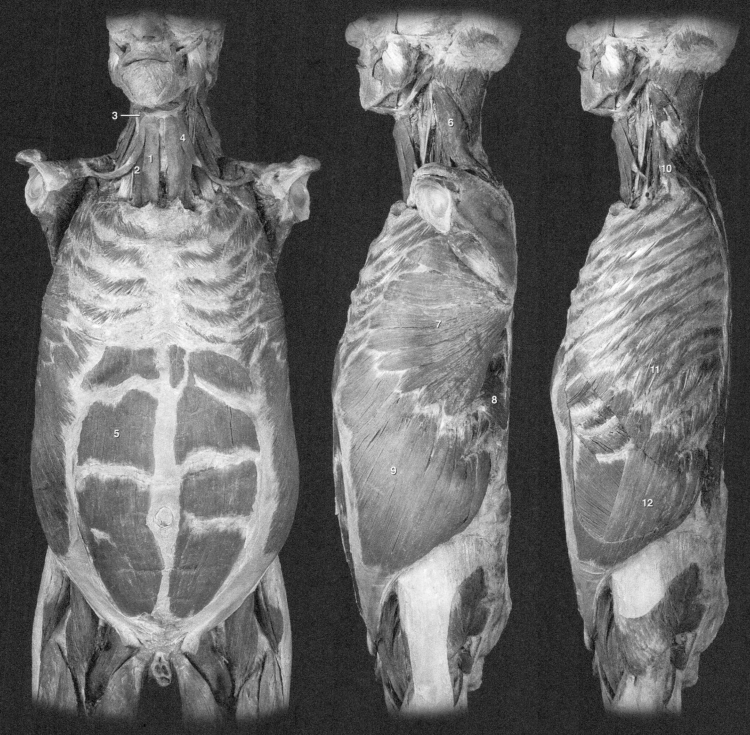

Ventral hypaxial muscles
Anterior view

Lateral supracostal hypaxial muscles
Lateral view

Lateral external hypaxial muscles
Lateral view

Ventral Musculature
1 Sternohyoid muscle
2 Sternothyroid muscle
3 Thyrohyoid muscle
4 Omohyoid muscle
5 Rectus abdominis muscle

Lateral Supracostal Musculature
6 Levator scapulae muscle
7 Serratus anterior muscle
8 Serratus posterior inferior muscle
9 External oblique muscle (superficial lamina)

Lateral External Musculature
10 Posterior scalene muscle
11 External intercostal muscle
12 External oblique muscle (deep lamina)

Lateral Middle Musculature
13 Middle scalene muscle
14 Internal intercostal muscle
15 Internal oblique muscle

Lateral Internal Musculature
16 Anterior scalene muscle

17 Innermost intercostal muscle
18 Transversus abdominis muscle

Subvertebral Musculature
19 Longus capitis muscle
20 Longus colli muscle
21 Psoas major muscle
22 Psoas minor muscle

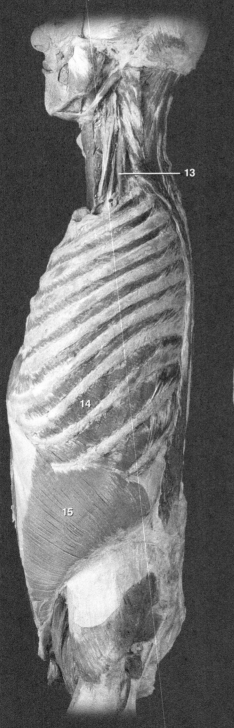

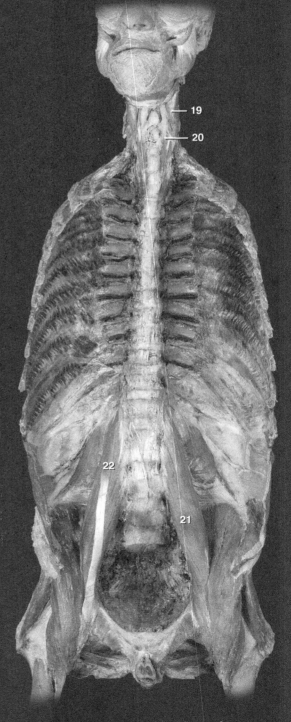

Lateral middle hypaxial muscles
Lateral view

Lateral internal hypaxial muscles
Lateral view

Subvertebral hypaxial muscles
Lateral view

Cervical Hypaxial Muscles

The muscular wall of the neck arises from the hypomeres of the cervical somites and develops in accordance with the anterior and lateral body wall muscle pattern. A close scrutiny of the cervical hypaxial muscles reveals a ventral muscle, which has split into numerous subdivisions, a four-layered lateral muscle wall where the muscles have lost their sheet-like structure, and a subvertebral muscle on the anterior surface of the neck vertebrae. The cervical trunk muscles have a variety of functions. Some of the muscles function to stabilize and move the cervical vertebral column. Some of the muscles assist in raising the upper ribs. Some are annexed by the upper limb to support the pectoral girdle. The strap-like ventral muscles, which run from sternum to larynx to hyoid bone to mandible, are active during mastication, swallowing, respiration, and sound production. These seemingly varied muscles are all innervated by the anterior rami of the cervical spinal nerves.

Cervical Hypaxial Muscles
1 Sternohyoid muscle
2 Sternothyroid muscle
3 Thyrohyiod muscle
4 Omohyoid muscle
5 Geniohyoid muscle
6 Anterior scalene muscle
7 Middle scalene muscle
8 Posterior scalene muscle
9 Levator scapulae muscle
10 Longus colli muscle

Other Muscles and Structures
11 Anterior digastricus muscle
12 Mylohyoid muscle
13 Sternocleidomastoid muscle
14 Trapezius muscle
15 Deltoid muscle
16 Pectoralis major muscle
17 Serratus anterior muscle
18 Cricothyroid muscle
19 Stylohyoid muscle
20 Posterior digastricus muscle

21 Subclavian artery
22 Root of brachial plexus
23 Common carotid artery
24 Vagus nerve
25 Thyroid cartilage
26 Thyroid gland
27 Trachea
28 External intercostal muscle
29 Internal intercostal muscle

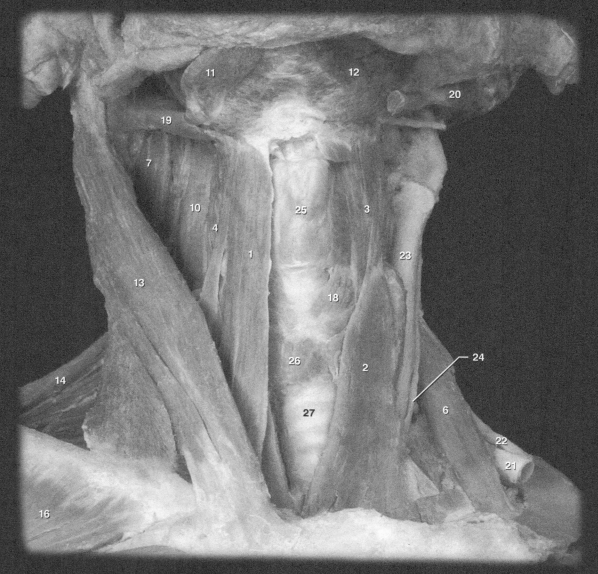

Dissection of neck muscles
Anterior view

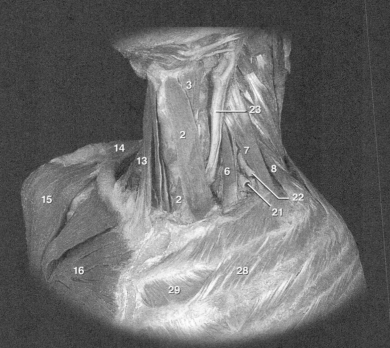

Dissection of cervical hypaxial muscles
Anterolateral view

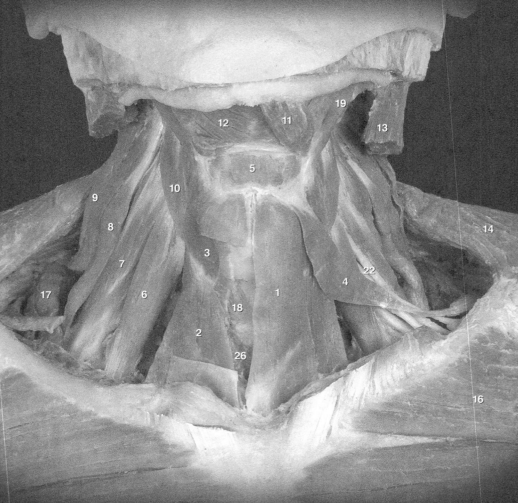

Dissection of cervical hypaxial muscles
Anterior view

Thoracic and Abdominal Hypaxial Muscles

The muscles of the thorax and abdomen develop from the hypomere of the thoracic and abdominal somites of the embryo. Like the neck they clearly demonstrate the muscle pattern of the vertebrate body wall. The thoracic body wall differs from the abdomen in having well-developed ribs that dominate the wall and limit the movements of the vertebral column. Because of the well-developed segmental ribs, the muscles of the thoracic wall retain their segmental origins. The uniquely mammalian diaphragm muscle is a member of this group that plays an important role in respiration. The outermost layer of the lateral muscle wall is well developed in the thorax. Some portions of this muscle layer remain associated with the ribs, while the rhomboid muscles (depicted in the upper limb chapter that follows) and large serratus anterior muscle migrate onto the scapula to become principal stabilizers of the upper limb. The ventral ramus of each of the thoracic and upper lumbar spinal nerves innervates these muscles.

Thoracic and Abdominal Musculature
1   Rectus abdominis muscle
2   Serratus anterior muscle
3   External intercostal muscle
4   External oblique muscle (superficial lamina)
5   External oblique muscle (deep lamina)
6   Internal intercostal muscle
7   Internal oblique muscle
8   Innermost intercostal muscle
9   Transversus abdominis muscle

Other Muscles and Structures
10   Platysma muscle
11   Sternohyoid muscle

12   Sternothyroid muscle
13   Omohyoid muscle
14   Sternocleidomastoid muscle
15   Trapezius muscle
16   Deltoid muscle
17   Pectoralis major muscle
18   Anterior scalene muscle
19   Middle scalene muscle
20   Posterior scalene muscle
21   Biceps brachii muscle
22   Tensor fasciae latae muscle
23   Gluteus medius muscle
24   Gluteus minimis muscle
25   Iliopsoas muscle

26   Pectineus muscle
27   Adductor longus muscle
28   External lamina of rectus sheath
29   Linea alba
30   Tendinous intersections
31   Internal lamina of rectus sheath
32   Semilunar line
33   Arcuate line
34   Transversalis fascia
35   Inguinal ligament
36   Spermatic cord
37   Inferior epigastric vessels
38   Cutaneous nerves

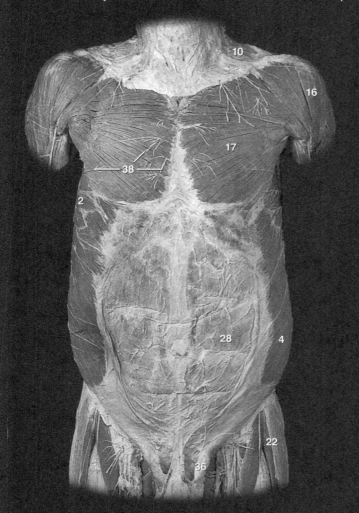
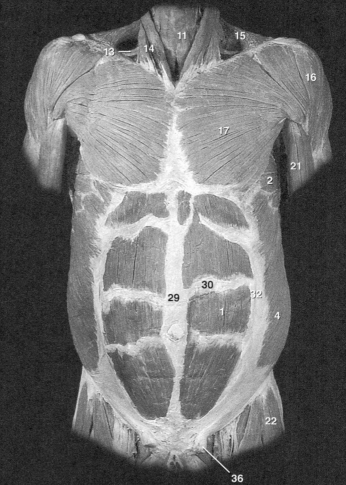

Dissections of thoracic and abdominal hypaxial muscles
Anterior view

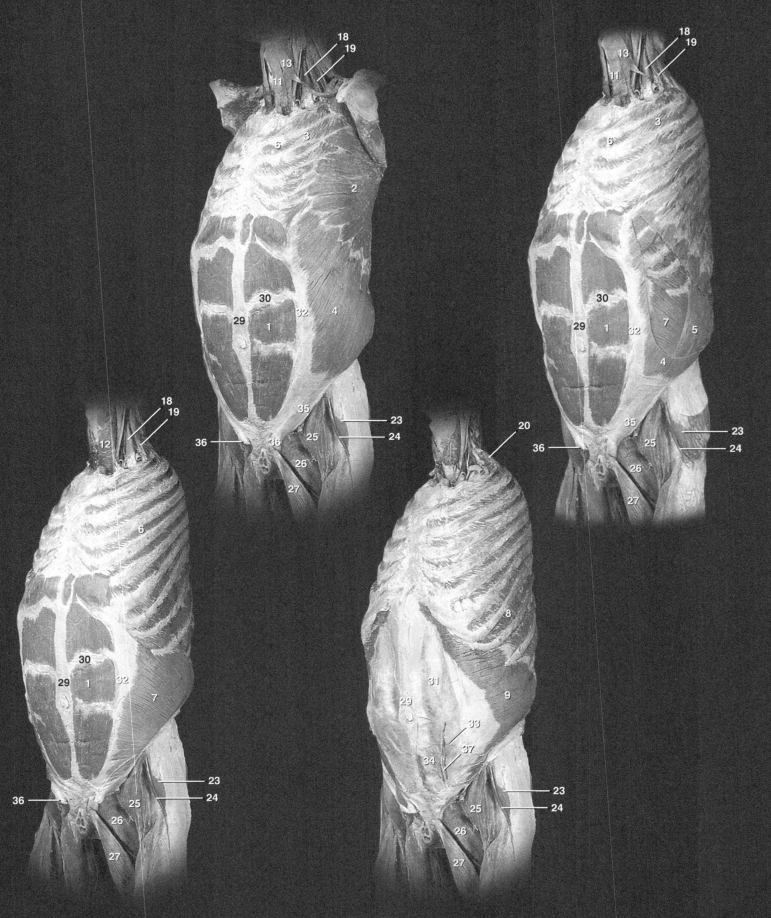

Dissections of thoracic and abdominal hypaxial muscles
Anteriolateral view

Again we would have you notice the rarely described deep lamina of the external oblique muscle. Notice its continuity with the external intercostal muscles, while the superficial lamina of the external oblique interdigitates with the serratus anterior muscle. Also note the similar fiber orientations of the intercostal muscles and their homologues in the abdominal wall. The photos of the diaphragm on the opposite page clearly reveal the continuity of this internal layer muscle with its internal homologue in the abdomen – the transversus abdominis muscle.

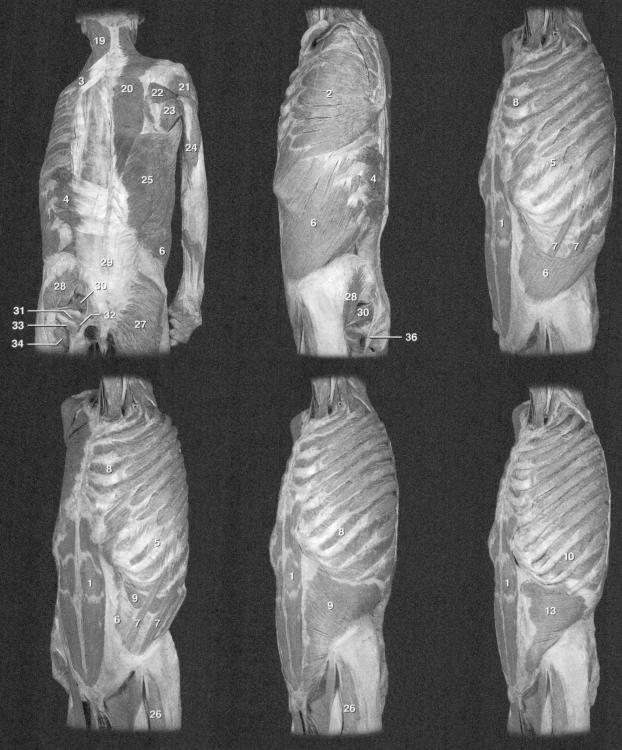

Dissections of lateral muscle layers of thoracic and abdominal wall
Posterior view upper left, Lateral view upper center, Posterolateral view all others

Thoracic and Abdominal Musculature
1 Rectus abdominis muscle
2 Serratus anterior muscle
3 Serratus posterior superior muscle
4 Serratus posterior inferior muscle
5 External intercostal muscle
6 External oblique muscle (superficial lamina)
7 External oblique muscle (deep lamina)
8 Internal intercostal muscle
9 Internal oblique muscle
10 Innermost intercostal muscle
11 Subcostal muscle
12 Diaphragm
13 Transversus abdominis muscle
14 Quadratus lumborum muscle
15 Psoas major muscle
16 Psoas minor muscle

Other Muscles and Structures
17 Longus capitis muscle
18 Longus colli muscle
19 Splenius capitis muscle
20 Trapezius muscle
21 Deltoid muscle
22 Infraspinatus muscle
23 Teres major muscle
24 Triceps brachii muscle
25 Latissimus dorsi muscle
26 Tensor fasciae latae muscle
27 Gluteus maximus muscle
28 Gluteus medius muscle
29 Gluteus minimis muscle
30 Piriformis muscle
31 Superior gemellus muscle
32 Obturator internus muscle
33 Inferior gemellus muscle
34 Quadratus femoris muscle
35 Iliacus muscle
36 Sacrotuberous ligament

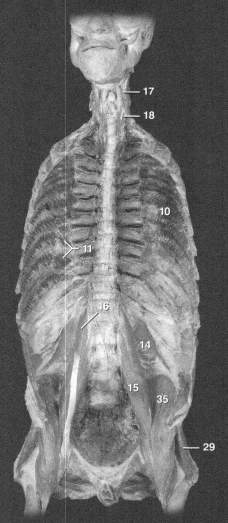

Dissection of hypaxial subvertebral muscles
Anterior view

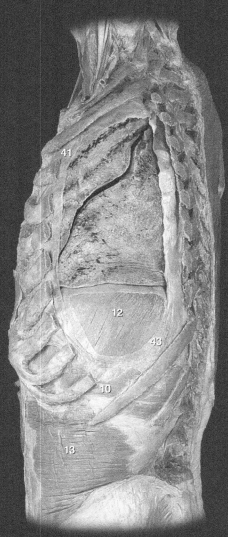

Dissection revealing diaphragm
Lateral view

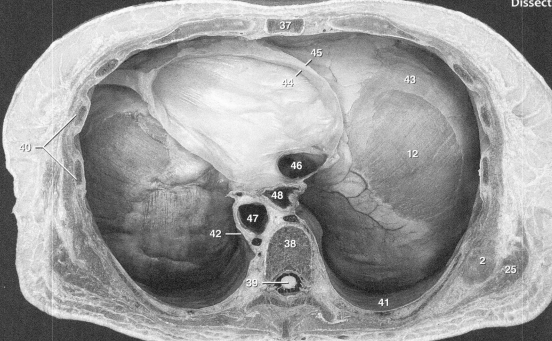

Dissection revealing diaphragm
Superior view

37 Sternum
38 Thoracic vertebra
39 Spinal cord
40 Ribs
41 Costal pleura
42 Mediastinal pleura
43 Diaphragmatic pleura
44 Parietal pericardium
45 Fibrous pericardium
46 Inferior vena cava
47 Thoracic aorta
48 Esophagus

173

Perineal Hypaxial Muscles

The ventral, subvertebral, and lateral supracostal muscles are either annexed by the lower limb or terminate above the pelvic region of the trunk. Therefore, the three inner layers of the lateral wall become the major contributors to the pelvic hypaxial wall. The three muscle layers from each side pass into the bottom of the pelvis where they meet in the midline to surround the urethra, vagina, and anus. This three-layered muscle floor at the bottom of the pelvis is called the pelvic diaphragm (internal layer) and the perineum (middle and external layers.) The pelvic diaphragm forms a basin-shaped floor that supports the pelvic viscera. The perineal muscles span the diamond-shaped pelvic outlet, and are divided into an anterior urogenital triangle and a posterior anal triangle. The perineal muscles support the pelvic viscera, form important sphincter muscles that surround the urethral and anal orifices, assist in erectile function, and propel the sperm from the male penis during ejaculation. Additional views of these muscles in both the male and female are depicted in the reproductive system chapter.

Perineal Musculature
1 Obturator externus muscle
2 Ischiocavernosus muscle
3 Bulbospongiosus muscle
4 Superficial transverse perinei muscle
5 Superficial external anal sphincter muscle

6 Deep external anal sphincter muscle
7 Deep transverse perinei muscle
8 Levator ani muscle
9 Ischiococcygeus muscle

Other Muscles and Structures
10 Gluteus maximus muscle
11 Penis (cut)
12 Obturator nerve
13 Ischial tuberosity
14 Coccyx
15 Perineal body

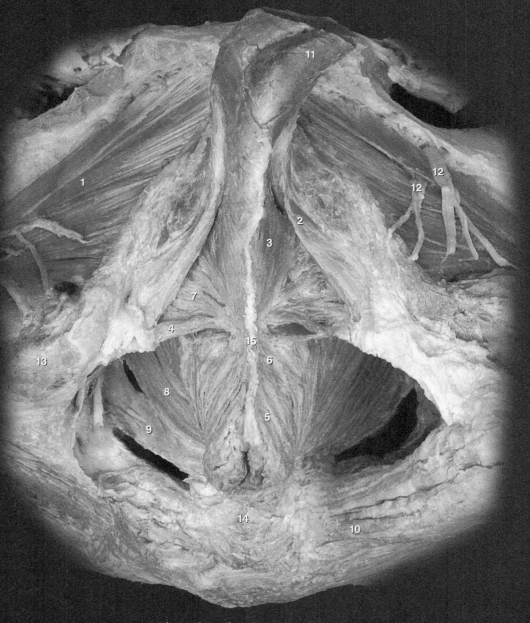

Dissection of male perineal muscles
Inferior view

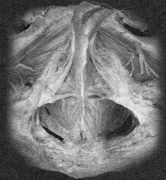

External perineal muscles

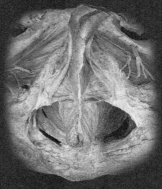

Middle perineal muscles

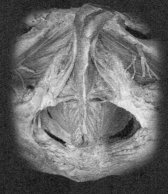

Internal perineal muscles

11 | Upper Limb Muscles

While the majority of the muscles of the upper limb arise as true limb muscles from the embryonic somites, some of the upper limb muscles are annexed from the body wall and head musculature to support and stabilize the scapula and suspend it from the trunk skeleton. The levator scapulae, rhomboideus major and minor, serratus anterior, pectoralis minor, and subclavius muscles are annexed lateral body wall muscles that help suspend the scapula, while the trapezius is an annexed branchial arch muscle that is also a part of the scapular group. Unlike these annexed body wall and head muscles, the true muscles of the limb arise from mesenchymal migrations of the somites into the developing limb bud. These migrations form two distinct muscle masses in the limb, an anterior muscle group and a posterior muscle group. As the limb develops, the two distinct muscle groups become separated by connective tissue septa and bones into anterior and posterior muscle compartments within the different sections of the limb. As the ventral rami of the associated spinal nerves grow into the developing upper limb bud, a nerve network, or plexus, develops. From this plexus posterior divisions of the network send branches into the posterior muscle compartments and anterior divisions of the network send branches into the anterior muscle compartments. At the proximal end of the limb, some of the true limb muscles from the anterior and posterior compartments increase in size and migrate back onto the trunk. As they spread onto the trunk, they cover the body wall muscles and attach to the axial skeleton. This muscular expansion of the proximal limb muscles increases their mechanical advantage at the shoulder joint. Because of this interesting arrangement of body wall muscles and true limb muscles at the shoulder end of the superior limb, a clear compartment organization is not evident. For this reason, we will group these muscles into groups that share some common feature, such as a common attachment or function. In the limb proper we group the muscles into their developmental anterior and posterior muscle compartments. This greatly simplifies the learning process because most of the muscles in a compartment share common attachments, actions, and nerves. Grouping things in this way can help to simplify the learning process.

Find more information about the muscles of the upper limb in
REAL ANATOMY

Upper Limb Muscles

This chapter depicts the interesting array of muscles of the upper limb. Because of its weak ligamentous association with the axial skeleton, the upper limb annexed muscles from the outer layer of the trunk wall and head to help suspend it from the axial skeleton. This scapular muscle sling, which has no homologous counterpart in the lower limb, is the major difference between the muscles of the upper and lower limbs. On the pages that follow we present the muscles of the upper limb and organize them primarily by developmental groups, with the exception of the muscles of the shoulder joint (see the outline below). The opposite page and the two pages that follow show anterior and posterior views of the upper limb muscles and their relationships to the trunk musculature.

Pectoral Girdle Muscles
(Annexed from head muscles (trapezius) and outermost layer of lateral trunk muscles to support and stabilize scapula)
    Trapezius
    Levator scapulae
    Rhomboideus major
    Rhomboiedus minor
    Serratus anterior
    Pectoralis minor
    Subclavius

Shoulder Joint Muscles
    Rotator cuff muscles
    (Muscles with a ligamentous role that function as stabilizers of the weakly ligamentous shoulder joint)
      Supraspinatus
      Infraspinatus
      Teres minor
      Subscapularis
    Intertubercular groove muscles
    (Muscles that share an insertion on the intertubercular groove and are prime movers of the shoulder joint)
      Pectoralis major
      Latissimus dorsi
      Teres major
    Deltoid

Anterior Brachial Muscles
(Nerve supply - musculocutaneous nerve; function as flexors of the shoulder and elbow)
    Coracobrachialis
    Brachialis
    Biceps brachii

Posterior Brachial Muscles
(Nerve supply - radial nerve, like all posterior compartment muscles; functions as extensor of shoulder and elbow)
    Triceps brachii

Anterior Antebrachial Muscles
(Nerve supply - median and ulnar nerves; function as flexors of wrist and digits)
    Superficial muscles
      Pronator teres
      Flexor carpi radialis
      Palmaris longus
      Flexor carpi ulnaris
      Flexor digitorum superficialis
    Deep muscles
      Flexor digitorum profundus
      Flexor pollicis longus
      Pronator quadratus

Posterior Antebrachial Muscles
(Nerve supply - radial nerve; function as extensors of the wrist and digits)
    Lateral muscles
      Brachioradialis
      Extensor carpi radialis longus
      Extensor carpi radialis brevis
      Extensor digitorum
      Extensor digiti minimi
      Extensor carpi ulnaris
      Anconeus
      Supinator
    Radial muscles
      Abductor pollicis longus
      Extensor pollicis longus
      Extensor pollicis brevis
      Extensor indicis

Hand Muscles
(All intrinsic hand muscles arise from anterior muscles of embryonic limb and are innervated by the median and ulnar nerve from the anterior divisions of the plexus)
    Thenar Muscles
    (All supplied by the median nerve except adductor pollicis)
      Abductor pollicis brevis
      Flexor pollicis brevis
      Opponens pollicis
      Adductor pollicis
    Hypothenar Muscles
    (All supplied by the ulnar nerve)
      Palmaris brevis
      Abductor digiti minimi
      Flexor digiti minimi
      Opponens digiti minimi
    Intermetacarpal Muscles
    (All supplied by the ulnar nerve except first two lumbricals)
      Lumbricales
      Palmar interossei
      Dorsal interossei

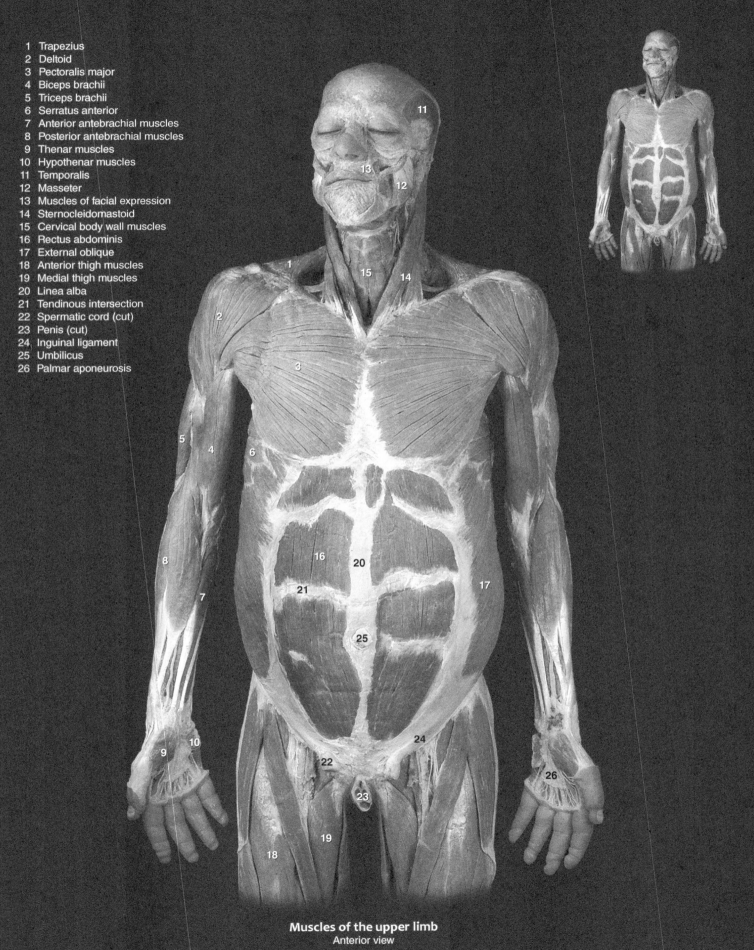

1 Trapezius
2 Deltoid
3 Pectoralis major
4 Biceps brachii
5 Triceps brachii
6 Serratus anterior
7 Anterior antebrachial muscles
8 Posterior antebrachial muscles
9 Thenar muscles
10 Hypothenar muscles
11 Temporalis
12 Masseter
13 Muscles of facial expression
14 Sternocleidomastoid
15 Cervical body wall muscles
16 Rectus abdominis
17 External oblique
18 Anterior thigh muscles
19 Medial thigh muscles
20 Linea alba
21 Tendinous intersection
22 Spermatic cord (cut)
23 Penis (cut)
24 Inguinal ligament
25 Umbilicus
26 Palmar aponeurosis

Muscles of the upper limb
Anterior view

Upper Limb Muscles

Upper Limb Muscles
 1 Trapezius
 2 Deltoid
 3 Pectoralis major
 4 Biceps brachii
 5 Triceps brachii
 6 Serratus anterior
 7 Teres major
 8 Infraspinatus
 9 Teres minor
10 Latissimus dorsi
11 Posterior antebrachial muscles
12 Anterior antebrachial muscles
13 Hypothenar muscles
14 Intermetacarpal muscle

Other Muscles and Structures
15 Muscles of mastication
16 Muscles of facial expression
17 Sternocleidomastoid
18 Rectus abdominis
19 External oblique
20 Gluteal muscles
21 Posterior thigh muscles
22 Thoracolumbar fascia
23 Antebrachial fascia
24 Iliotibial tract

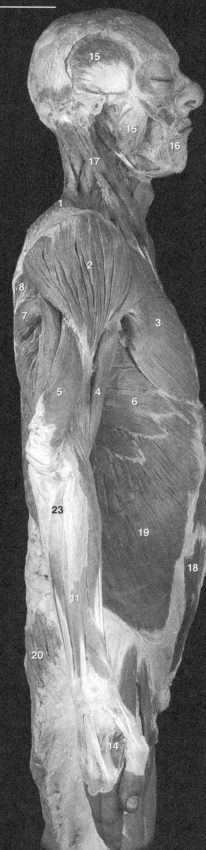

Muscles of the upper limb
Right lateral view

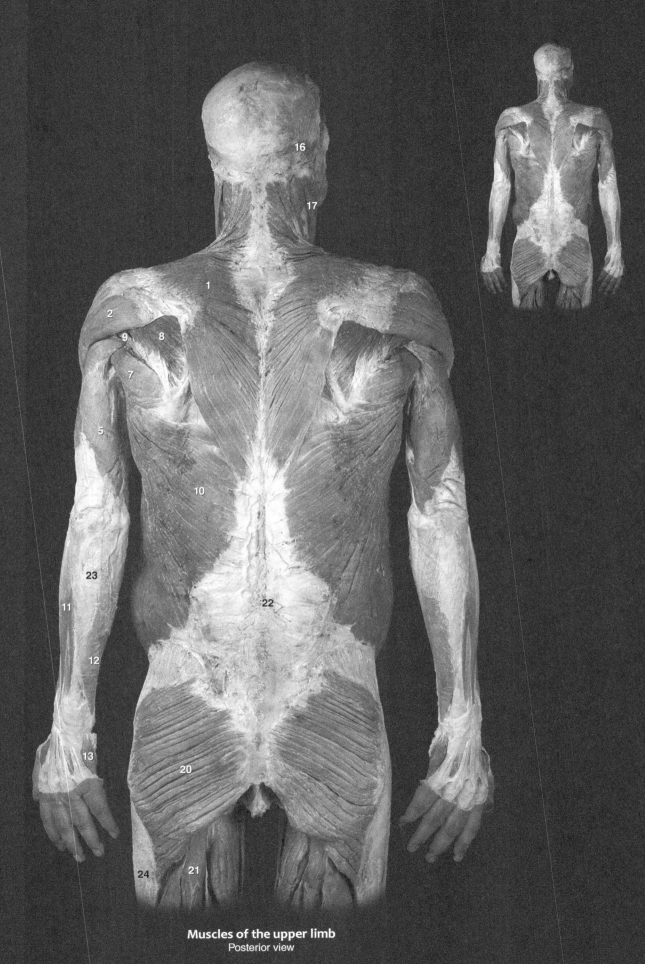

Muscles of the upper limb
Posterior view

179

Scapular Muscles

The muscles that insert on the scapula and anchor it to the trunk form an extensive muscular sling. During development the upper limb annexes these muscles from the head and trunk wall. They share the common functional goal of moving the scapula, stabilizing it, and anchoring it to the axial skeleton. These muscles are some of the larger muscles of the upper limb, yet produce visibly minor movements of the skeleton. Realize, however, that their major role is to stabilize and anchor the scapula to the axial skeleton. With the exception of the pectoralis minor, the nerves that supply these muscles arise from the roots of the brachial plexus.

Scapular Musles
1 Trapezius
2 Levator scapulae
3 Rhomboideus minor
4 Rhomboideus major
5 Serratus anterior
6 Pectoralis minor
7 Subclavius

Other Muscles and Structures
8 Sternocleidomastoid
9 Omohyoid
10 Clavicle
11 Deltoid
12 Coracobrachialis
13 Pectoralis major (cut)
14 External intercostal
15 Internal intercostal
16 Biceps brachii
17 Brachialis
18 Triceps brachii
19 Latissimus dorsi
20 Supraspinatus
21 Infraspinatus
22 Teres major
23 External oblique
24 Rectus abdominis
25 Brachioradialis
26 Extensor carpi radialis longus
27 Serratus posteror inferior
28 Teres minor
29 External oblique aponeurosis
30 Trachea
31 Spine of scapula
32 Greater tubercle of humerus
33 Rib

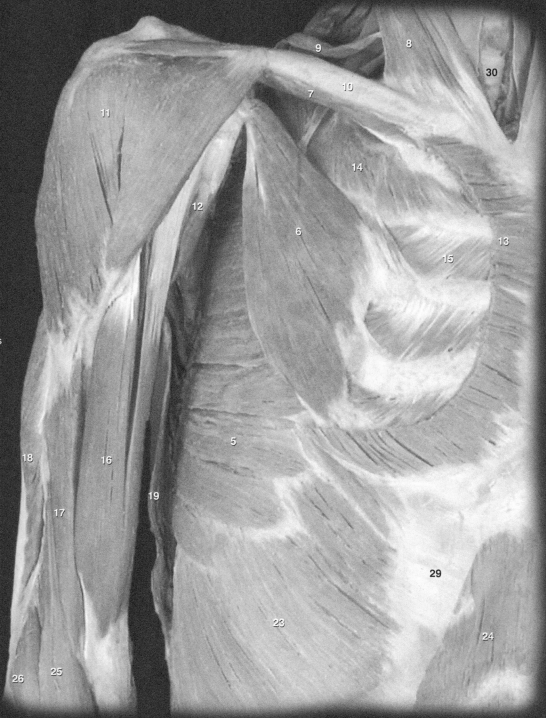

Muscles of right brachium, shoulder, and chest
Anterior view

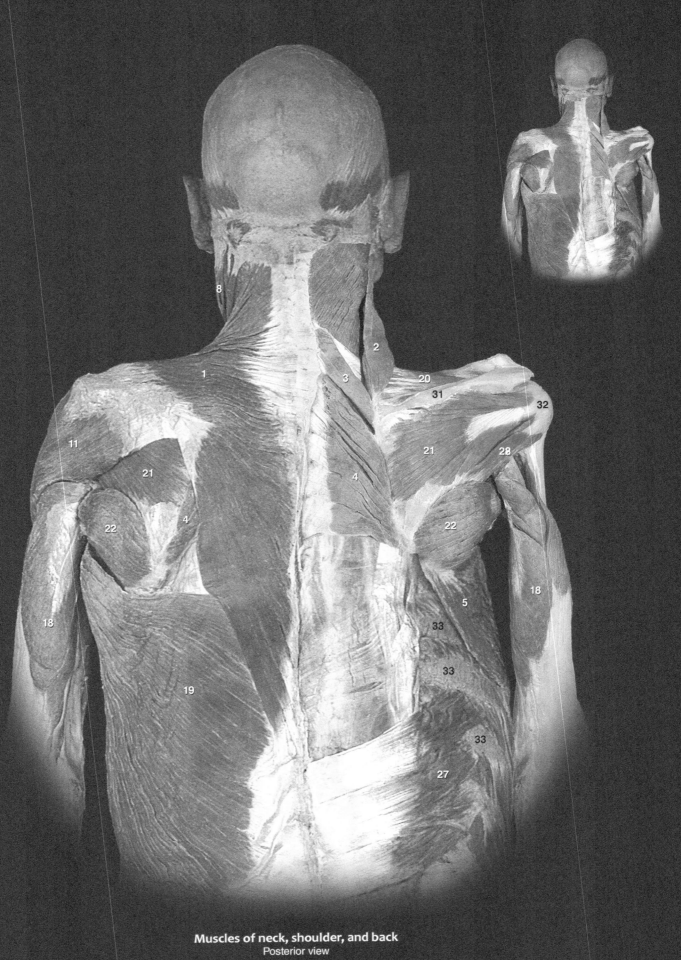

Muscles of neck, shoulder, and back
Posterior view

Shoulder Muscles - Rotator Cuff

The rotator cuff muscles are an important muscle group that play a critical role in stabilizing the shoulder joint. The four muscles (supraspinatus, infraspinatus, teres minor, and subscapularis) have thick, flat tendons of insertion that form a strong musculotendinous cuff around all but the inferior aspect of the glenohumeral joint. These tendons are intimately applied to the fibrous membrane of the joint capsule. Individually each muscle contributes little to the total range of motion of the humerus at the glenohumeral joint. However, they play a prominent role in stabilizing the joint and positioning and stabilizing the head of the humerus in the glenoid cavity. When the rotator cuff muscles are compromised by injury, the shoulder joint loses stability and becomes highly susceptible to dislocation.

Rotator Cuff Muscles
1 Supraspinatus
2 Infraspinatus
3 Teres minor
4 Subscapularis

Other Muscles and Structures
5 Biceps brachii
6 Coracobrachialis
7 Triceps brachii
8 Teres major
9 Coracoid process of scapula
10 Superior angle of scapula
11 Inferior angle of scapula
12 Spine of scapula
13 Medial border of scapula
14 Greater tubercle of humerus

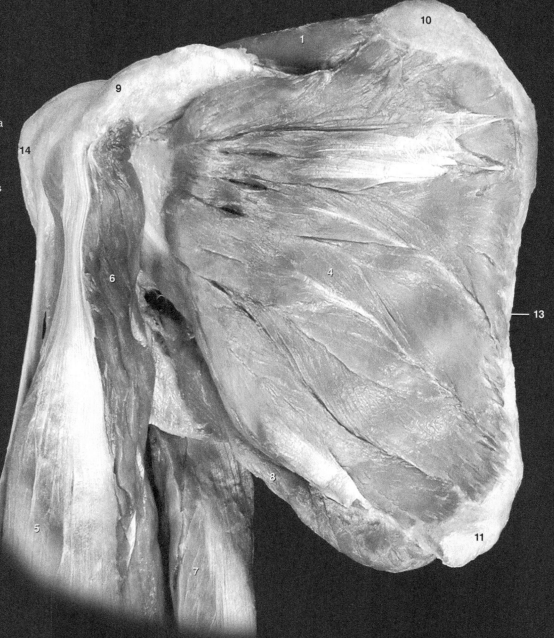

Deep dissection of the right shoulder muscles
Anterior view

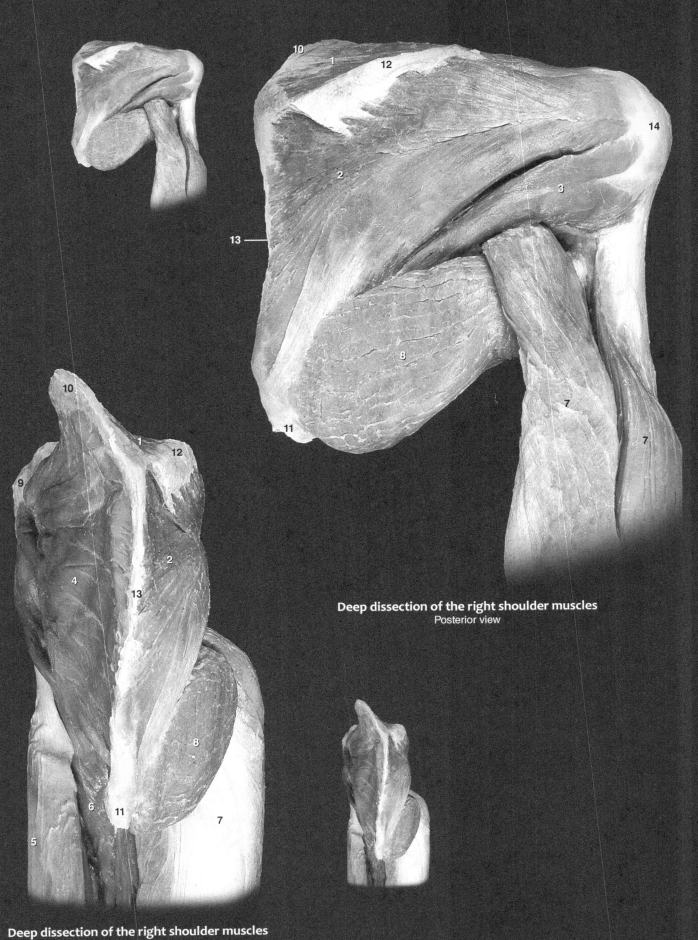

Deep dissection of the right shoulder muscles
Posterior view

Deep dissection of the right shoulder muscles
Medial view

Shoulder Muscles - Prime Movers

The prime movers of the shoulder joint are the muscles that share a common attachment on the intertubercular groove (pectoralis major, teres major, and latissimus dorsi) and the deltoid muscle. These large muscles are superficial to the muscles of the rotator cuff and form extensive attachments on the pectoral girdle and axial skeleton. Inserting more distally on the humerus then the muscles of the rotator cuff, they have a better mechanical advantage and produce the major movements of the shoulder joint. The intertubercular groove muscles also form the anterior and posterior walls of the axilla. The large pectoralis major forms the anterior wall of the axilla, while the sheet-like latissimus dorsi and thick, round teres major form the posterior axillary wall.

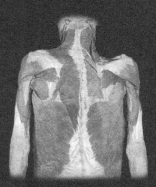

Shoulder Prime Movers
1 Deltoid
2 Pectoralis major
3 Teres major
4 Latissimus dorsi

Other Muscles and Structures
5 Levator scapulae
6 Rhomboideus minor
7 Rhomboideus major
8 Supraspinatus
9 Infraspinatus
10 Teres minor
11 Triceps brachii
12 Trapezius
13 Spleneus capitis
14 Serratus anterior
15 Pectoralis minor
16 External intercostal
17 Internal intercostal
18 Rectus abdominis
19 Coracobrachialis
20 Biceps brachii
21 Brachialis
22 Posterior scalene
23 Middle scalene
24 Anterior scalene
25 Omohyoid
26 Sternohyoid
27 Sternothyroid
28 Thyrohyoid
29 Sternocleidomastoid
30 External oblique
31 Brachioradialis
32 Clavicle
33 Humerus
34 Spine of scapula
35 Thoracolumbar fascia
36 Linea alba
37 Common carotid artery

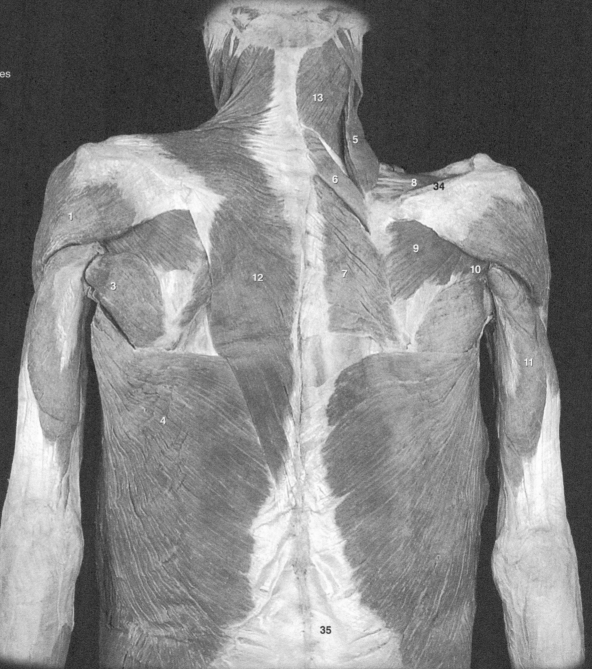

Muscles of neck, shoulder, brachium, and back
Posterior view

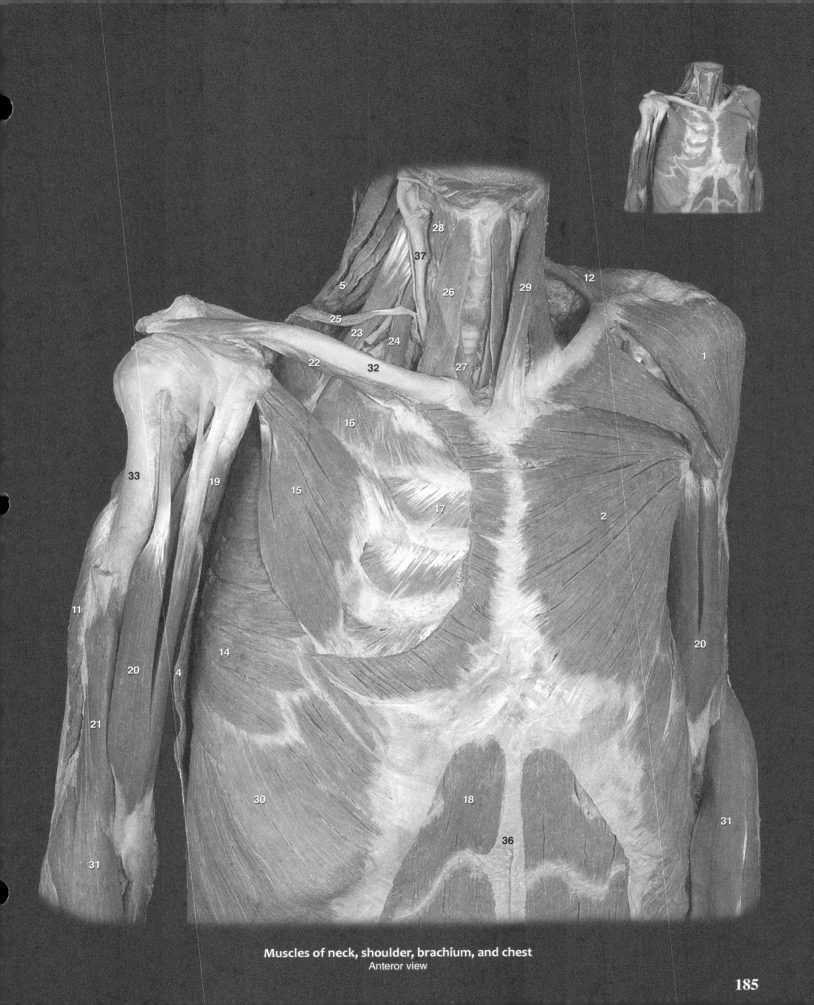

Muscles of neck, shoulder, brachium, and chest
Anteror view

Anterior Brachial Muscles

The anterior muscle compartment of the brachium consists of three muscles — the coracobrachialis, brachialis, and biceps brachii. The coracobrachialis and brachialis each cross a single joint, the shoulder joint and elbow joint respectively. The biceps brachii crosses three joints, the shoulder, and the humero-ulnar and radio-ulnar joints of the elbow. The muscles share in common the actions of flexion of the shoulder and elbow. All three muscles are innervated by the musculocutaneous nerve.

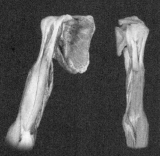

Anterior Brachial Muscles
1 Coracobrachialis
2 Brachialis
3 Biceps brachii - long head
4 Biceps brachii - short head
5 Triceps brachii

Other Muscles and Structures
6 Supraspinatus
7 Subscapularis
8 Teres major
9 Brachioradialis
10 Pronator teres
11 Coracoid process
12 Superior angle
13 Inferior angle
14 Greater tubercle
15 Lesser tubercle

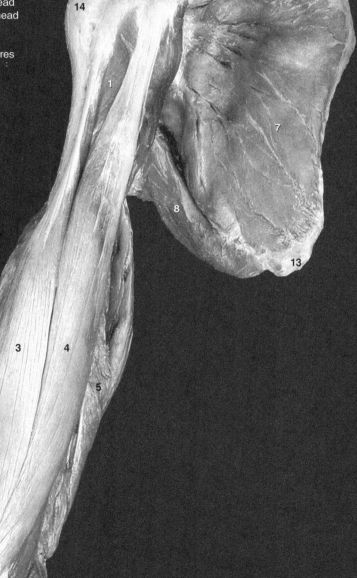

Muscles of the right brachium and scapula
Anterior view

Deep muscles of the right brachium
Anterior view

Posterior Brachial Muscles

The three headed triceps brachii muscle is the sole muscle of the posterior compartment of the brachium. This large muscle extends the shoulder and elbow joints and is innervated by the radial nerve.

Posterior Brachial Muscles
1 Triceps brachii - medial head
2 Triceps brachii - lateral head
3 Triceps brachii - long head
4 Biceps brachii - long head
5 Beceps brachii - short head
6 Brachialis

Other Muscles and Structures
7 Supraspinatus
8 Infraspinatus
9 Teres minor
10 Teres major
11 Humerus
12 Greater tubercle
13 Spine of scapula
14 Brachail artery

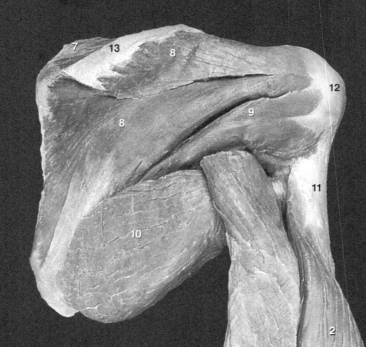

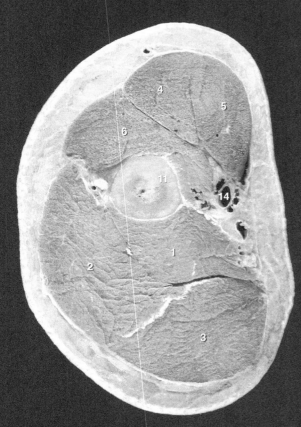

Transverse section of right midbrachim
Inferior view

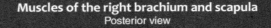

Muscles of the right brachium and scapula
Posterior view

Anterior Antebrachial Muscles

The muscles of the anterior antebrachium form three distinct muscle layers. The superficial group has four superficial muscles (pronator teres, flexor carpi radialis, palmaris longus, and flexor carpi ulnaris) covering the intermediate flexor digitorum superficialis. All five of these muscles share a common attachment on the medial epicondyle of the humerus. The three deep muscles (flexor digitorum profundus, flexor pollicis longus, and pronator quadratus) do not cross the elbow joint. Other than the two pronators, all the muscles are flexors of either the wrist or digits. The median nerve innervates all but the flexor carpi ulnaris and the ulnar half of the flexor digitorum profundus, both of which are supplied by the ulnar nerve.

Anterior Antebrachial Muscles
1 Pronator teres
2 Flexor carpi radialis
3 Palmaris longus
4 Flexor carpi ulnaris
5 Flexor digitorum superficialis
6 Flexor digitorum profundus
7 Flexor pollicis longus
8 Pronator quadratus

Other Muscles and Structures
9 Brachialis
10 Palmar aponeurosis
11 Brachial artery
12 Radial artery
13 Ulnar artery
14 Anterior interosseous artery
15 Interosseous membrane
16 Abductor pollicis brevis
17 Flexor pollicis brevis
18 Lumbricals
19 Adductor pollicis
20 Flexor digiti minimi brevis
21 Abductor digiti minimi
22 Palmaris brevis
23 Supinator
24 Superficial transverse metacarpal ligament

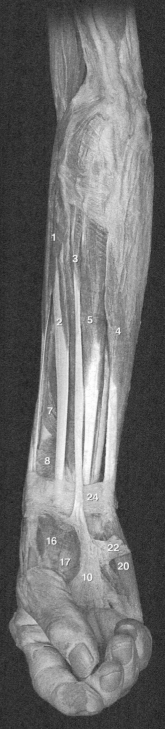

Superficial muscles of the right antebrachium
Anterior view, hand pronated

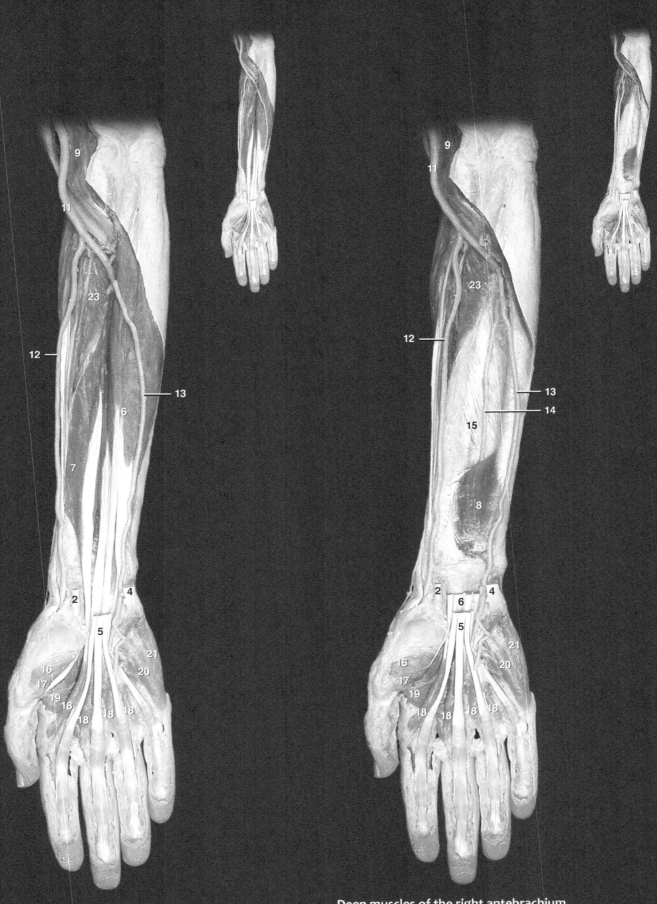

Deep muscles of the right antebrachium
Anterior view, superficial muscles removed and hand pronated

Deep muscles of the right antebrachium
Anterior view, muscles removed to expose pronator quadratus

There are two muscle groups in the posterior antebrachium — the eight muscles of the lateral group that share a common attachment on or near the lateral epicondyle of the humerus and the four muscles of the radial group that course along the distal aspect of the radius to insert on the thumb and first finger. Like the triceps of the posterior brachial compartment, all the muscles of the posterior antebrachium receive innervation via the radial nerve. With a few exceptions, the muscles are extensors of either the elbow, wrist, or digits.

Posterior Antebrachial Muscles
1 Brachioradialis
2 Anconeus
3 Supinator
4 Extensor carpi radialis longus
5 Extensor carpi radialis brevis
6 Extensor digitorum
7 Extensor digiti minimi
8 Extensor carpi ulnaris
9 Abductor pollicis longus
10 Extensor pollicis longus
11 Extensor pollicis brevis
12 Extensor indicis

Other Muscles and Structures
13 Biceps brachii
14 Brachialis
15 Triceps brachii
16 Flexor carpi radialis
17 Pronator teres
18 Flexor pollicis longus
19 Abductor digiti minimi
20 Dorsal interossei

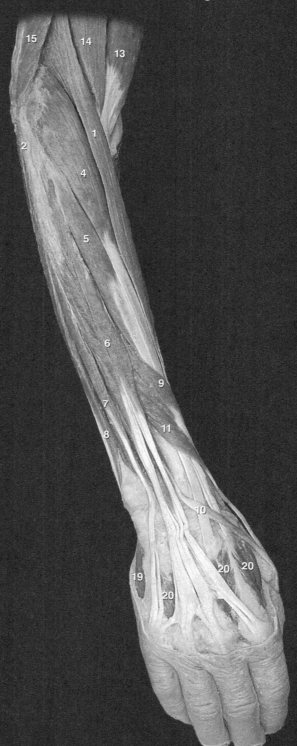

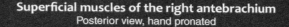

Superficial muscles of the right antebrachium
Posterior view, hand pronated

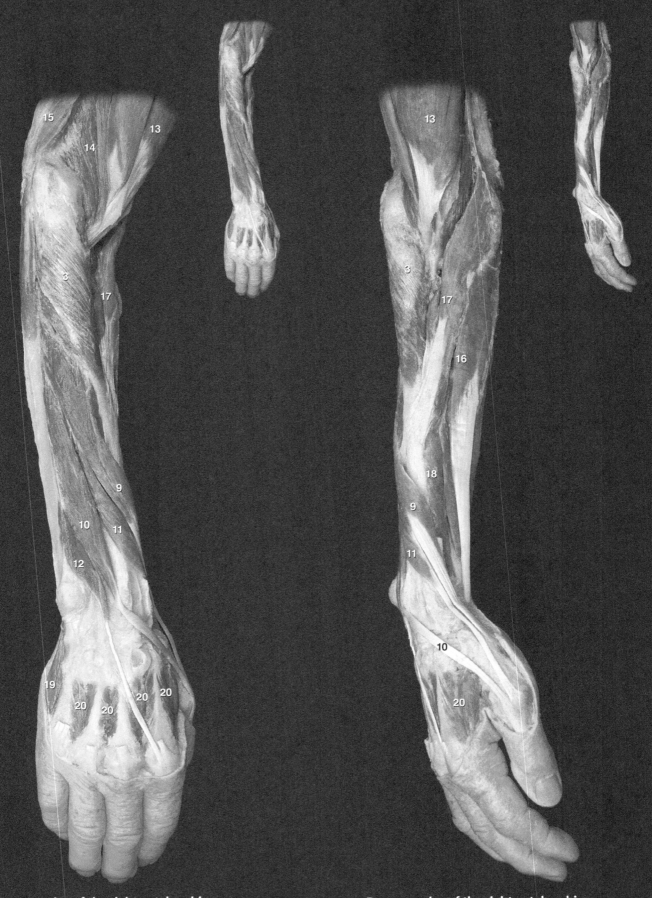

Deep muscles of the right antebrachium
Posterior view, lateral group muscles removed and hand pronated

Deep muscles of the right antebrachium
Anterolateral view, lateral group muscles removed and hand pronated

Hand Muscles

There are three muscle groups in the hand — the muscles of the thenar eminence at the base of the thumb, the muscles of the hypothenar eminence at the base of the little finger, and the three layers of intermetacarpal muscles that occupy the spaces between the metacarpal bones. All of these muscles arise from the anterior muscles of the embryonic limb bud and receive anterior division nerve supply from the median and ulnar nerves as they pass from the anterior antebrachium into the hand. While the median nerve supplies the majority of the muscles of the anterior antebrachium, the ulnar nerve supplies all but three of the muscles in the hand.

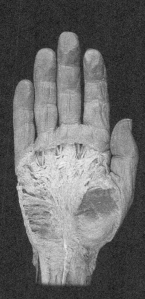

**Muscles of the
thenar eminence**

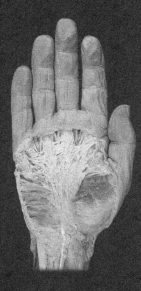

**Muscles of the
hypothenar eminence**

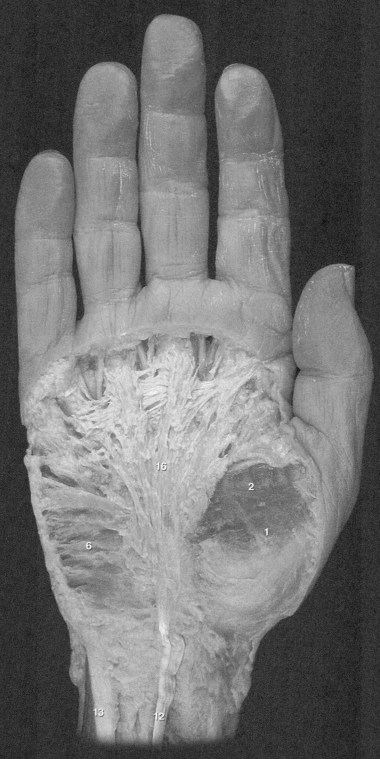

Superficial muscles of the right hand
Anterior view

Hand Muscles
1 Abductor pollicis brevis
2 Flexor pollicis brevis
3 Adductor pollicis
4 Abductor digiti minimi
5 Flexor digiti minimi brevis
6 Palmaris brevis

7 Lumbricals
8 Palmar interossei
9 Dorsal interossei

Other Muscles and Structures
10 Flexor digitorum superficialis
11 Flexor digitorum profundus

12 Palmaris longus
13 Flexor carpi ulnaris
14 Flexor pollicis longus
15 Flexor carpi radialis
16 Palmar aponeurosis
17 Flexor retinaculum
18 Ulna

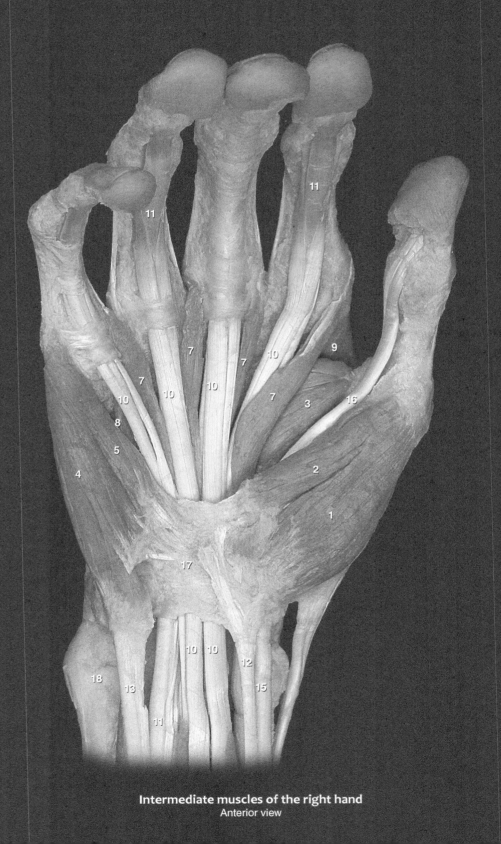

Intermediate muscles of the right hand
Anterior view

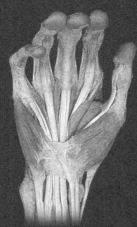

Muscles of the thenar eminence

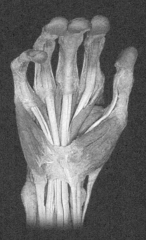

Muscles of the hypothenar eminence

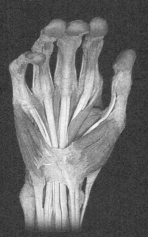

Intermetacarpal muscles

Hand Muscles

Hand Muscles
1 Abductor pollicis brevis (cut)
2 Flexor pollicis brevis (cut)
3 Opponens pollicis
4 Adductor pollicis
5 Abductor digiti minimi
6 Flexor digiti minimi brevis
7 Opponens digiti minimi
8 Palmaris brevis

9 Lumbricals (cut)
10 Palmar interossei
11 Dorsal interossei

Other Muscles and Structures
12 Flexor digitorum superficialis
13 Flexor digitorum profundus
14 Carpal tunnel

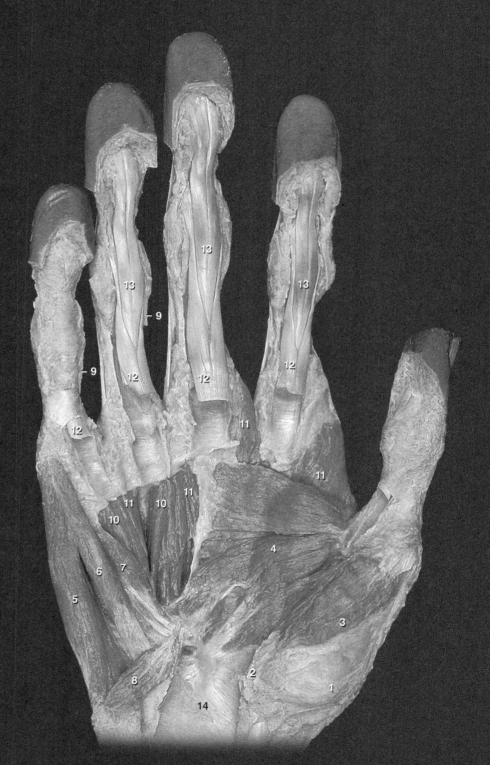

Deep muscles of the right hand
Anterior view

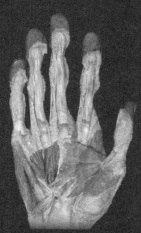

Muscles of the thenar eminence

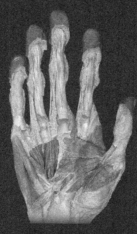

Muscles of the hypothenar eminence

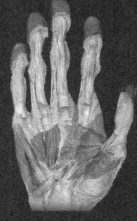

Intermetacarpal muscles

12 | Lower Limb Muscles

The design of the inferior limb musculature is similar to that of the true limb muscles of the superior limb. The major difference between the two limbs is that the proximal end of the lower limb forms a direct skeletal attachment to the vertebral column via the strong sacro-iliac joint, unlike the unattached scapula of the superior limb. Because of this difference, the inferior limb does not require body wall muscles to support, stabilize, and suspend it from the axial skeleton. There are two additional features that are important to keep in mind when studying this powerful locomotor limb. First, during development of the lower limb the embryonic posterior muscles rotate and reposition themselves to the anterior aspect of the limb. For this reason the knee and ankle move directly opposite the elbow and wrist. The second notable feature is that there are three muscle compartments in the thigh and leg, as compared to just two in the brachium and antebrachium. One of the two original compartments in each lower limb segment (thigh and leg) splits to give rise to an additional compartment. The thigh has an anterior compartment and a posterior compartment, but the posterior compartment is subdivided into posterior and medial compartments. The leg has a large posterior compartment and a smaller anterior compartment and the anterior compartment is subdivided into anterior and lateral compartments. As with the upper limb, we present the muscles of the lower limb proper in their muscle compartments. Again, this greatly simplifies the learning process because most of the muscles in a compartment share similar attachments, perform common actions, and have a common nerve supply. Unlike the compartmental muscles of the lower limb proper, the proximal muscles of the lower limb that surround the hip joint are a more diverse group of muscles. Some are true limb muscles, while others are annexed muscles from the trunk wall. We organize these hip muscles into three groups — the deep hip rotator muscles, the gluteal muscles, and the hip flexors.

Find more information about the muscles of the lower limb in

REAL ANATOMY

Lower Limb Muscles

The muscles of the lower limb share similarities with their upper limb counterparts, yet have important differences. As you will notice in the groups below there are no homologues in the lower limb to the scapular muscles of the upper limb. Like the shoulder muscles, the muscles surrounding the hip joint are a varied group of muscles, with some annexed from the body wall of the abdominopelvic region. In the limb proper the muscles develop in muscular compartments as they do in the upper limb; however, the embryonic posterior aspect of the limb rotates to an anterior position. As a result, the nerves that arise from the posterior divisions of the lumbosacral plexus innervate the anterior muscle compartments, and the nerves from the anterior divisions of the plexus innervate the posterior muscle compartments. The developmental groups of muscles and their nerve supply are outlined below.

Hip Muscles
    Gluteal muscles
    (Nerve supply - gluteal nerves, superior to maximus and inferior to the other three; arise from lateral aspect of ilium and are prime movers and stabilizers of hip joint)
        Gluteus maximus
        Gluteus medius
        Gluteus minimus
        Tensor fasciae latae
    Deep hip rotator muscles
    (All are lateral rotators of the hip joint and insert on the medial aspect of greater trochancter)
        Piriformis
        Obturator internus
        Obturator externus
        Superior gemellus
        Inferior gemellus
        Quadratus femoris
    Hip flexor muscles
        Psoas major
        Iliacus

Anterior Thigh Muscles
(Nerve supply - femoral nerve; major extensor group of the knee)
    Sartorius
    Quadriceps femoris
        Rectus femoris
        Vastus lateralis
        Vastus intermedius
        Vastus medialis
    Articularis genu

Medial Thigh Muscles
(Nerve supply - obturator nerve with exception of pectineus, which is supplied by femoral nerve and condylar head of adductor magnus, which is supplied by tibial nerve)
    Pectineus
    Adductor brevis
    Adductor longus
    Adductor magnus
    Adductor minimis
    Gracilis

Posterior Thigh Muscles
(Nerve supply - Tibial nerve with exception of short head of biceps femoris, which is supplied by common fibular nerve)
    Biceps femoris
    Semitendinosus
    Semimembranosus

Anterior Leg Muscles
(Nerve supply - deep fibular nerve)
    Tibialis anterior
    Extensor digitorum longus
    Extensor hallucis longus
    Peroneus tertius

Lateral Leg Muscles
(Nerve supply - superficial fibular nerve)
    Peroneus longus
    Peroneus brevis

Posterior Leg Muscles
(Nerve supply - tibial nerve)
    Triceps surae
        Gastrocnemius
        Soleus
    Plantaris
    Popliteus
    Tibialis posterior
    Flexor digitorum longus
    Flexor hallucis longus

Dorsal Foot Muscles
(Nerve supply - deep fibular nerve)
    Extensor hallucis brevis
    Extensor digitorum brevis

Plantar Foot Muscles
(Nerve supply - tibial nerve via its terminal branches, medial plantar nerve supplies first lumbrical, abductor hallucis, flexor hallucis brevis, and flexor digitorum brevis; lateral plantar nerve supplies all the others)
    First layer
        Abductor hallucis
        Flexor digitorum brevis
        Abductor digiti minimi
    Second layer
        Quadratus plantae
        Lumbricales
    Third layer
        Flexor halluci brevis
        Adductor hallucis
        Flexor digiti minimi brevis
    Fourth layer
        Plantar interossei
        Dorsal interossei

1	Tensor fasciae latae	9	Vastus lateralis	17	Soleus
2	Iliacus	10	Vastus medialis	18	Tibialis anterior
3	Psoas major	11	Gluteus maximus	19	Fibularis longus
4	Pectineus	12	Adductor magnus	20	Fibularis brevis
5	Adductor longus	13	Biceps femoris	21	Iliotibial tract
6	Gracilis	14	Semitendinosus	22	Calcaneal tendon
7	Sartorius	15	Semimembranosus	23	Quadriceps tendon
8	Rectus femoris	16	Gastrocnemius	24	Flexor digitorum longus

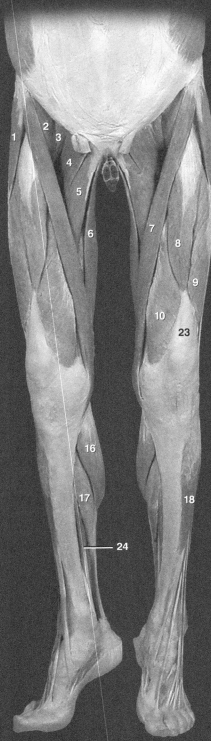

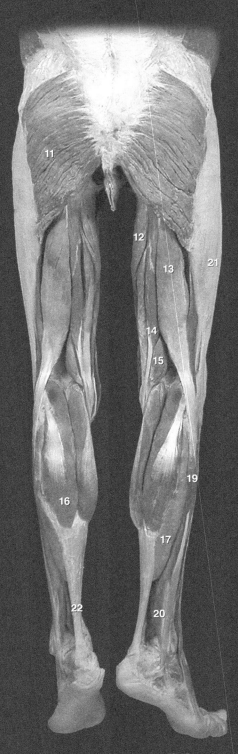

Muscles of the lower limb
Anterior view

Muscles of the lower limb
Posterior view

Hip Muscles

The muscles that surround the hip joint form three groups. The gluteal muscles arise from the posterior musculature of the embryonic limb bud and are prime movers of the hip joint. They create the characteristic profile of the human buttocks. The deep hip rotator muscles are closely associated with the body wall of the pelvic region. Five of the six muscles sit deep to the gluteal musculature on the posterior aspect of the hip joint. The hip flexors are deep body wall muscles of the abdominal wall that have been annexed by the lower limb during development. These muscles, the psoas major and iliacus, form a pulley over the superior ramus of the pubis on their descent onto the lesser trochanter of the femur.

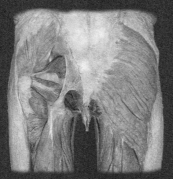

Gluteal muscles

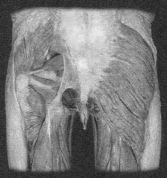

Deep hip rotator muscles

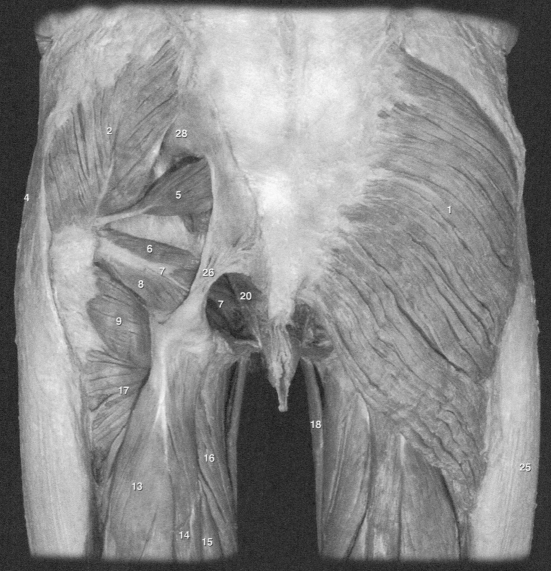

Muscles of the gluteal region, gluteus maximus removed on left
Posterior view

Gluteal Muscles
1 Gluteus maximus
2 Gluteus medius
3 Gluteus minimis
4 Tensor fasciae latae

Deep Hip Rotator Muscles
5 Piriformis
6 Superior gemellus

7 Obturator internus
8 Inferior gemellus
9 Quadratus femoris
10 Obturator externus

Hip Flexor Muscles
11 Psoas major
12 Iliacus

Other Muscles and Structures
13 Biceps femoris
14 Semitendinosus
15 Semimembranosus
16 Adductor magnus
17 Adductor minimus
18 Gracilis
19 Vastus intermedius
20 Pelvic diaphragm

21 Transversus abdominis
22 Quadratus lumborum
23 Psoas minor
24 Pectineus (cut)
25 Iliotibal tract
26 Sacrotuberous ligament
27 Penis (cut)
28 Ilium
29 Femur

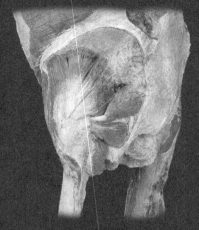

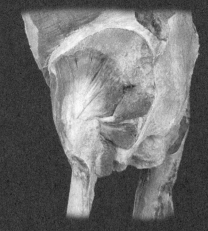

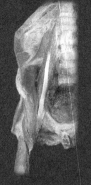

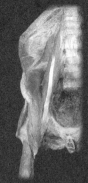

Gluteal muscles

Hip flexor muscles

Deep hip rotator muscles

Gluteal muscles

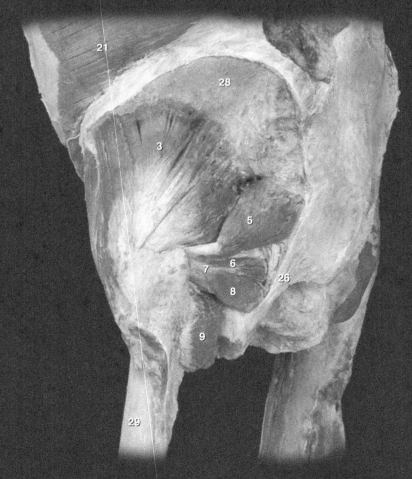

Muscles of gluteal region, gluteus maximus and medius removed
Posterolateral view

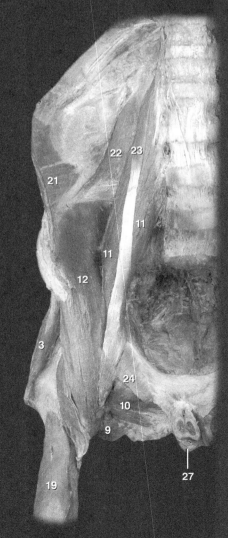

Deep dissection of iliopsoas muscles
Anterior view

Anterior Thigh Muscles

The four major muscles of the anterior compartment form the quadriceps femoris muscle group. The four muscles of this group converge to form the strong quadriceps tendon that surrounds all but the posterior surface of the patella. As the sole extensors of the knee, the quadriceps are essential for running, jumping, and kicking. The sartorius, which is the longest muscle in the body, is a knee flexor. The small articularis genus raises the suprapatellar bursa during extension of the knee. All of the muscles in this compartment receive their innervation via the femoral nerve from the posterior divisions of the lumbar plexus.

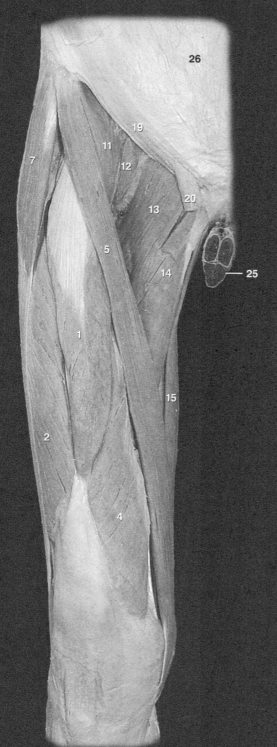

Muscles of the thigh
Anterior view, left thigh

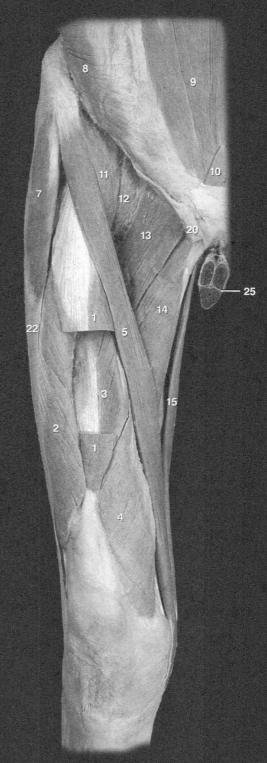

Muscles of the thigh, rectus femoris cut
Anterior view. left thigh

Anterior Thigh Muscles
1 Rectus femoris
2 Vastus lateralis
3 Vastus intermedius
4 Vastus medialis
5 Sartorious
6 Articularis genus

Other Muscles and Structures
7 Tensor fasciae latae
8 Transversus abdominis
9 Rectus abdominis
10 Pyramidalis
11 Iliacus
12 Psoas major

13 Pectineus
14 Adductor longus
15 Gracilis
16 Gluteus minimis
17 Obturator externus
18 Quadratus femoris
19 Inguinal ligament

20 Spermatic cord
21 Linea alba
22 Iliotibial tract
23 Femur
24 Inferior epigastric vessels
25 Penis (cut)
26 Rectus sheath

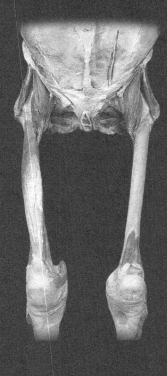

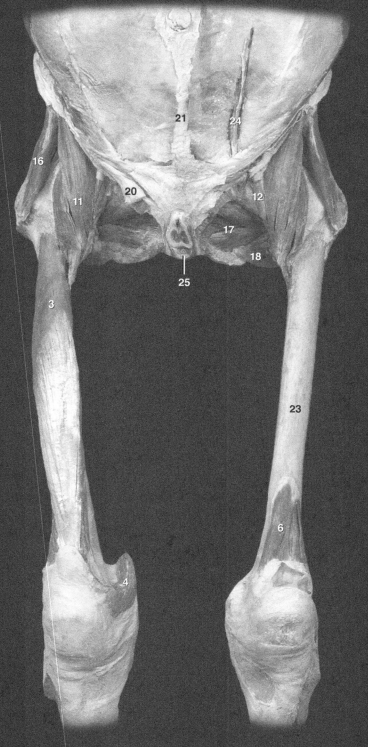

Deep muscles of the thigh
Anterior view

201

Medial Thigh Muscles

The six muscles of the medial compartment are all capable of adducting the hip joint. The pectineus and four adductor muscles all originate from a medial position on the pubis and ischium and project laterally to insert on the posterior surface of the femur. The gracilis muscle differs from the others in the group by crossing the knee joint in addition to the hip. It courses with the sartorius muscle as a flexor of the knee. With the exception of the pectineus and condylar part of the adductor magnus, all the muscles are innervated by the obturator nerve, which arises from the anterior divisions of the lumbar plexus.

Medial Thigh Muscles
1 Pectineus
2 Adductor longus
3 Adductor brevis
4 Adductor magnus
5 Adductor minimis
6 Gracilis

Other Muscles and Structures
7 Sartorius
8 Iliacus
9 Psoas major
10 Tensor fasciae latae
11 Rectus femoris
12 Obturator externus
13 Vastus lateralis
14 Articularis genus
15 Gluteus medius
16 Piriformis
17 Superior gemellus
18 Obturator internus
19 Inferior gemellus
20 Quadratus femoris
21 Biceps femoris (short head)
22 Gastrocnemius
23 Plantaris
24 Soleus
25 Pelvic diaphragm
26 Transversus abdominis
27 Rectus abdominis
28 Spermatic cord
29 Sacrotuberous ligament
30 Femur
31 Penis (cut)

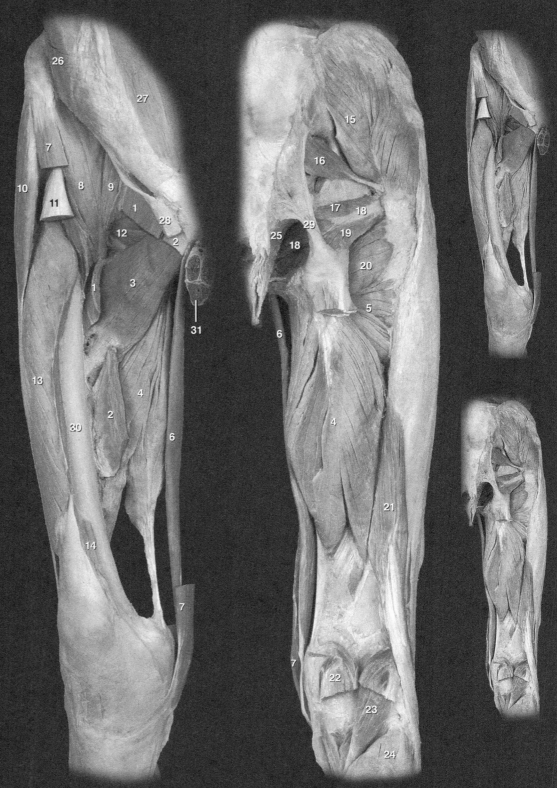

Dissection of medial thigh muscles
Anterior view, right thigh

Dissection of medial thigh muscles
Posterior view, right thigh

Posterior Thigh Muscles

Like the medial compartment of the thigh, the biceps femoris, semimembranosus, and semitendinosus arise from the embryonic anterior, or flexor, musculature. The muscles of this compartment, the smallest of the three thigh compartments, are long, two-joint muscles that share much in common. All three muscles arise from the ischial tuberosity, extend the hip and flex the knee, and receive their nerve supply via the tibial branch of the sciatic nerve (with the exception of the short head of the biceps femoris, which is innervated by the common fibular branch of the sciatic nerve). Often referred to as the hamstring muscles, these muscles work with the sartorius and gracilis as the strong flexors of the knee joint.

Posterior Thigh Muscles
1   Biceps femoris (long head)
2   Biceps femoris (short head)
3   Semitendinosus
4   Semimembranosus

Other Muscles and Structures
5   Gluteus maximus
6   Gluteus medius
7   Piriformis
8   Superior gemellus
9   Obturator internus
10 Inferior gemellus
11 Quadratus femoris
12 Adductor minimus
13 Adductor magnus
14 Pelvic diaphragm
15 Gracilis
16 Gastrocnemius
17 Sacrotuberous ligament
18 Iliotibial tract
19 Ilium

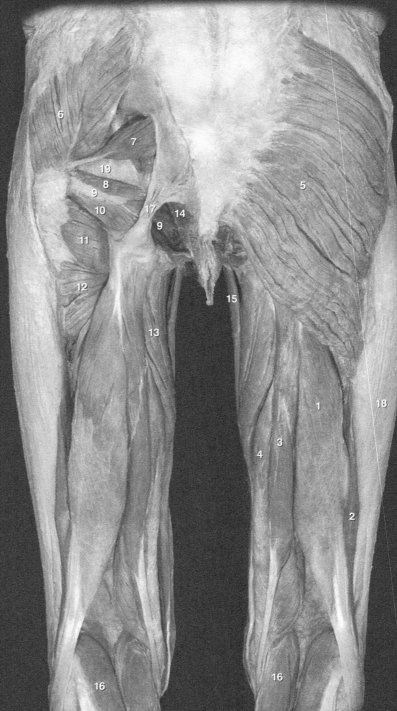

Muscles of the gluteal region and thigh
Posterior view, gluteus maximus removed on left

Thigh Muscles

1 Rectus femoris
2 Vastus lateralis
3 Vastus intermedius
4 Vastus medialis
5 Sartorious
6 Gracilis
7 Adductor longus
8 Adductor magnus
9 Biceps femoris
10 Semitendinosus
11 Semimembranosus
12 Femoral artery
13 Femoral vein
14 Hypodermis
15 Femur
16 Yellow bone marrow
17 Sciatic nerve
18 Saphenous nerve

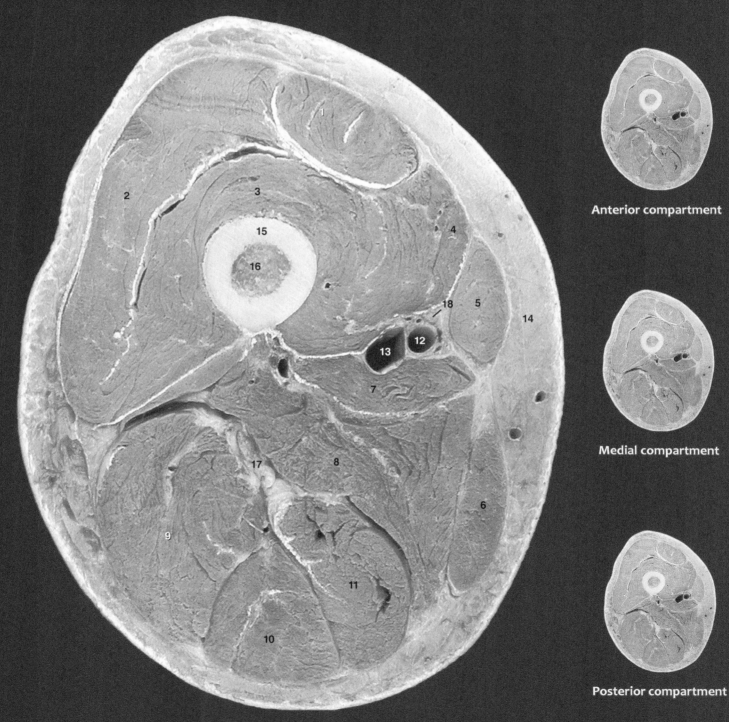

Anterior compartment

Medial compartment

Posterior compartment

Transverse section of right thigh
Inferior view, level at mid thigh

Anterior Leg Muscles

The anterior compartment of the leg consists of four muscles, all of which dorsal flex the ankle joint and are innervated by the deep fibular nerve from the posterior divisions of the sacral plexus. These muscles sit in a tight fascial compartment anterior to the interosseous membrane and between the tibia and fibula. As their tendons cross the ankle joint they are held firmly in place between the tibial and fibular malleoli by two strong retinacular bands. Two of the muscles, the tibialis anterior and fibularis tertius, insert on the ankle. The other two muscles, the extensor digitorum longus and extensor hallucis longus, reach the ends of the digits and also function as digital extensors.

Anterior Leg Muscles
 1 Tibialis anterior
 2 Extensor digitorum longus
 3 Extensor hallucis longus
 4 Fibularis tertius

Other Muscles and Structures
 5 Vastus lateralis
 6 Fibularis longus
 7 Fibularis brevis
 8 Gastrocnemius
 9 Soleus
10 Extensor hallucis brevis
11 Extensor digitorum brevis
12 Interosseous membrane
13 Anterior tibial vessels
14 Extensor retinaculum
15 Tibia
16 Patellar ligament

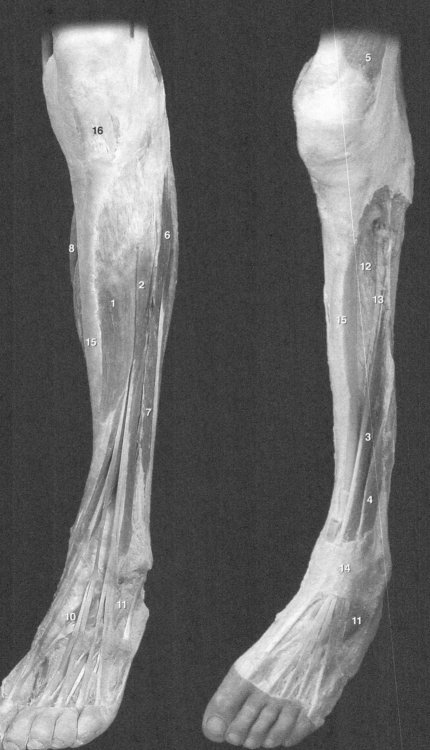

Superficial muscles of the anterior crus
Anterior view

Deep muscles of the anterior crus
Anterolateral view

Lateral Leg Muscles

The small lateral compartment, like the anterior compartment, arises from the embryonic dorsal limb muscles. The two muscles within this compartment, the fibularis longus and fibularis brevis, are similar. They both arise from the lateral aspect of the fibula. They both pursue a pulley-like course behind the lateral malleolus, under the cover of a retinaculum, in their passage to the bottom of the foot. They both plantar flex and evert the foot. The superficial fibular nerve, from the posterior divisions of the sacral plexus, supplies both muscles.

Lateral Leg Muscles
1 Fibularis longus
2 Fibularis brevis

Other Muscles and Structures
3 Gastrocnemius
4 Soleus
5 Fibularis tertius
6 Extensor digitorum longus
7 Tibialis anterior
8 Extensor hallucis longus
9 Extensor digitorum brevis
10 Interosseous membrane
11 Calcaneal tendon
12 Femur
13 Tibia
14 Fibula
15 Lateral malleolus
16 Patellar ligament

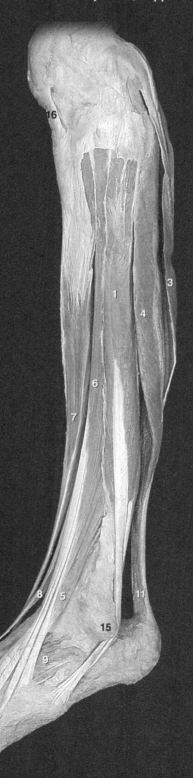

Muscles of the crus
Lateral view

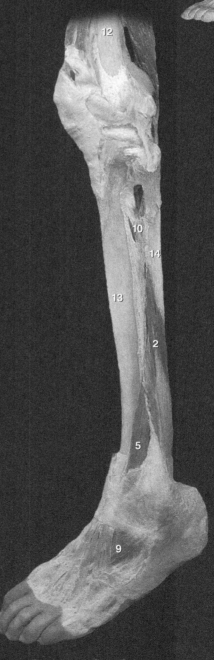

Deep muscles of the crus
Lateral view

Posterior Leg Muscles

The posterior compartment of the leg comprises the large muscle mass on the back of the leg that is often referred to as the calf. This compartment has two distinct muscle groups – a large superficial group and a smaller deep group. The superficial group, the gastrocnemius, the soleus, and the plantaris, each insert on the calcaneus. The gastrocnemius and soleus combine to form the large tendocalcaneus, or Achilles tendon. The smaller, deep group consists of four muscles, three of which form a pulley-like arrangement around the medial malleolus. These are the flexor hallucis longus, flexor digitorum longus, and tibialis anterior. The fourth muscle in the group is the deeply situated popliteus that occupies the floor of the popliteal fossa.

Posterior Leg Muscles
1   Tibialis posterior
2   Flexor digitorum longus
3   Flexor hallucis longus
4   Popliteus
5   Plantaris
6   Soleus
7   Gastrocnemius

Other Muscles and Structures
8   Fibularis brevis
9   Fibularis longus (tendon)
10 Flexor digitorum brevis
11 Abductor hallucis
12 Flexor hallucis brevis
13 Abductor digiti minimi
14 Calcaneal tendon
15 Fibula

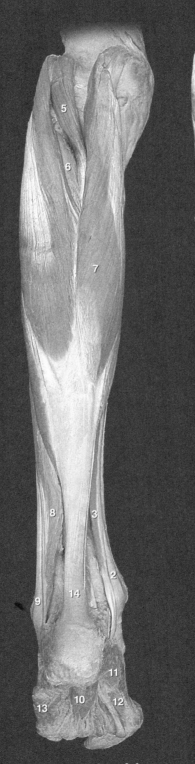

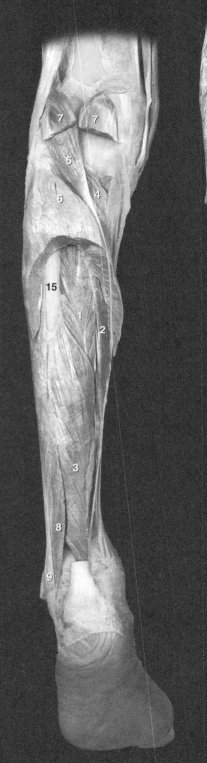

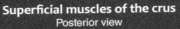

Superficial muscles of the crus
Posterior view

Deep muscles of the crus
Posterior view

Foot Muscles

Situated on the dorsal surface of the foot are two short digital extensor muscles, the extensor hallucis brevis and extensor digitorum brevis. These thin muscle sheets help the long digital extensors of the anterior compartment extend the digits. Like the anterior compartment muscles, they are innervated by the deep fibular nerve. The plantar muscles of the foot are much more substantial than the thin dorsal muscles of the foot. These muscles sit beneath the thick subcutaneous fat pad on the bottom of the foot. From superficial to deep, the plantar muscles form four layers.

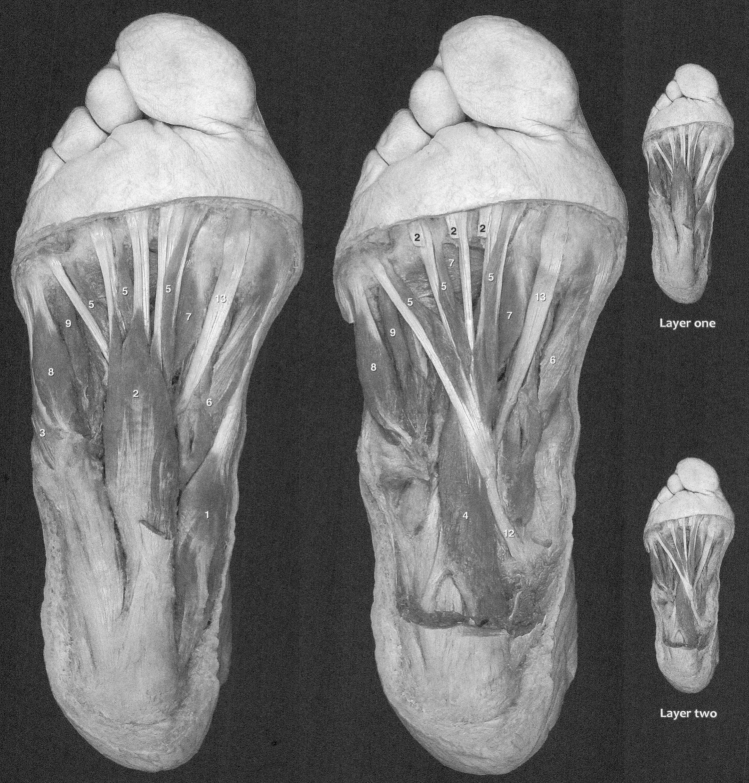

Dissection of foot, plantar aponeurosis removed
Plantar view

Dissection of foot, first muscle layer removed
Plantar view

Layer one

Layer two

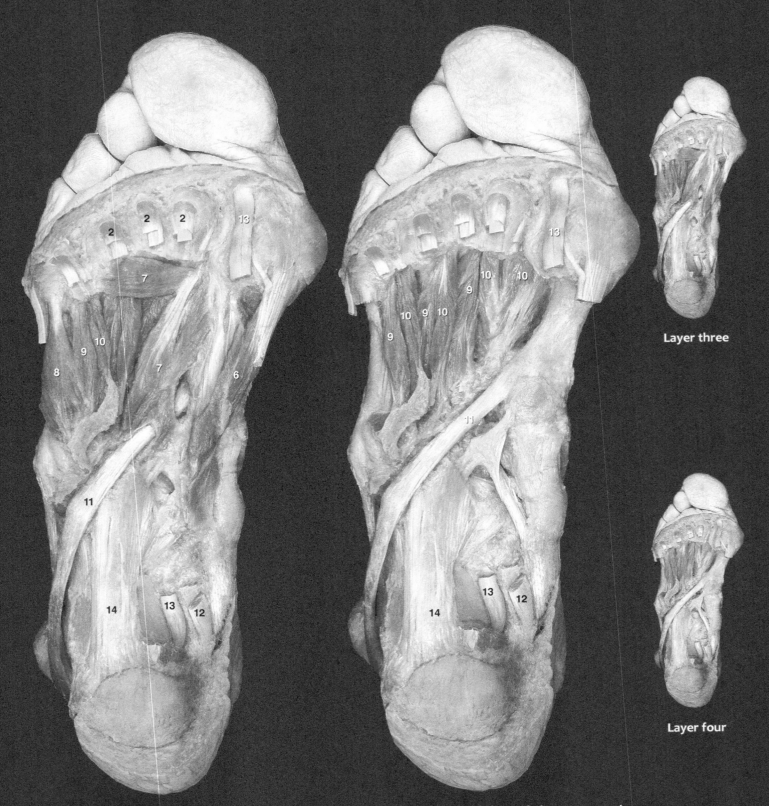

Foot Muscles
1 Abductor hallucis
2 Flexor digitorum brevis
3 Abductor digiti minimi
4 Quadratus plantae
5 Lumbricals
6 Flexor hallucis brevis
7 Adductor hallucis
8 Flexor digiti minimi brevis
9 Plantar interossei
10 Dorsal interossei

Other Muscles and Structures
11 Fibularis longus (tendon)
12 Flexor digitorum longus (tendon)
13 Flexor hallucis longus (tendon)
14 Long plantar ligament

Layer three

Layer four

Dissection of foot, second muscle layer removed
Plantar view

Dissection of foot, third muscle layer removed
Plantar view

209

Foot Muscles

Foot Muscles
1. Extensor hallucis brevis
2. Extensor digitorum brevis

Other Muscles and Structures
3. Tibialis anterior (tendon)
4. Extensor hallucic longus (cut)
5. Extensor digitorum longus (cut)
6. Fibularis longus (tendon)
7. Fibularis brevis (tendon)
8. Deep fibular nerve
9. Dorsalis pedis artery

Dissection of left foot
Dorsal view

Dorsal foot muscles

13 | Peripheral Nervous System

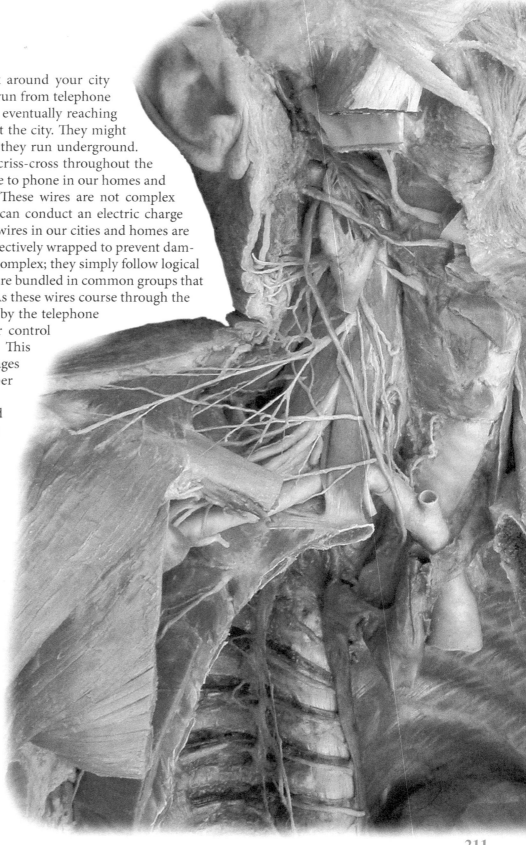

Look around your city or town and notice the telephone wires that run from telephone pole to telephone pole along the city streets, eventually reaching the homes and places of business throughout the city. They might not always be visible because in some cities they run underground. Regardless of where they occur, these wires criss-cross throughout the city distributing electrical current from phone to phone in our homes and places of school, work, and entertainment. These wires are not complex structures; they are simply metal wires that can conduct an electric charge from one phone to another. These telephone wires in our cities and homes are typically insulated from one another and protectively wrapped to prevent damage. Their pathways through the city are not complex; they simply follow logical routes to different parts of the city. The wires are bundled in common groups that follow shared pathways to similar locations. As these wires course through the city they relay to telephone centers operated by the telephone companies. At these centers the wires enter control rooms where they form complex circuits. This complex circuitry allows the electrical messages to be processed and directed to the proper phones.

Like the telephone wires of our cities and homes, the nerves of the peripheral nervous system are really rather simple structures. They consist of long, insulated axons bundled together in protective collagenous wrappings. These axons pass in bundled groups that follow logical routes to the different regions of the body where they communicate with receptor (sensory receptors) or effector strucutres (muscles or glands). Like telephone wires, these neuronal wires conduct electrical messages to and from the central processing center (brain and spinal cord). This chapter will depict the basic design of the structures called nerves and demonstrate the pathways of the nerves throughout the body.

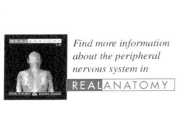

Find more information about the peripheral nervous system in
REAL ANATOMY

Structure of a Nerve

Nerves are bundles of axons running between the central nervous system and the peripheral tissues of the body. While all nerves have a similar basic structure, they vary in the types and numbers of neurons bundled within. The basic design of a nerve consists of neurons wrapped by neurolemmocytes to form the nerve fiber. The fibers are protectively wrapped and nourished by a vascular loose connective tissue, the endoneurium. Many endoneurial wrapped fibers are surrounded by a collagenous perineurium to form the fasciculus of the nerve, and all the fasciculi are wrapped in a collagenous sheath, the epineurium, to form the nerve.

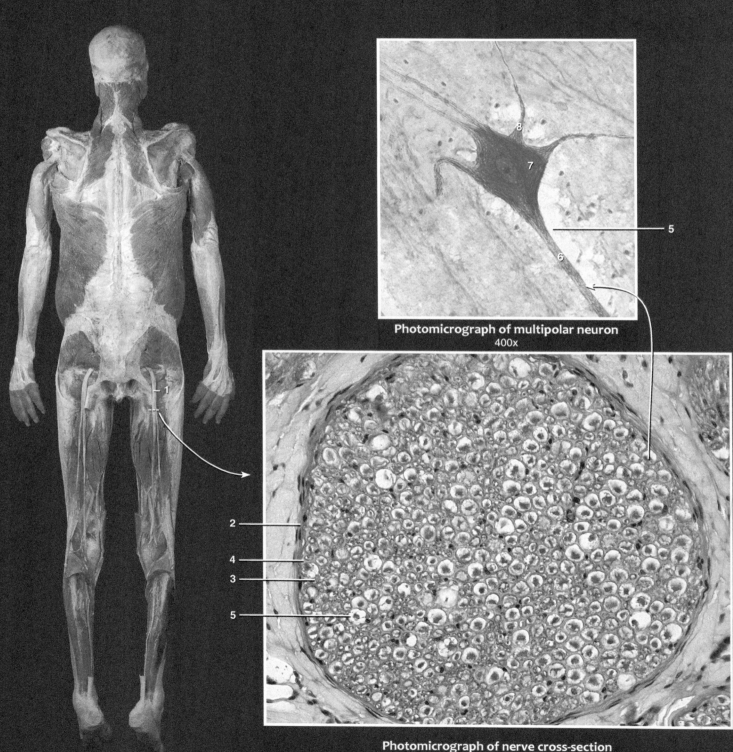

Photomicrograph of multipolar neuron
400x

Photomicrograph of nerve cross-section
200x

Dissection of sciatic nerve
Posterior view

Spinal Nerve Structure

The spinal nerves arise from the spinal cord as a series of small neuronal bundles called rootlets — ventral (motor)rootlets and dorsal (sensory) rootlets. Each series of ventral rootlets converges to form larger ventral roots. Likewise each series of dorsal rootlets converges to form larger dorsal roots. The dorsal and ventral roots project laterally and converge to form the spinal nerve trunk. A ganglion, the dorsal root ganglion, is present on the dorsal root just prior to the spinal nerve trunk. Branching from the trunk are two large branches and a variable series of smaller branches. Each branch follows a specific course to different peripheral regions. The two largest branches, the ventral ramus and dorsal ramus, are somatic branches that run in the musculoskeletal wall of the body. Smaller visceral branches, the meningeal nerve, the white and gray communicating rami, and the parasympathetic splanchnic nerves form the autonomic pathways to smooth muscle and glandular tissue.

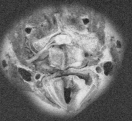

Structure of a Nerve
1. Sciatic nerve
2. Epineurium
3. Perineurium
4. Endoneurium
5. Myelin sheath
6. Axon
7. Cell body
8. Dendrite

Spinal Nerve Structures
9. Ventral rootlets
10. Dorsal rootlets
11. Dorsal root
12. Dorsal root ganglion
13. Ventral root
14. Spinal nerve trunk
15. Ventral ramus
16. Dorsal ramus

Other Structures
17. Spinal cord
18. Cervical vertebra
19. Vertebral artery
20. Common carotid artery
21. Internal jugular vein
22. Laryngopharynx
23. Larynx
24. Thyroid cartilage
25. Cricoid cartilage
26. Vocalis muscle

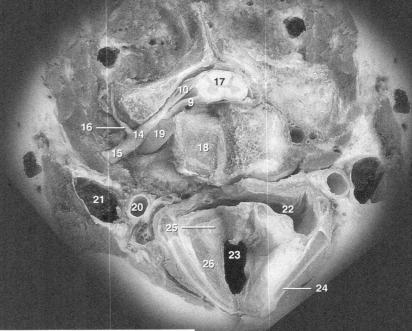

Dissection of cervical spinal cord
Superior view

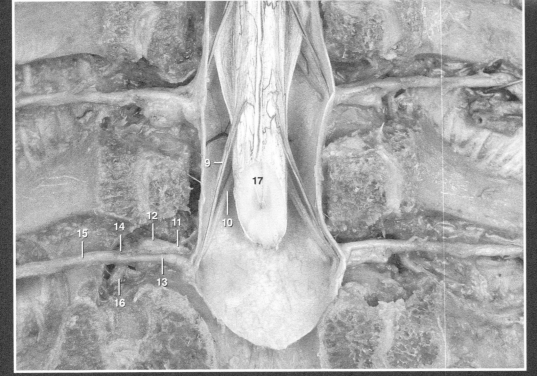

Dissection of spinal cord, thoracic vertebral bodies removed
Anterior view

Spinal Nerves

With slight variation, the basic pattern of the spinal nerve repeats itself thirty-one times along the entire length of the spinal cord. With the exception of the first spinal nerve, each spinal nerve level emerges from within the vertebral column to pass peripherally between successive vertebrae. Because of the developmental differences in the growth rate of the vertebral column and associated spinal cord, the lower roots of the spinal nerves are dragged downward by the lengthening vertebral column. With each succeeding spinal nerve level the roots become longer and more oblique in their course, eventually extending beyond the end of the spinal cord as the vertically oriented cauda equina.

Spinal Nerves
1 Spinal nerve
2 Cervical dorsal rootlets
3 Thoracic dorsal rootlets
4 Lumbosacral dorsal rootlets
5 Dorsal rami
6 Cauda equina
7 Filum terminale

Other Structures
8 Cerebrum
9 Cerebellum
10 Medulla oblongata
11 Spinal cord

12 Dura mater
13 Superior sagittal sinus
14 Transverse sinus
15 Opening of straight sinus
16 Confluence of sinuses

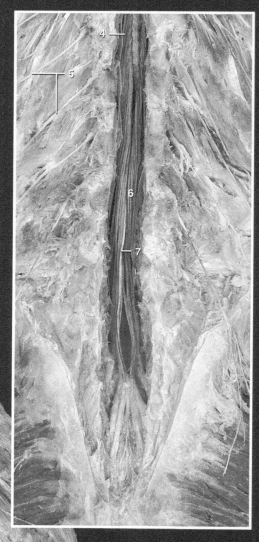

Dissection exposing cauda equina
Posterior view

Dissection revealing spinal cord and brain
Posterior view

Dorsal Rami

The dorsal rami of the spinal nerves arise at all spinal levels and pursue a posterior course into the muscles, connective tissue, and skin of the back. They innervate all the epaxial muscles comprising the extensors of the vertebral column. The cutaneous distribution of the dorsal rami spans from the top of the head, down the posterior trunk, to the superior half of the gluteal region. With the exception of levels C1, S4, S5, and the coccygeal, the dorsal rami split into lateral and medial branches as they course posteriorly into the back.

Dorsal Rami
1 Greater occipital nerve
2 Least occipital nerve
3 Dorsal ramus
4 Medial branch
5 Lateral branch

Other Structures
6 Rectus capitis posterior major muscle
7 Rectus capitis posterior minor muscle
8 Obliquus superioris muscle
9 Obliquus inferioris muscle
10 Posterior digastricus muscle

11 Semispinalis cervicis muscle
12 Intertransversarii thoracic muscle
13 Levatores costarum muscles
14 External intercostal muscle
15 External oblique muscle
16 Internal oblique muscle

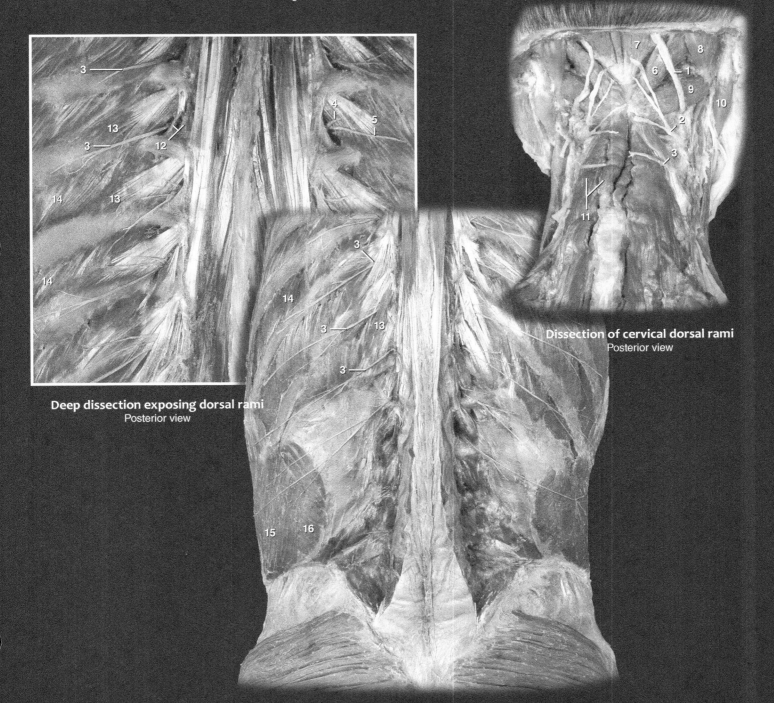

Deep dissection exposing dorsal rami
Posterior view

Dissection of cervical dorsal rami
Posterior view

Erector spinae muscle removed to expose dorsal rami
Posterior view

This next series of pages illustrates the ventral rami of the spinal nerves. The ventral rami innervate the majority of the skeletal muscles (all hypaxial and limb muscles). The cervical plexus forms from the ventral rami of the first four cervical spinal nerves. As these ventral rami pass laterally between the middle and internal layers of the lateral cervical body wall, they form ascending and descending branches that communicate to form the cervical plexus. Emerging from this plexus are the nerves that innervate the muscles of the hypaxial cervical wall, as well as cutaneous branches that serve the overlying skin of the lateral head, neck and upper thorax.

Cervical Plexus Nerves
1 Lesser occipital nerve
2 Great auricular nerve
3 Transverse cutaneous nerve
4 Supraclavicular nerve
5 Phrenic nerve
6 Ansa cervicalis
7 Nerve to geniohyoid muscle
8 Nerve to thyrohyoid muscle
9 Nerve to superior omohyoid muscle
10 Nerve to sternohyoid muscle
11 Nerve to sternothyroid muscle
12 Nerve to inferior omohyoid muscle

Other Nerves and Structures
13 Hypoglossal nerve
14 Vagus nerve
15 Superior trunk of brachial plexus
16 Common carotid artery
17 Carotid sinus
18 Internal carotid artery
19 External carotid artery
20 Parotid gland
21 Sternocleidomastoid muscle
22 Thyrohyoid muscle
23 Omohyoid muscle
24 Sternohyoid muscle
25 Sternothyroid muscle
26 Anterior scalene muscle
27 Middle scalene muscle
28 Levator scapulae muscle

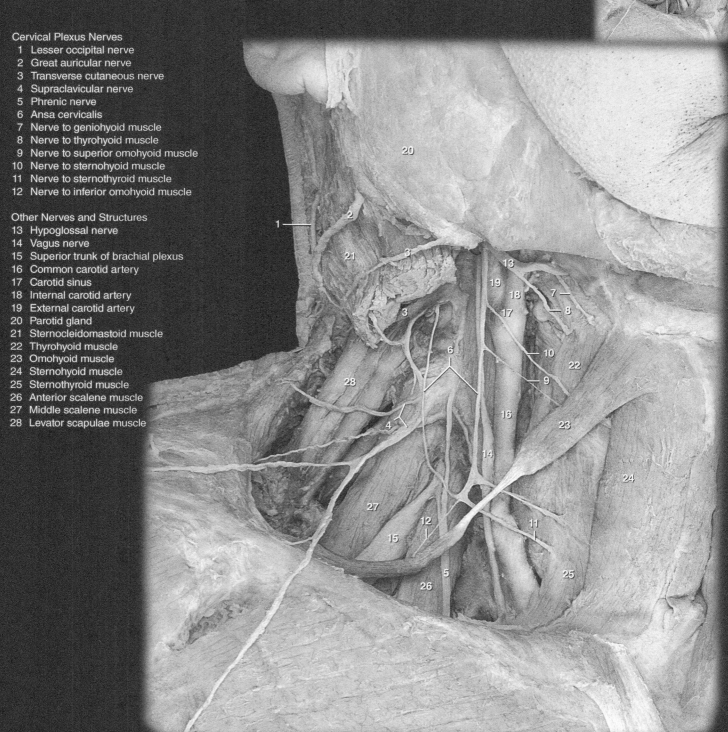

Dissection of cervical plexus
Anterior view

Brachial Plexus

The brachial plexus arises from the last four cervical ventral rami and the first thoracic ventral ramus. The four cervical ventral rami pass laterally between the middle and internal layers of the lateral cervical body wall, the middle and anterior scalene muscles, respectively. As they emerge through the scalenes, they connection with with one another as well as with the ascending branch of the first thoracic ventral ramus. This is the beginning of the nerve plexus that will innervate almost all the muscles and associated skin of the upper limb.

Brachial Plexus Nerves
1 Dorsal scapular nerve
2 Suprascapular nerve
3 Nerve to the subclavius muscle
4 Lateral pectoral nerve
5 Upper subscapular nerve
6 Musculocutaneous nerve
7 Axillary nerve
8 Radial nerve
9 Median nerve
10 Ulnar nerve
11 Lower subscapular nerve
12 Thoracodorsal nerve

13 Long thoracic nerve
14 Medial pectoral nerve
15 Superior trunk
16 Middle trunk
17 Inferior trunk
18 Lateral cord
19 Posterior cord
20 Medial cord

Other Nerves and Strucures
21 Phrenic nerve
22 Anterior scalene muscle
23 Middle scalene muscle

24 Levator scapulae muscle
25 Subclavius muscle
26 Pectoralis minor muscle
27 Pectoralis major muscle
28 Deltoid muscle
29 Biceps brachii muscle
30 Subscapularis muscle
31 Teres major muscle
32 Latissimus dorsi muscle
33 Serratus anterior muscle
34 Clavicle

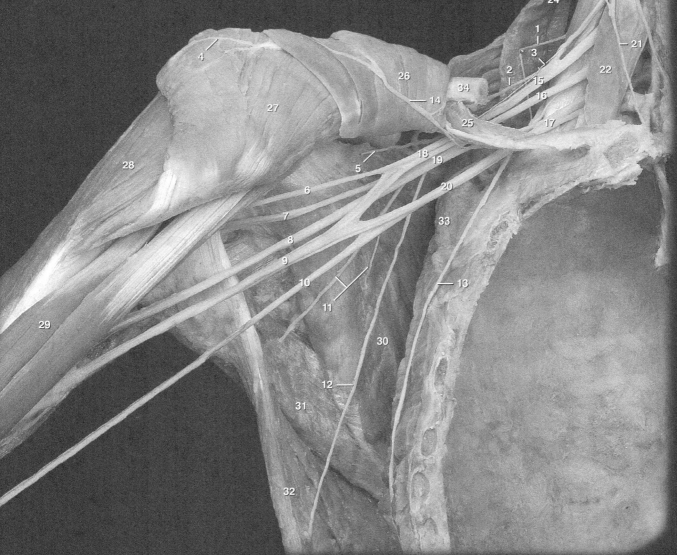

Dissection of brachial plexus
Anterior view

Lumbar Plexus

The lumbar plexus arises from the ventral rami of the first four lumbar spinal nerves. The plexus emerges laterally through the intervertebral foramina to pass anterolateral between the two heads of the psoas major muscle. The more superior branches of the plexus enter the abdominal body wall to innervate the abdominal muscles. The lower nerves of the plexus course into the lower limb as the lateral femoral cutaneous, femoral, and obturator nerves. The lumbar plexus is a transitory plexus that begins as a series of body wall nerves and eventually transitions into limb innervation. The first ventral ramus of the plexus is basically a segmental nerve that follows the basic segmental nerve pattern in the ventral body wall. The second lumbar ventral ramus forms segmental branches in the body wall and other branches that contribute to limb innervation. The third and fourth ventral rami contribute solely to innervation of the lower limb anatomy.

Lumbar Plexus Nerves
1 Subcostal nerve
2 Iliohypogastric nerve
3 Ilioinguinal nerve
4 Genitofemoral nerve
5 Genital branch of genitofemoral nerve
6 Femoral branch of genitofemoral nerve
7 Lateral femoral cutaneous nerve
8 Femoral nerve
9 Obturator nerve
10 Lumbosacral trunk

Sacral Plexus Nerves
11 Superior gluteal nerve
12 Inferior gluteal nerve
13 Posterior femoral cutaneous nerve
14 Nerve to the obturator internus muscle
15 Pudendal nerve
16 Perforating cutaneous nerve
17 Inferior cluneal nerve
18 Sciatic nerve
19 Upper bands of sacral plexus

Other Structures
20 Diaphragm
21 Psoas major muscle
22 Psoas minor muscle
23 Quadratus lumborum muscle
24 Iliacus muscle
25 Obturator externus muscle
26 Sartorius muscle
27 Tensor fasciae latae muscle
28 Gluteus maximus muscle
29 Gluteus medius muscle
30 Gluteus minimis muscle
31 Piriformis muscle
32 Superior gemellus muscle
33 Obturator internus muscle
34 Inferior gemellus muscle
35 Sacrotuberous ligament
36 Penis

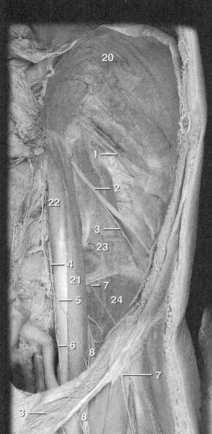

Abdominal dissection of lumbar plexus
Anterior view

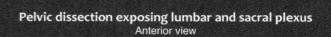

Pelvic dissection exposing lumbar and sacral plexus
Anterior view

Sacral Plexus

The sacral plexus forms from the ventral rami of the last two lumbar and the first four sacral spinal nerves. The fourth and fifth lumbar spinal nerves form a descending communication, the lumbosacral trunk, that joins with the upper sacral spinal nerves as they exit the anterior foramina of the sacrum. On the anterior surface of the sacrum the large roots of the plexus are noticeable before they exit through the greater sciatic notch on their course into the pelvic wall and lower limb. This plexus forms the total nerve supply to the pelvic body wall, and, along with the limb branches from the lumbar plexus, is the nerve supply for the lower limb.

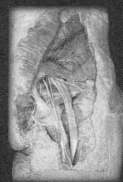

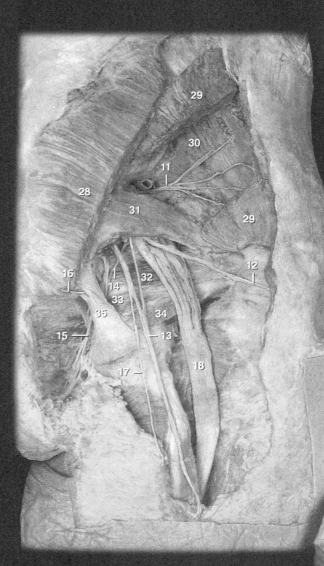

Dissection of sacral plexus nerves
Posterior view

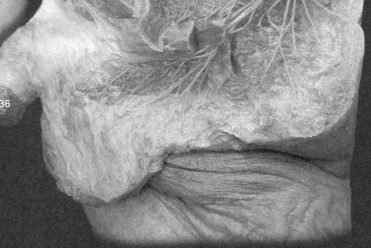

Dissection of pudendal nerves and vessels
Lateral view

Intercostal Nerves

Unlike the ventral rami in the cervical, lumbar, and sacral regions, which form plexuses, most of the thoracic ventral rami remain segmental like their dorsal counterparts. These thoracic ventral rami, called the intercostal and subcostal nerves, emerge from the spinal nerve trunk and enter the intercostal space just inferior to each of the twelve ribs. Each of these segmental nerves has a similar structural design. The main trunk of the nerve runs through the intercostal space, with the segmental arteries and veins, between the middle and internal muscle layers of the body wall. Accompanying the main branch is a smaller collateral branch, which emerges from the main branch near the angle of the rib, and runs inferior to the main branch through the intercostal space. The main branch also gives rise to lateral and anterior cutaneous branches that supply the skin, or dermatome, of each segment.

Intercostal Nerves
1   Main trunk
2   Collateral branch

Other Nerves and Structures
3   Subcostal nerve
4   Iliohypogastric nerve
5   Posterior intercostal vein
6   Posterior intercostal artery
7   Innermost intercostal muscle
8   Transversus abdominis muscle
9   Gluteus medius muscle
10 Piriformis muscle
11 Iliocostalis muscles
12 Rib 12

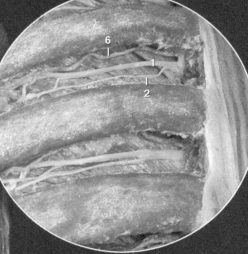

Dissection of intercostal space
Lateral view

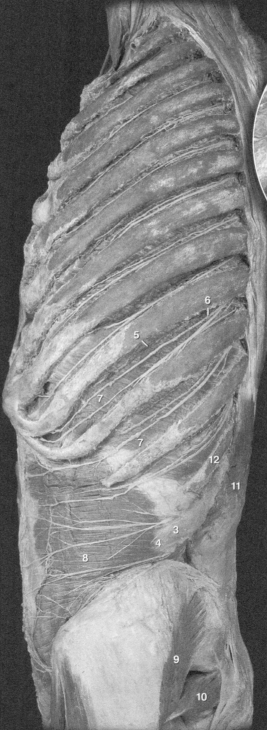

Dissection of intercostal nerves
Lateral view

Cutaneous Nerves

Many small nerves, named cutaneous nerves, branch from the spinal and cranial nerves and course through and between muscles to emerge into the integumentary covering of the body. These detailed dissections reveal all the cutaneous nerves of the body.

1 Greater occipital nerve
2 Transverse cervical nerves
3 Supraclavicular nerves
4 Medial cutaneous branches (dorsal rami)
5 Lateral cutaneous branches (dorsal rami)
6 Anterior cutaneous branches (ventral rami)
7 Lateral cutaneous branches (ventral rami)

8 Superior lateral brachial cutaneous nerves
9 Posterior brachial cutaneous nerves
10 Inferior lateral brachial cutaneous nerves
11 Posterior antebrachial cutaneous nerve
12 Lateral antebrachial cutaneous nerve
13 Medial antebrachial cutaneous nerve
14 Femoral branch of genitofemoral nerve
15 Lateral cutaneous branch of subcostal nerve
16 Anterior cutaneous branch of femoral nerve
17 Lateral femoral cutaneous nerve
18 Superior cluneal nerves
19 Inferior cluneal nerve
20 Posterior femoral cutaneous nerve
21 Saphenous nerve
22 Lateral sural cutaneous nerve

Dissections exposing cutaneous nerves
Anterior view to left, Posterior view to right

Autonomic Nerves

In contrast to the somatic branches of the spinal nerve, the visceral branches leave the body wall to form nerve pathways that enter the body cavities. Within the cavities these nerves form the autonomic nerve pathways, sympathetic and parasympathetic, to the viscera. The autonomic nerves relay input signals from the wall of the tubular gut and other viscera, while carrying output signals to smooth muscle,

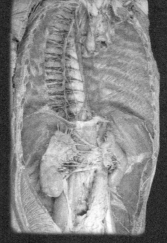

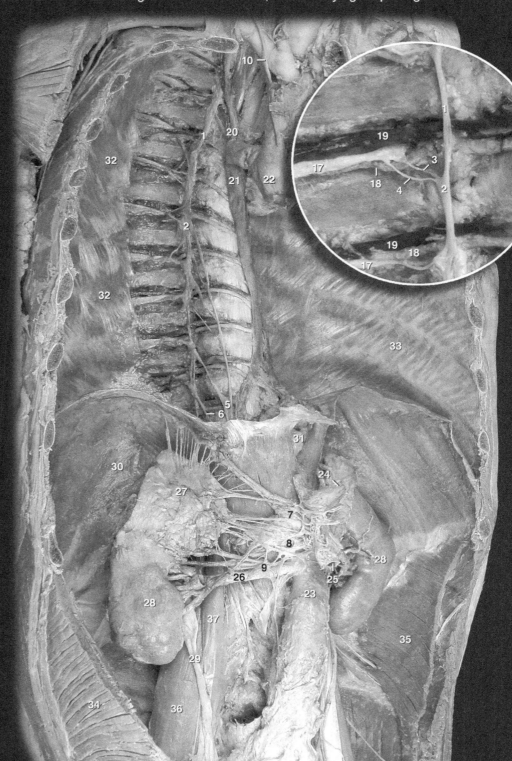

Autonomic Nerves
1 Sympathetic trunk nerve
2 Sympathetic trunk ganglion
3 White communicating ramus
4 Gray communicating ramus
5 Greater splanchnic nerve
6 Lesser splanchnic nerve
7 Coeliac ganglion
8 Superior mesenteric ganglion
9 Aorticorenal ganglion
10 Vagus nerve
11 Recurrent laryngeal nerve
12 Anterior vagal trunk
13 Posterior vagal trunk
14 Inferior cardiac plexus
15 Pulmonary plexus
16 Esophageal plexus

Other Structures
17 Intercostal nerve
18 Posterior intercostal artery
19 Posterior intercostal vein
20 Right superior intercostal vein
21 Azygous vein
22 Superior vena cava
23 Aorta
24 Celiac trunk
25 Superior mesenteric artery
26 Renal artery
27 Suprarenal gland
28 Kidney
29 Ureter
30 Diaphragm
30 Esophageal hiatus
32 Subcostal muscle
33 Innermost intercostal muscle
34 Internal oblique muscle
35 Transversus abdominis muscle
36 Psoas major muscle
37 Psoas minor muscle

Deep dissection of sympathetic nerves, callout of communicating rami
Anterolateral view

cardiac muscle, and glands. Some of the autonomic nerves even rejoin the somatic pathways to supply the blood vessels and glands of the body wall. The sympathetic pathways are primarily associated with vascular smooth muscle control, and the parasympathetic pathways are principally responsible for the regulation and control of gut tube smooth muscle and glands. The sympathetic nerves are depicted on the opposite page, while the vagus nerve, which carries 75% of the parasympathetic output, is shown below as it follows the derivatives of the gut tube.

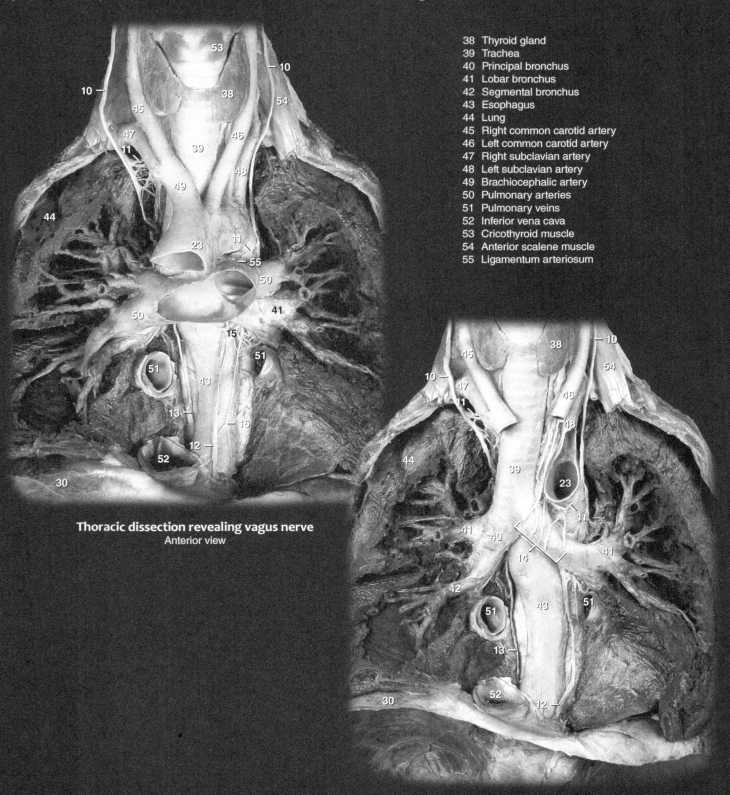

38 Thyroid gland
39 Trachea
40 Principal bronchus
41 Lobar bronchus
42 Segmental bronchus
43 Esophagus
44 Lung
45 Right common carotid artery
46 Left common carotid artery
47 Right subclavian artery
48 Left subclavian artery
49 Brachiocephalic artery
50 Pulmonary arteries
51 Pulmonary veins
52 Inferior vena cava
53 Cricothyroid muscle
54 Anterior scalene muscle
55 Ligamentum arteriosum

Thoracic dissection revealing vagus nerve
Anterior view

Deeper thoracic dissection revealing vagus nerve
Anterior view

Cranial Nerves

Cranial nerves segregate into three distinct groups based on associations they form during development. In number there are twelve cranial nerves, which originate in pairs from a rostral to caudal sequence from the brain. The first category, the special sensory cranial nerves, are afferent pathways established between the the brain and the special sensory structures of the nose, eye, and ear. The second category, the ventral or somitic motor cranial nerves, are homologous with the ventral roots of the spinal nerves. They originate from the brainstem as efferent pathways to somitic skeletal muscles within the head. The final category, comprising the largest of the

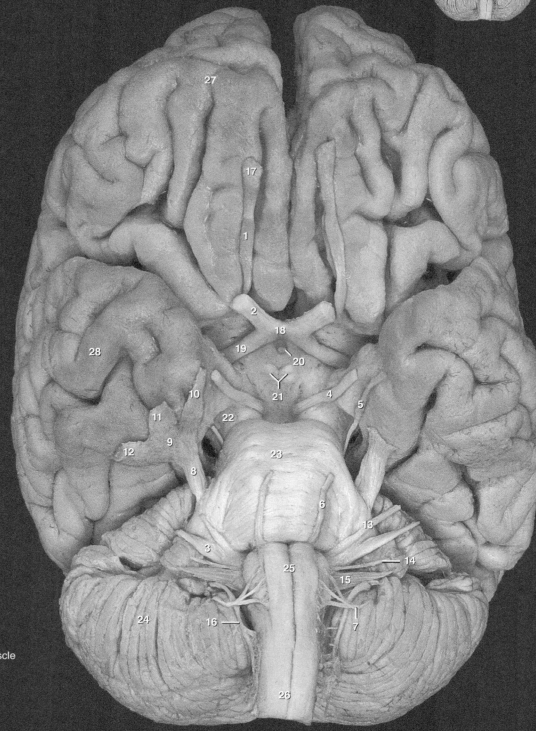

Special Sensory Nerves
  1 Olfactory nerve
  2 Optic nerve
  3 Vestibulocochlear nerve

Somitic Motor Nerves
  4 Occulomotor nerve
  5 Trochlear nerve
  6 Abducens nerve
  7 Hypoglossal nerve

Pharyngeal Arch Nerves
  8 Trigeminal nerve
  9 Trigeminal ganglion
 10 Opthalmic branch
 11 Maxillary branch
 12 Mandibular branch
 13 Facial nerve
 14 Glossopharyngeal nerve
 15 Vagus nerve
 16 Accessory nerve

Other Structures
 17 Olfactory bulb
 18 Optic chiasm
 19 Optic tract
 20 Infundibulum
 21 Mammillary bodies
 22 Cerebral peduncle
 23 Pons
 24 Cerebellum
 25 Medulla oblongata
 26 Spinal cord
 27 Frontal lobe
 28 Temporal lobe
 29 Insular lobe
 30 Parietal lobe
 31 Occipital lobe
 32 Right lateral ventricle
 33 Choroid plexus
 34 Falx cerebri
 35 Falx cerebelli
 36 Straight sinus
 37 Superior sagittal sinus
 38 Corpora quadrigemina
 39 Pineal gland
 40 Third ventricle
 41 Fourth ventricle
 42 Geniculate ganglion
 43 Anterior cerebral artery
 44 Internal carotid artery
 45 Levator palpebrae superioris muscle
 46 Superior rectus muscle
 47 Lateral rectus muscle
 48 Superior oblique muscle
 49 Nasociliary nerve
 50 Long ciliary nerve
 51 Ciliary ganglion
 52 Eye

Base of brain with cranial nerves
Inferior view

cranial nerves, are those cranial nerves associated with the pharyngeal arches. The dorsal or pharyngeal arch cranial nerves are developmentally similar to the dorsal roots of the spinal nerves. These five dorsal cranial nerves form the general sensory afferent pathways from the peripheral tissues of the head. However, because these nerve pathways coursed through the specialized arches forming the pharyngeal wall of the foregut, they established parasympathetic efferent pathways to the glandular tissue of the gut wall, along with motor efferent pathways to the skeletal muscles derived from the pharyngeal arch tissues.

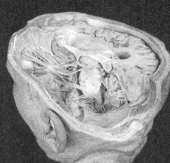

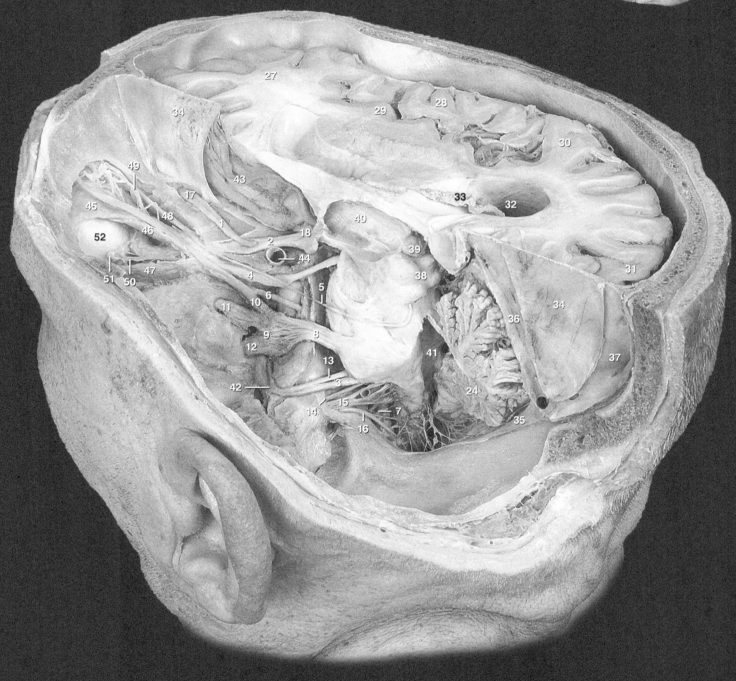

Intracranial dissection of cranial nerves
Posterolateral view

Cranial Nerves

Cranial nerves V and VII, the trigeminal and facial nerves respectively, have the most extensive distribution to the tissues of the head. This page and the three pages that follow depict the peripheral distribution of many of the branches of the trigeminal and facial nerves.

Trigeminal Nerve
1 Auriculotemporal nerve
2 Supraorbital nerve
3 Infraorbital nerve
4 Mental nerve
5 Maxillary branch
6 Nerve of the pterygoid canal
7 Pterygopalatine ganglion
8 Nasopalatine nerve (cut)

9 Superior posterior lateral nasal branch
10 Inferior posterior lateral nasal branch
11 Pharyngeal branch
12 Lesser palatine nerve
13 Greater palatine nerve

Facial Nerve
14 Temporal branches
15 Zygomatic branches

16 Buccal branches
17 Mandibular branches
18 Cervical branch

Other Nerves and Structures
19 Greater occipital nerve
20 Lesser occipital nerve
21 Great auricular nerve
22 Auricularis posterior muscle

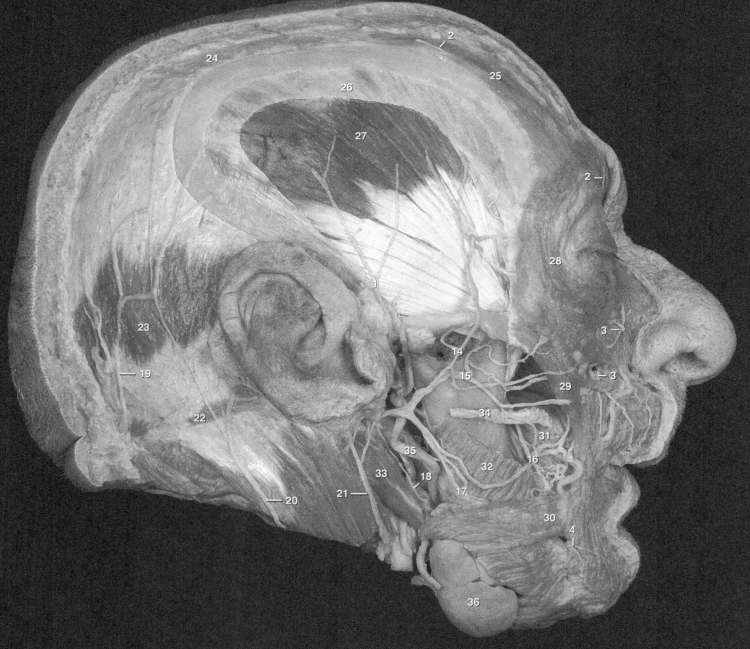

Dissection of head exposing branches of the facial nerve
Lateral view

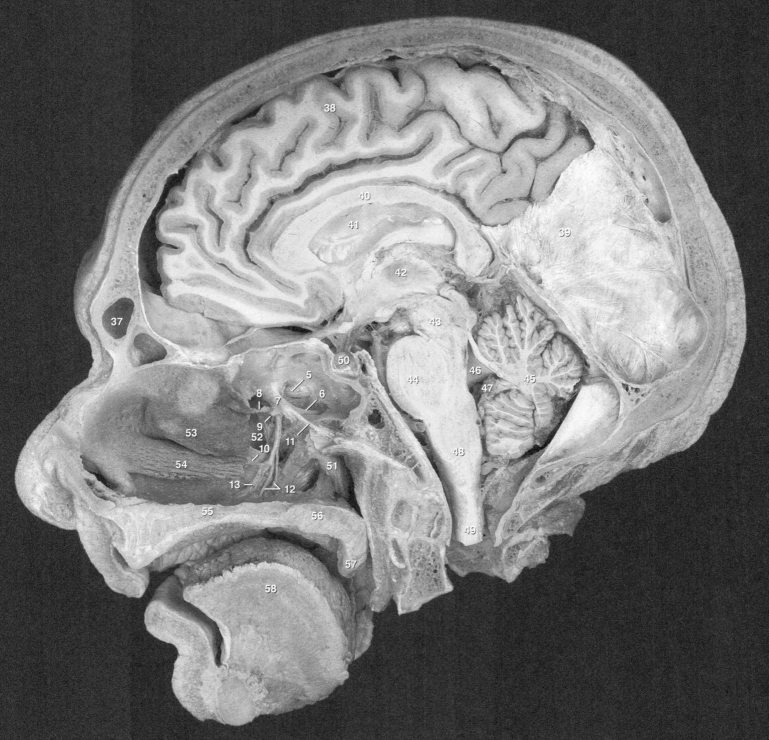

23	Occipital belly of epicranius muscle	32	Masseter muscle	41	Septum pellucidum	50	Pituitary gland
24	Galia aponeurotica	33	Posterior digastricus muscle	42	Thalamus	51	Torus tubarius
25	Frontal belly of epicranius muscle	34	Parotid duct	43	Midbrain	52	Maxillary sinus
26	Temporal fascia	35	External carotid artery	44	Pons	53	Middle nasal concha
27	Temporalis muscle	36	Submandibular gland	45	Cerebellum	54	Inferior nasal concha
28	Orbicularis oculi muscle	37	Frontal sinus	46	Fourth ventricle	55	Hard palate
29	Zygomaticus major muscle	38	Cerebrum	47	Choroid plexus	56	Soft palate
30	Risorius muscle	39	Falx cerebri	48	Medulla oblongata	57	Uvula
31	Buccinator muscle	40	Corpus callosum	49	Spinal cord	58	Tongue

Parasagittal section and dissection of head exposing branches of the trigeminal and facial nerve
Medial view

Cranial Nerves

1 Nerve to temporalis muscle
2 Buccal nerve
3 Middle superior alveolar nerve
4 Posterior superior alveolar nerve
5 Lingual nerve
6 Chorda tympani nerve

7 Inferior alveolar nerve
8 Nerve to mylohyoid muscle
9 Pterygopalatine ganglion
10 Infraorbital nerve
11 Hypoglossal nerve
12 Submandibular ganglion

13 Superior laryngeal nerve

Other Structures
14 Orbicularis oculi muscle
15 Temporal fascia
16 Temporalis muscle

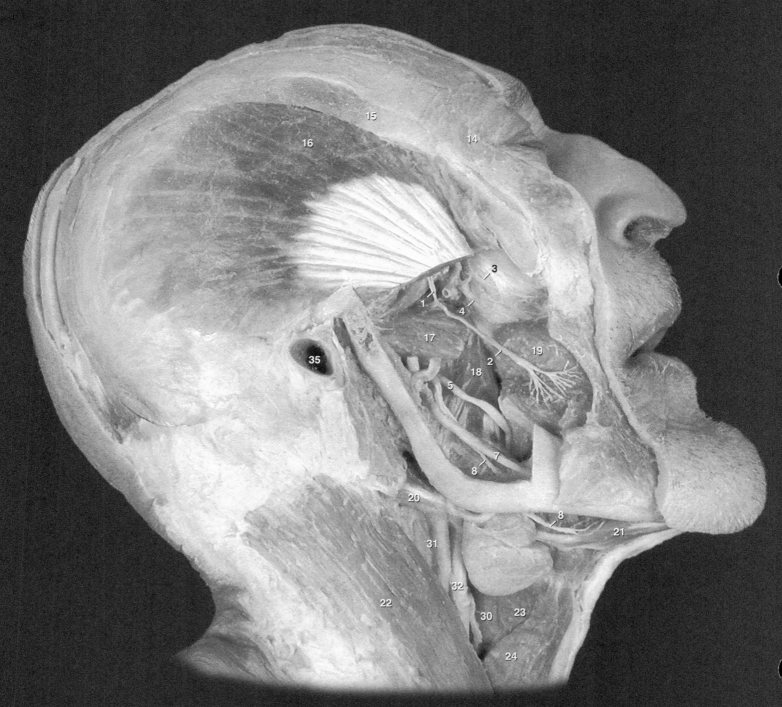

Dissection of head exposing branches of the trigeminal nerve
Lateral view

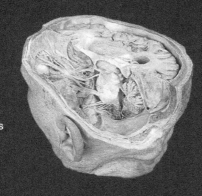

17 Lateral pterygoid muscle
18 Medial pterygoid muscle
19 Buccinator muscle
20 Posterior digastricus muscle
21 Anterior digastricus muscle
22 Sternocleidomastoid muscle
23 Thyrohyoid muscle

24 Omohyoid muscle
25 Styloglossus muscle
26 Stylohyoid muscle
27 Geniohyoid muscle
28 Mylohyoid muscle
29 Superior pharyngeal constrictor
30 Inferior pharyngeal constrictor

31 Internal jugular vein
32 Common carotid artery
33 Dura mater
34 Cerebrum
35 External acoustic meatus
36 Tongue

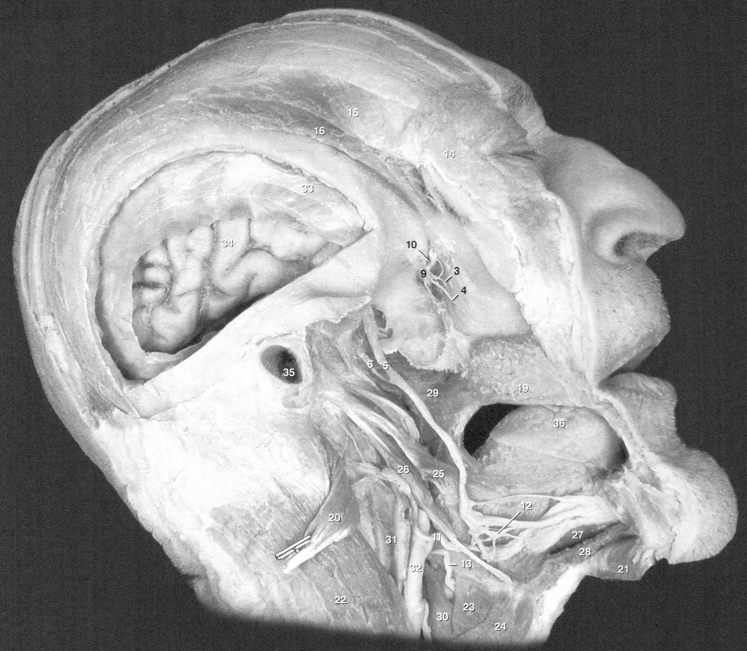

Dissection of head with mandible removed
Lateral view

Sensory Receptors

Sensory receptors are the transducers of the nervous system; that is, they convert the different types of energy we experience such as mechanical energy (touch, pressure, sound waves, etc.), thermal energy (heat), chemical energy (taste, smell), and electromagnetic energy (light) into the electrical energy of the nervous impulse. They do this by facilitating the depolarization of the peripheral terminals of the sensory neurons. This initiates the nervous impulse along the sensory neuron, and this input is carried by the sensory neuron to the processing centers of the brain and spinal cord, which will be the topic of the next chapter.

1 Epidermis
2 Corpuscle of touch (Meissner's)
3 Dermis
4 Dermal papilla
5 Neuron
6 Lamellated corpuscle
7 Taste bud
8 Taste pore
9 Gustatory hair
10 Gustatory receptor cell
11 Supporting cell
12 Basal cell

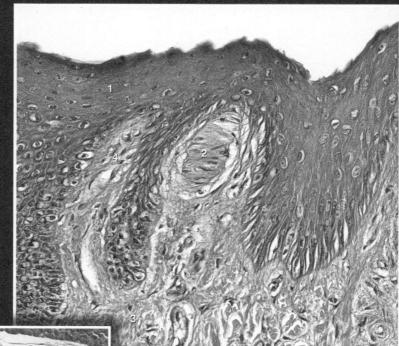

Photomicrograph of corpuscle of touch
200x

Photomicrograph of lamellated corpuscle
100x

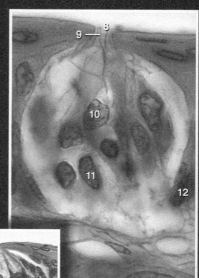

Photomicrographs of taste bud
200x (left), 700x (right)

14 | Central Nervous System

While the neuronal circuitry of the central nervous system is awe inspiring to say the least, the basic concepts behind this complex integration and control center have a simple design. At its simplest, the fundamental design of the central nervous system involves two features: gray matter and white matter. The gray matter centers represent the synaptic integration and control circuits; that is, these centers contain numerous highly dendritic interneurons along with the cell bodies of efferent neurons and axon terminals of incoming afferent neurons, all forming a myriad of synaptic circuits. In these gray centers input is integrated, compared, sensed, and stored to give rise to coordinated, controlled output. The white matter, on the other hand, represents conduction tracts between the synaptic gray centers. These white tracts consist mainly of the myelinated axons of interneurons relaying signals from one gray center to another.

A second simple concept to keep in mind is that the complexity of the central nervous system increases from a caudal to cranial direction. There is logic to this pattern because in the spinal cord the gray centers primarily function as integration networks that regulate input and output for their specific spinal nerve levels. In other words, they are segmental control centers. Input entering a spinal nerve level initiates reflexive output back to the peripheral tissues at that same spinal level. Connecting these segmental gray centers via interneuronal tracts leads to greater association between neighboring levels, therefore improving integration and control. If one segmental gray center can relay information received from its center to neighboring centers, then there can be a greater spread of control generated in response to local segmental input. Now take this a step further by relaying information via white tracts from each of the segmental control centers to higher centers. These higher centers receive input from all the lower segmental centers, integrating the input to gain a full body perspective, while generating the necessary output signals to exert coordinated full body control. Because of this added circuitry the cranial or brain end of the central nervous system increases in size. This additive accumulation of interconnected gray centers accounts for the structure of the brain and its amazing functional properties.

Because much of the central nervous system circuitry is of a more microscopic nature and beyond the scope of this book. In this chapter we attempt to depict the basic gross anatomy of the central nervous system and its protective coverings.

Find more information about the central nervous system in REALANATOMY

Spinal Cord

Extending from the brainstem is a long slender rod of nerve tissue, the spinal cord. The cord exits the foramen magnum of the skull and descends within the vertebral canal of the boney vertebral column. It is about 45 cm long (18 inches) and ends between the first and second lumbar vertebrae. Although there are some slight regional variations, the cross-sectional anatomy of the spinal cord is generally the same throughout its length. The gray matter of the spinal cord forms a butterfly-shaped region in the center of the cord that is surrounded by the white matter. As is the theme throughout the central nervous system, gray matter consists primarily of neuronal cell bodies and their dendrites, short interneurons, and glial cells. The white matter is organized into tracts, which are bundles of myelinated nerve fibers (axons of long interneurons and sensory neurons) that communicate between the gray circuit centers at all levels of the spinal cord and brain.

Each side of the H-shaped gray matter of the spinal cord has a dorsal horn and a ventral horn sandwiching an intermediate gray region. Entering the dorsal horns from the dorsal rootlets are the axons of the afferent neurons, which synapse with small interneuron pools to form segmental integration centers for that level of the body. The dorsal horn and intermediate gray matter contain numerous small interneurons. The intermediate gray also contains, at certain levels, the preganglionic efferent neurons of the autonomic output. The ventral horns are primarily populated by the efferent neurons to the skeletal muscles of their respective spinal levels. The white matter tracts are grouped into columns of myelinated axons that extend the length of the cord. Each of these tracts begins or ends within a particular area of the cord and brain, and each is specific in the type of information that it transmits. Some are ascending tracts that carry signals derived from sensory input. For example, one tract carries information derived from pain and temperature receptors, whereas another carries information regarding touch. Other tracts are descending tracts that relay messages from the brain to motor neurons in the ventral horn.

Both the white and gray matter exhibit regional differences throughout the length of the spinal cord. There is relatively more white matter at the cranial end of the spinal cord than at the caudal end. Notice that the gray matter, especially the ventral horn, is the largest at lower cervical levels and at lower lumbar-upper sacral levels. These levels correspond to upper and lower limb anatomy respectively, where large amounts of muscle tissue require motor innervtion from the ventral horn motor neuron pools.

1 Dorsal horn of gray matter	7 Central canal	13 Conus medullaris
2 Lateral horn of gray matter	8 Dorsolateral fasciculus	14 Cauda equina
3 Ventral horn of gray matter	9 Dorsal root of spinal nerve	15 Dorsal rami of spinal nerve
4 Posterior funiculus of white matter	10 Dorsal root ganglion	16 Cerebrum
5 Lateral funiculus of white matter	11 Ventral root of spinal nerve	17 Cerebellum
6 Anterior funiculus of white matter	12 Spinal cord	18 First lumbar vertebra

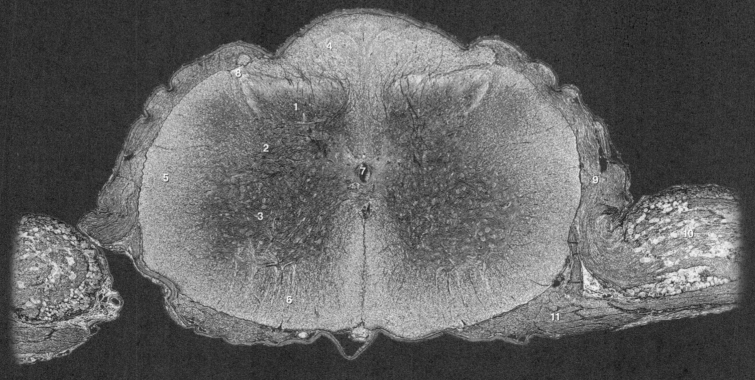

Photomicrograph of spinal cord
50x

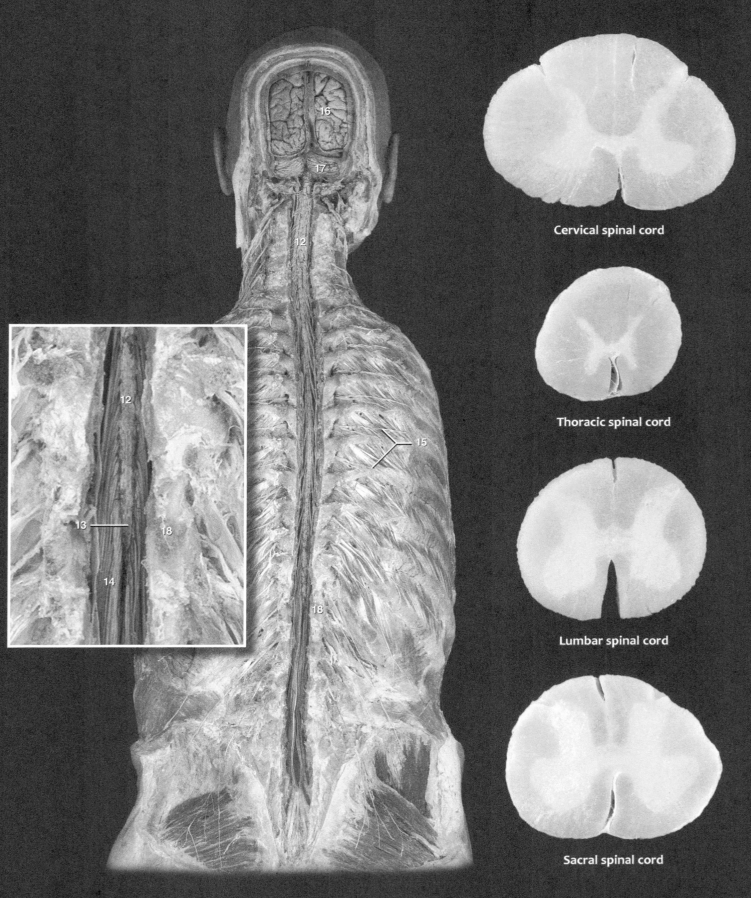

Cervical spinal cord

Thoracic spinal cord

Lumbar spinal cord

Sacral spinal cord

Dissection of vertebral column and skull revealing brain and spinal cord
Posterior view, with call-out of terminal end of cord

Brain

The brain is the large, anterior-expansion of the neural tube situated within the cranium. Rapid development of the rostral e nd of the neural tube forms three expanded regions — the prosencephalon, mesencephalon, and rhombencephalon. The prosencephalon undergoes further development to form the telencephalon and diencephalon, and the rhombencephalon continues to develop to form a metencephalon and myelencephalon. These five embryonic regions give rise to the brain. The telencephalon becomes the cerebrum, the diencephalon becomes the thalamic regions, the mesencephalon becomes the midbrain, the metencephalon becomes the cerebellum and pons, and the myelencephalon becomes the medulla oblongata. A variety of views of the full brain are depicted on this and the facing page.

1 Spinal cord
2 Medulla oblongata
3 Pons
4 Cerebellum
5 Midbrain
6 Diencephalon
7 Frontal lobe of cerebrum
8 Parietal lobe of cerebrum

9   Occipital lobe of cerebrum
10 Temporal lobe of cerebrum
11 Longitudinal fissure
12 Transverse fissure
13 Lateral cerebral sulcus
14 Anterior median fissure
15 Gyrus
16 Sulcus

17 Central sulcus
18 Precentral gyrus
19 Postcentral gyrus
20 Precentral sulcus
21 Postcentral sulcus
22 Inferior frontal gyrus
23 Superior temporal gyrus

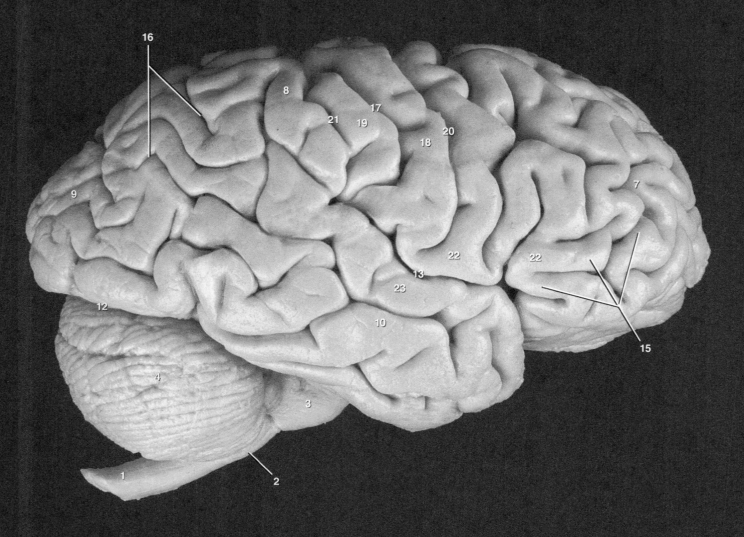

Brain
Lateral view

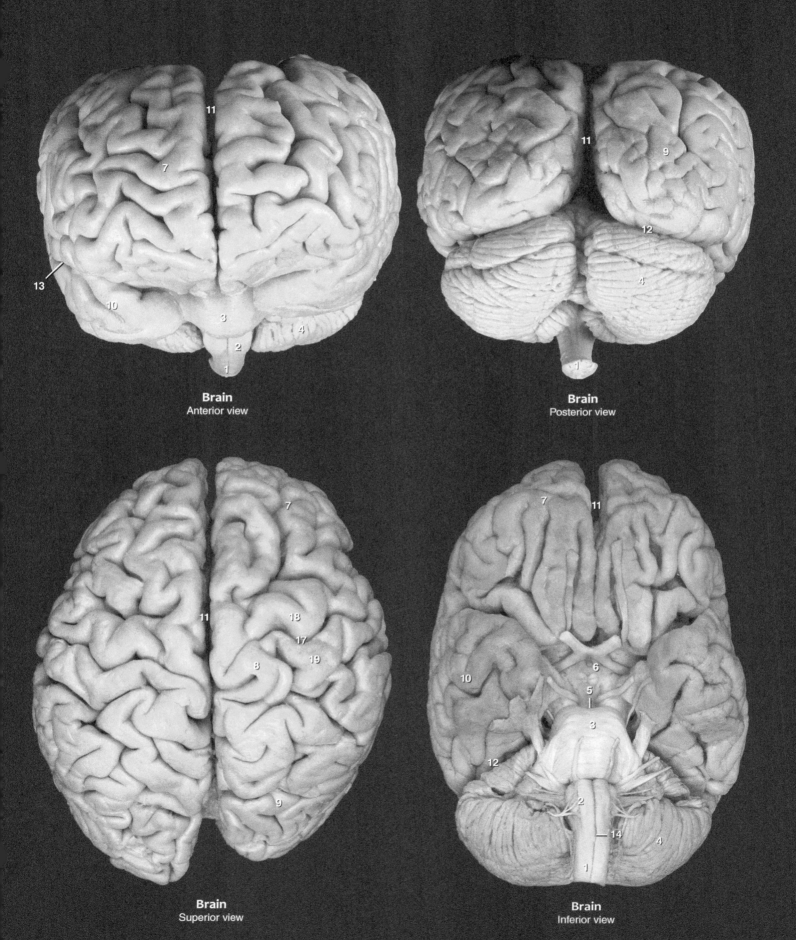

Brain
Anterior view

Brain
Posterior view

Brain
Superior view

Brain
Inferior view

Brain Regions

As the spinal cord ascends through the foramen magnum to enter the skull, the cranial central nervous system gradually expands in size to form the large central processing circuitry we call the brain. The increasing size of the brain results from the addition of more and more gray processing centers to the basic cord-like brain stem. The caudal part of the brain, called the brain stem, consists of the medulla oblongata, pons, and midbrain. Though all of these structural regions exhibit their own specializations, they have certain fiber tracts in common and all have nuclei for the cranial nerves. Added to the brain stem are the more rostral portions of the brain — the cerebellum, diencephalon, and cerebral hemispheres. These large processing centers greatly increase the size of the brain. The images on the facing page show the principal parts of the brain.

1 Spinal cord
2 Medulla oblongata
3 Pons
4 Cerebellum
5 Fourth ventricle
6 Midbrain
7 Inferior colliculus

8 Superior colliculus
9 Thalamus of diencephalon
10 Hypothalamus of diencephalon
11 Interthalamic adhesion
12 Pineal gland
13 Mammillary body
14 Optic tract

15 Frontal lobe of cerebrum
16 Parietal lobe of cerebrum
17 Occipital lobe of cerebrum
18 Temporal lobe of cerebrum
19 Corpus callosum
20 Lateral ventricle
21 Fornix

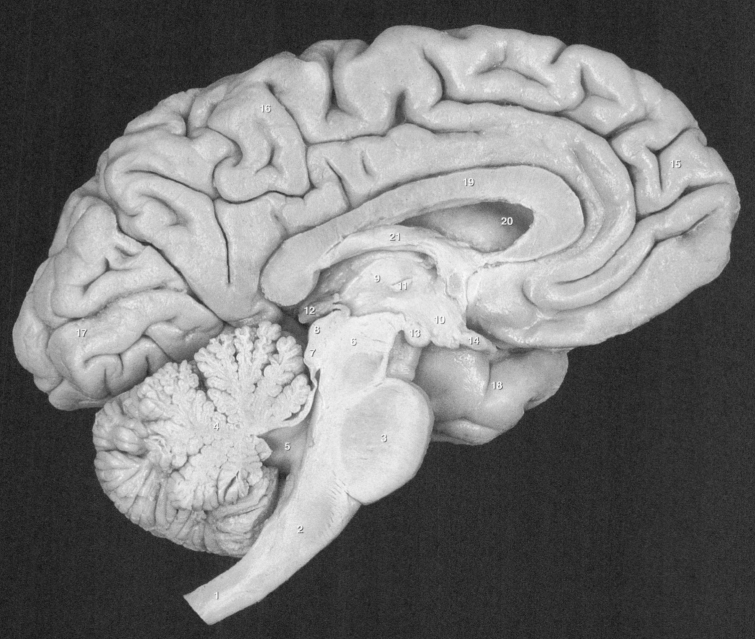

Sagittal section of the brain
Medial view

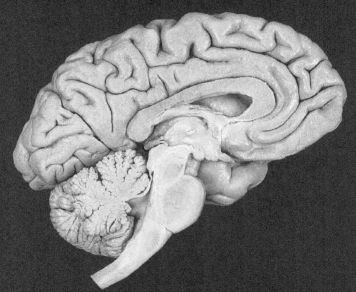

Medulla oblongata

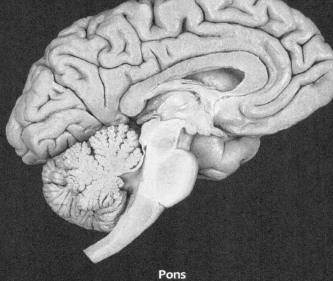

Pons

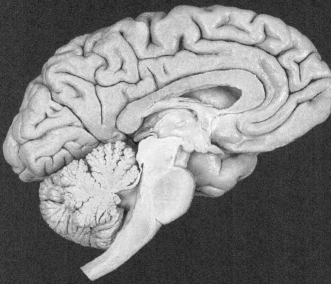

Midbrain

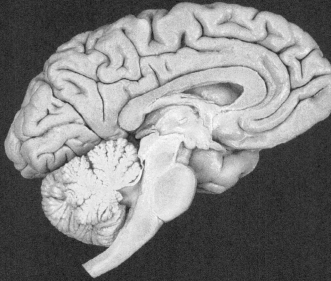

Cerebellum

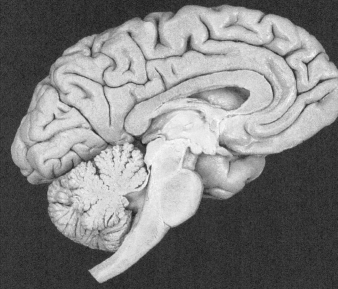

Diencephalon – epithalamus, thalamus, hypothalamus

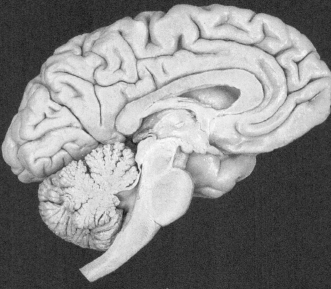

Cerebrum

Cerebrum

The cerebrum, by far the largest part of the human brain, consists of the cerebral hemispheres and the basal nuclei. The large, obvious cerebrum is divided into two halves, the right and left cerebral hemispheres. Each cerebral hemisphere has an outer layer of gray matter, the cerebral cortex, covering deeper networks of interconnecting white tracts that connect different areas of the cortex with one another and with lower brain centers. The amount of cortex is greatly increased by a complex folding of the cerebral surface. The folds produce hills, gyri (singular gyrus), and depressions, sulci (singular sulcus). This cortical surface forms the highest level of processing circuitry in the brain. The two hemispheres are connected to each other by the corpus callosum, a thick band consisting of an estimated 300 million neuronal axons traversing between the two hemispheres. Located deep within the cerebrum is another region of gray matter, the basal nuclei, which form key integration centers between the cortex and lower brain centers.

1 Central sulcus
2 Precentral gyrus
3 Postcentral gyrus
4 Precentral sulcus
5 Postcentral sulcus
6 Parieto-occipital sulcus
7 Transverse occipital sulcus

8 Calcarine sulcus
9 Superior temporal gyrus
10 Middle temporal gyrus
11 Inferior temporal gyrus
12 Inferior frontal gyrus
13 Middle frontal gyrus
14 Superior frontal gyrus

15 Short gyri
16 Long gyrus
17 Limen
18 Pons
19 Cerebellum
20 Medulla oblongata
21 Spinal cord

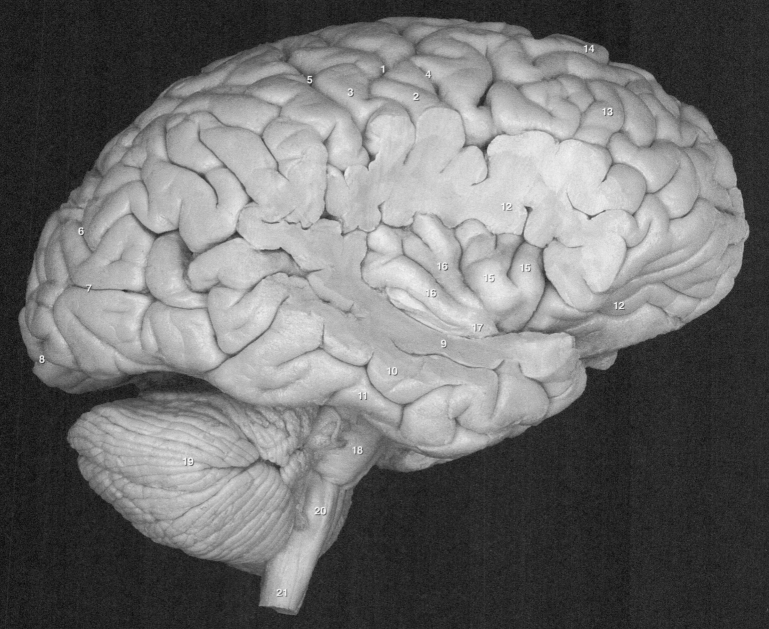

Brain dissection revealing insular lobe
Lateral view

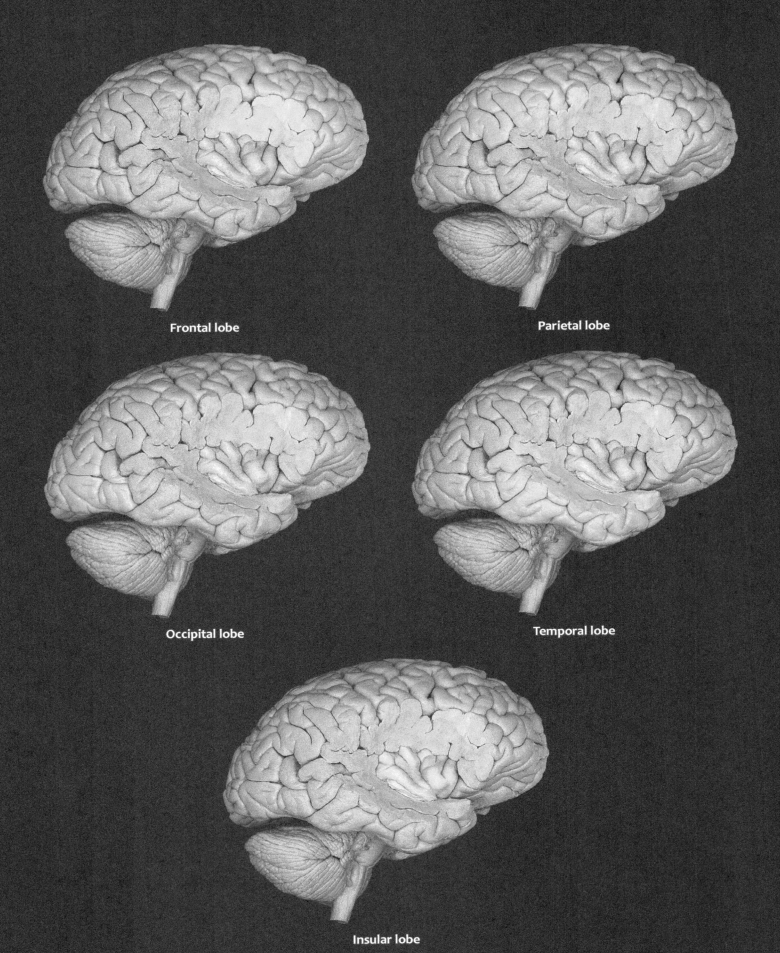

Frontal lobe

Parietal lobe

Occipital lobe

Temporal lobe

Insular lobe

Cerebellum

Immediately above the medulla oblongata the central nervous system expands dorsally to form the cerebellum, which means little brain. The cerebellum, like the cerebrum, has a highly folded surface that greatly increases the surface area of its outer gray matter cortex. It is estimated that the cerebellum has in the neighborhood of 10 billion neurons, which have a variety of functional roles. The cerebellum processes input received from the cerebral cortex, various brain stem nuclei, and peripheral sensory receptors to smooth and coordinate complex, skilled movements. It plays an important role in posture and balance and functions in cognition and language processing.

1 Folia of cerebellum
2 Anterior lobe of cerebellum
3 Posterior lobe of cerebellum
4 Superior vermis
5 Inferior vermis
6 Postlunate fissure
7 Posterior cerebellar notch
8 Tonsil
9 Quadrangular lobe of anterior

10 Primary fissure
11 Flocculus
12 Lingula
13 Central lobule
14 Culmen
15 Declive
16 Folium
17 Tuber
18 Pyramid

19 Uvula
20 Nodulus
21 Midbrain
22 Superior medullary velum
23 Fourth ventricle
24 Median aperture
25 Cerebral aqueduct
26 Pons
27 Medulla oblongata

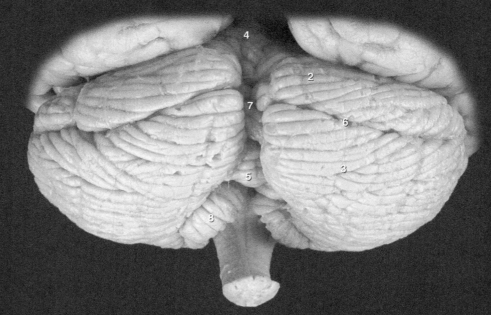

Cerebellum
Posterior view

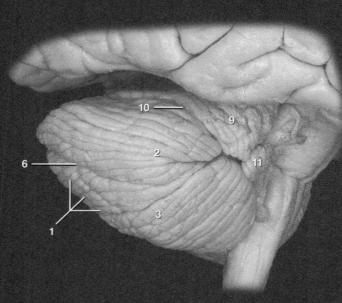

Cerebellum
Lateral view

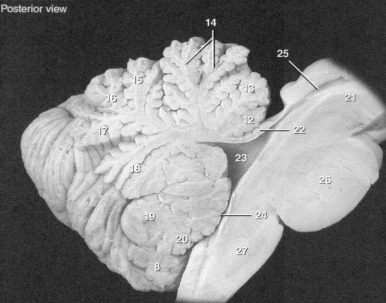

Sagittal section of cerebellum
Medial view

Diencephalon and Brainstem

The diencephalon, rostral to the midbrain and almost completely surrounded by the cerebral hemispheres, consists of four major parts — the thalamus, subthalamus, epithalamus, and hypothalamus. Projecting from the hypothalamus is the hypophysis, or pituitary gland. The brainstem consists of the medulla oblongata, pons, and midbrain. The medulla resembles the spinal cord in many ways. Like the cord it gives rise to many nerve roots; however, these are the roots of cranial nerves rather than spinal nerves. The pons is the bridge between the two cerebellar hemispheres. The ventral portion of the pons forms a large synaptic relay station consisting of scattered gray centers called the pontine nuclei. The dorsal portion of the pons is more like the other regions of the brainstem, the medulla and midbrain. The midbrain sits just above the pons and is obscured by the large, overlapping cerebral hemispheres. It contains nuclei for cranial nerves III and IV, as well as ascending and descending fiber tracts from the cerebrum.

1 Infundibulum
2 Anterior perforated substance
3 Tuber cinereum
4 Mammillary body
5 Posterior perforated substance
6 Pulvinar of thalamus
7 Pineal gland
8 Superior colliculus
9 Inferior colliculus
10 Medial geniculate ganglion
11 Pons
12 Superior cerebellar peduncle
13 Middle cerebellar peduncle
14 Inferior cerebellar peduncle
15 Medial eminence
16 Facial colliculus
17 Locus ceruleus
18 Trigeminal tubercle
19 Hypoglossal tubercle
20 Vestibular area
21 Sulcus limitans
22 Lateral recess
23 Obex
24 Olive
25 Pyramid
26 Third ventricle
27 Fourth ventricle
28 Cerebral crus
29 Superior medullary vellum
30 Flocculus of cerebellum
31 Caudate nucleus
32 Optic tract
33 Optic chiasm
34 Optic nerve
35 Oculomotor nerve
36 Trochlear nerve
37 Abducens nerve
38 Trigeminal nerve
39 Facial nerve
40 Vestibulocochlear nerve
41 Glossopharyngeal nerve
42 Vagus nerve
43 Accessory nerve
44 Hypoglossal nerve

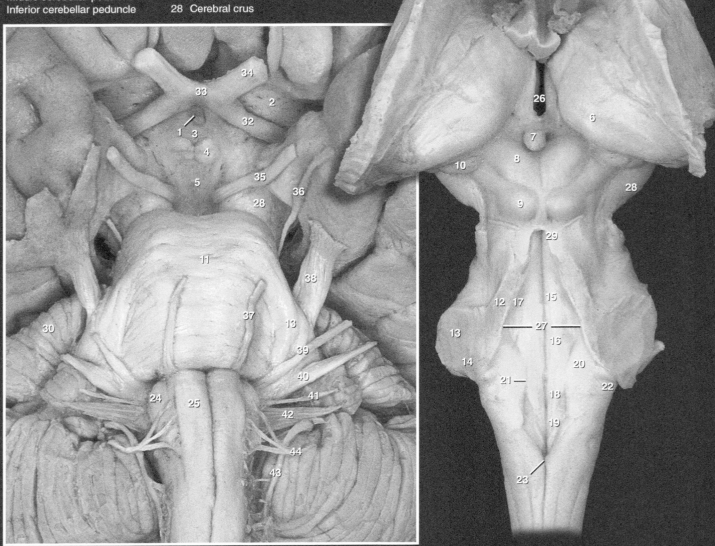

Brainstem
Ventral view

Brainstem
Posterior view

Brain Sections

The brain sections on this and the following page depict aspects of brain anatomy that are not evident on the external views of the brain, and the association of the brain with surrounding structures of the head. Each section is approximately 2 centimeters thick and is an anterior view of three sections in succession. The first section begins at the anterior aspect of the ear and the last section is just posterior to the ear.

1	Frontal lobe	14	Flocculus	27	Globus pallidus	40	Sigmoid sinus
2	Parietal lobe	15	Superior vermis	28	Medial thalamic nucleus	41	Internal jugular vein
3	Temporal lobe	16	Superior cerebellar peduncle	29	Lateral thalamic nucleus	42	Tympanic cavity
4	Insular lobe	17	Cerebral peduncle	30	Dentate gyrus	43	Cochlea
5	Lateral ventricle	18	Pituitary gland	31	Circular gyrus	44	Sphenoid sinus
6	Third ventricle	19	Pons	32	Optic chiasm	45	Mastoid air cells
7	Cerebral aqueduct	20	Olive	33	Facial nerve	46	Mandibular condyle
8	Fourth ventricle	21	Corpus callosum	34	Vestibulocochlear nerve	47	Occipital condyle
9	Septum pellucidum	22	Caudate nucleus	35	Vertebral artery	48	Atlas
10	Falx cerebri	23	Internal capsule	36	Middle cerebral artery	49	Axis
11	Tentorium cerebelli	24	Putamen	37	Internal carotid artery	50	Lateral pterygoid muscle
12	Anterior lobe of cerebellum	25	External capsule	38	Anterior cerebral artery	51	Medial pterygoid muscle
13	Posterior lobe of cerebellum	26	Body of fornix	39	Superior sagittal sinus	52	Sternocleidomastoid muscle

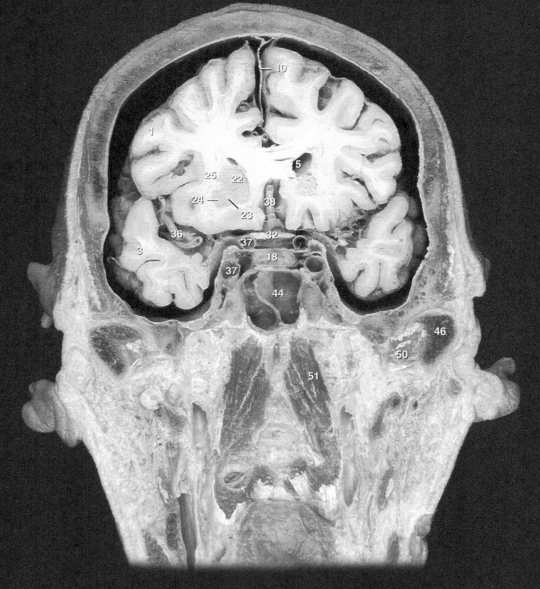

Frontal section of head at anterior aspect of auricle
Anterior view

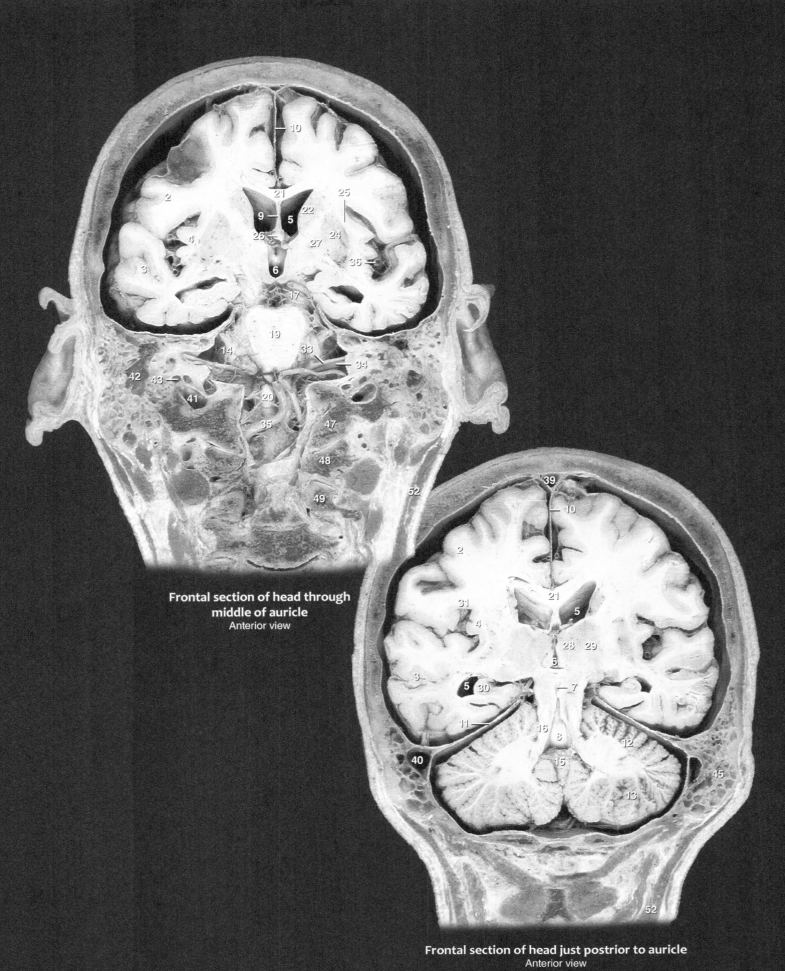

Frontal section of head through middle of auricle
Anterior view

Frontal section of head just postrior to auricle
Anterior view

Ventricular System

Developmentally the entire central nervous system forms from the hollow neural tube. As development proceeds and the wall of the neural tube becomes increasingly thicker, the hollow lumen of the tube undergoes changes in relative size and shape throughout different regions of the changing central nervous system. As a result of this developmental history, there remains a hollow interconnected center throughout the entire central nervous system. This hollow core forms the ventricular system. Beginning within the cerebral hemispheres are the large paired lateral ventricles. Each lateral ventricle has a C-shape like its corresponding hemisphere. The lateral ventricles communicate via the interventricular foramina with a midline cavity, the third ventricle. The third ventricle sits within the core of the diencephalon where the right and left thalamus form its lateral walls. From the third ventricle a narrow channel, the aqueduct of the midbrain or cerebral aqueduct, passes through the core of the midbrain. This narrow channel expands in the region of the pons and cerebellum to form the fourth ventricle. The fourth ventricle tapers through the medulla to enter the spinal cord as the central canal. Within the four ventricles of the brain convoluted aggregations of capillaries, called a choroid plexus, project into the cavity of the ventricle. These capillary projections are the principal site for the production of cerebrospinal fluid.

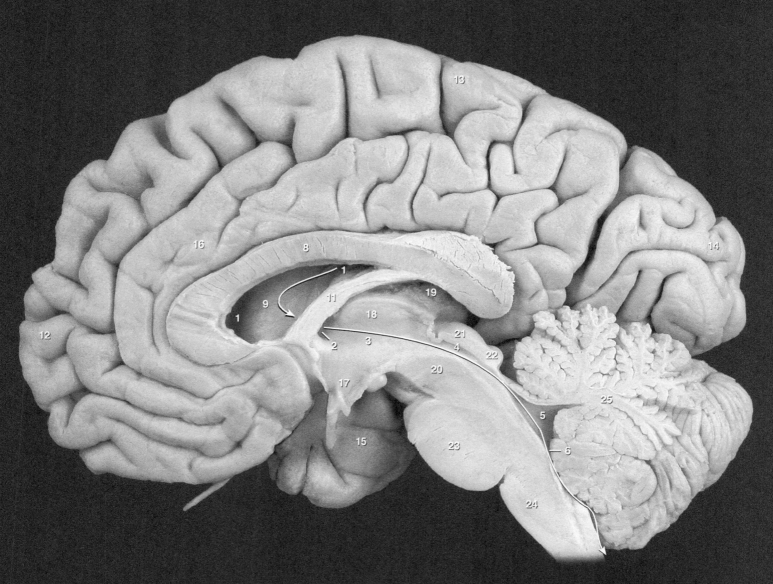

Sagittal section of braining revealing the ventricular system
Medial view, arrows show path of cerebrospinal fluid

1	Lateral ventricle	9	Caudate nucleus	17	Hypothalamus	25	Cerebellum
2	Interventricular foramen	10	Septum pellucidum	18	Thalamus	26	Falx cerebri
3	Third ventricle	11	Fornix	19	Pineal gland	27	Internal carotid artery
4	Cerebral aqueduct	12	Frontal lobe	20	Midbrain	28	Middle cerebellar peduncle
5	Fourth ventricle	13	Parietal lobe	21	Superior colliculus	29	Trochlear nerve
6	Median aperture	14	Occipital lobe	22	Inferior colliculus	30	Vestibulocochlear nerve
7	Choroid plexus	15	Temporal lobe	23	Pons	31	Vagus nerve
8	Corpus callosum	16	Cingulate gyrus	24	Medulla oblongata	32	Accessory nerve

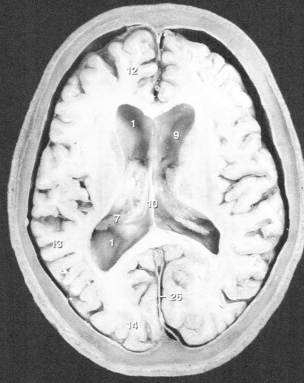

Floor of lateral ventricles
Superior view

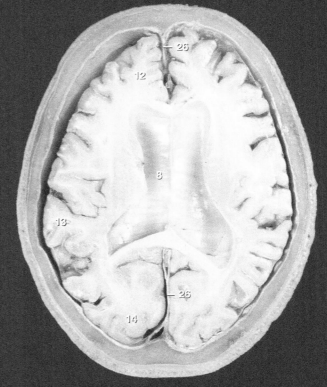

Roof of lateral ventricles
Inferior view

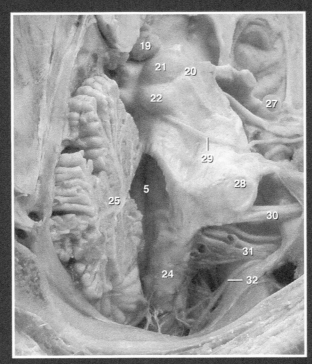

Fourth ventricle
Posterolateral view

Meninges

Within the cranium and vertebral column, the meninges form a protective encasement for the tissue of the brain and spinal cord. There are three meningeal membranes, the tough outer connective tissue pachymenix, the dura mater, and the epithelial inner leptomeninges, the arachnoid mater and pia mater. Between the leptomeningeal layers there is a fluid compartment called the subarachnoid space. Cerebrospinal fluid, secreted from the choroid plexuses of the ventricles, exits the ventricles to fill this compartment. The cerebrospinal fluid forms a hydraulic shock absorber and suspension system for the brain and spinal cord. In addition to protecting the central nervous system, the meninges support many of the blood vessels that are associated with the brain. Within the cranium the subdivisions of the dura mater split to form large venous channels, the dural venous sinuses, which drain all the tissues of the cranial vault, and these splits also form strong, fibrous septa that separate different parts of the brain.

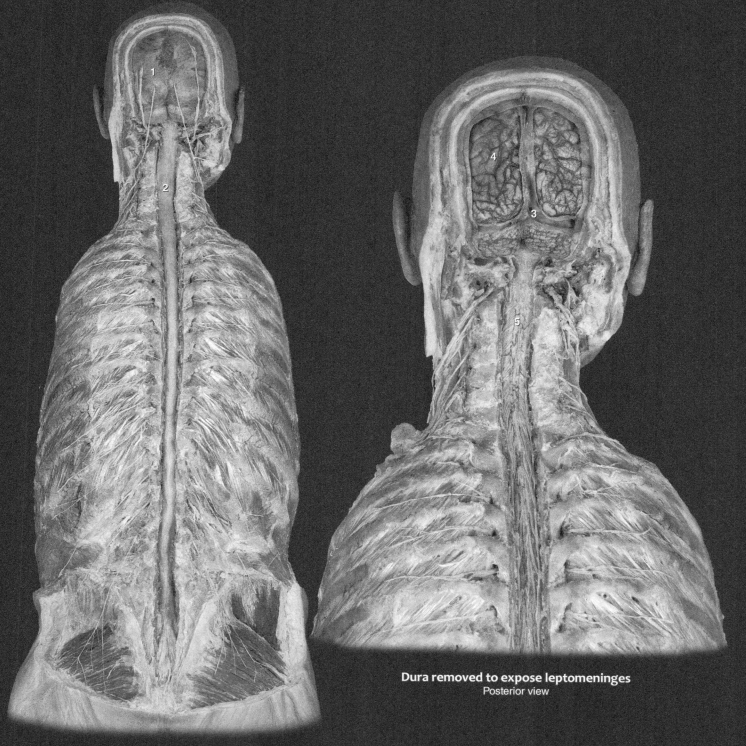

Dura removed to expose leptomeninges
Posterior view

Dissection of cranial and spinal dura mater
Posterior view

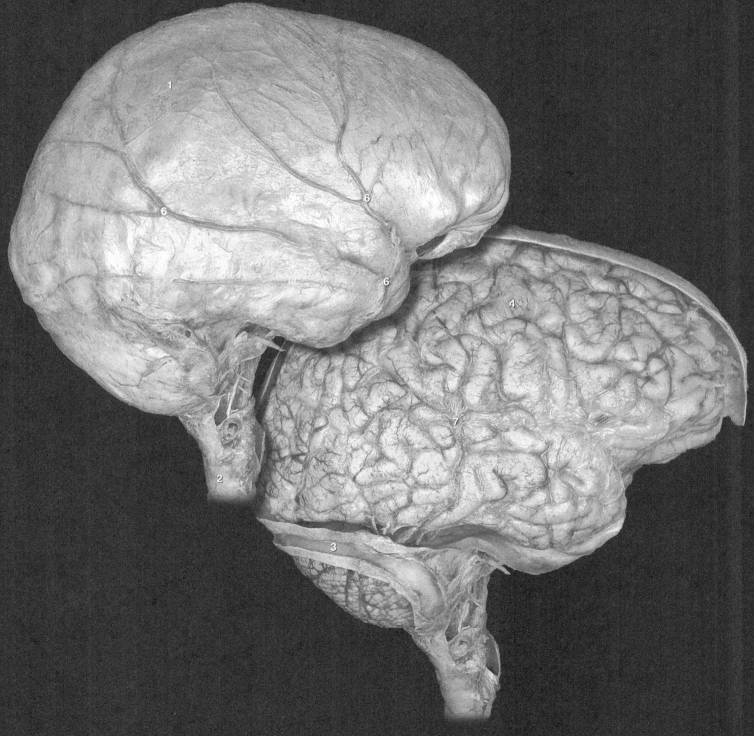

1 Cranial dura mater
2 Spinal dura mater
3 Dural venous sinus
4 Cranial leptomeninges - arachnoid is superficial to and covering pia mater
5 Spinal leptomeninges - arachnoid is superficial to and covering pia mater
6 Middle meningeal artery and branches in dura mater
7 Superficial middle cerebral vein and tributaries in subarachnoid space

Dural sac (above), Leptomeninges (below)
Lateral views

Meninges

1 Falx cerebri
2 Tentorium cerebelli (cut)
3 Superior sagittal sinus
4 Straight sinus
5 Transverse sinus
6 Lateral ventricle
7 Septum pellucidum
8 Third ventricle
9 Fourth ventricle
10 Cerebrum
11 Cerebellum
12 Corpus callosum
13 Choroid plexus
14 Optic chiasm
15 Trigeminal nerve

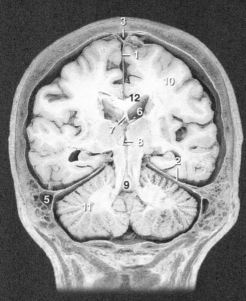

Head frontal section revealing dural septa
Anterior view

Dissection of cranium
Superoposterior view

Dissection of cranium
Superoposterior view

15 | Endocrine System

Like the nervous system, the endocrine system is a control system within the body. The nervous system administers its control over the body tissues via long wirelike cells that originate form complex circuits in the central nervous system. This circuitry receives sensory input, processes this input, and generates regulatory output. Endocrine control works in a much different fashion. The endocrine system consists of a number of different glands that function like radio transmitting stations. Just as different radio stations send radio signals of different wavelengths into the air, endocrine glands distribute different types of small molecules called hormones throughout the body via the circulatory system. These small molecules travel through the blood stream and are detected by effector organs in different parts of the body, much like radio waves are detected by radios in different parts of a city. Effector organs have receptor sites that are specific to specific hormones. This results in a "lock and key" function at the effector cell. When the hormone binds to the receptor site, it initiates a regulatory effect on the cell.

Because the hormones are distributed by the circulatory system, the speed of endocrine regulation is slower than that of nervous regulation, many minutes compared to milliseconds. Also, because of the distribution of the hormones via the circulatory system, endocrine effects can be experienced anywhere there are cells with the appropriate receptor site. In comparison to the nervous system, endocrine distribution is potentially very widespread. Because the hormone can lock into the receptor site and not be degraded instantly, the duration can be longer lasting than that initiated by a single nervous impulse.

Find more information about the endocrine system in
REAL ANATOMY

249

Hypothalamus

The hypothalamus occupies the area of the brain between the third ventricle and the subthalamus. It is a major intersection between the thalamus, cerebral cortex, and ascending fiber systems from the spinal cord and brainstem. It is the control center of the autonomic nervous system and regulates the function of numerous endocrine glands. The posterior pituitary gland, or neurohypophysis, is an outgrowth of the hypothalamus. Many factors influence the hypothalamus and dictate its controlling influence over tissues in the body. These factors include the nervous input that enters it, temperature, osmotic pressure, and levels of hormones in the circulating blood that pass through its capillaries.

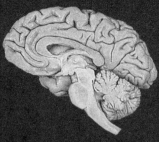

1	Hypothalamus	9	Midbrain
2	Pineal gland	10	Pons
3	Frontal lobe of cerebrum	11	Cerebellum
4	Parietal lobe of cerebrum	12	Medulla oblongata
5	Occipital lobe of cerebrum	13	Lateral ventricle
6	Temporal lobe of cerebrum	14	Fourth ventricle
7	Corpus callosum	15	Mammilary body
8	Thalamus	16	Spinal cord

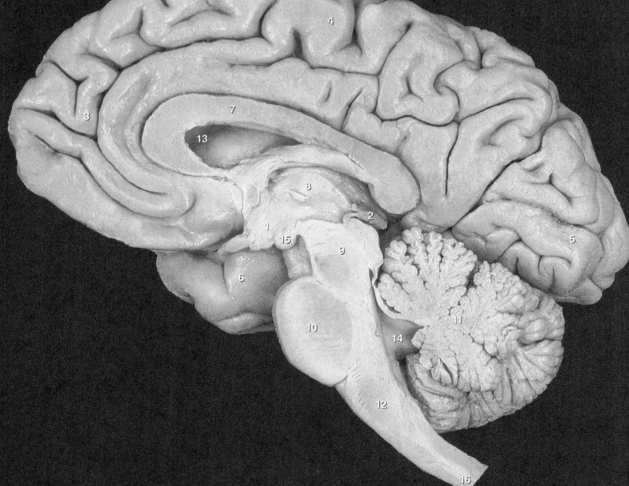

Sagittal section of brain
Medial view

Pituitary Gland

The pituitary gland, or hypophysis, "hangs" from the base of the brain via a connecting stalk, the infundibulum, which connects it to the hypothalamus. The infundibulum contains numerous nerve fibers that relay from the hypothalamus to the posterior portion of the pituitary gland. In addition to this nervous pathway between the hypothalamus and the pituitary, numerous small blood vessels pass between the two organs. The pituitary gland has two anatomically and functionally distinct lobes, the neurohypophysis (posterior lobe) and the adenohypophysis (anterior lobe). The posterior lobe arises as an outgrowth of the embryonic brain. It is composed of nervous tissue and forms a neural link with the hypothalamus through the infundibulum. The anterior lobe arises from the epithelial lining of the embryonic pharynx. It consists of glandular epithelial tissue and forms a vascular link with the hypothalamus via the small blood vessels that pass between the two regions.

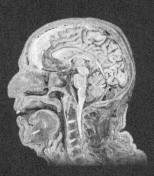

1 Pituitary gland
2 Infundibulum
3 Adenohypophysis
4 Neurohypophysis
5 Parenchyma consisting of acidophils, basophils, and chromophobes
6 Capillary with red blood cells
7 Parenchyma consisting of axons and pituicytes
8 Hypothalamus
9 Cerebrum
10 Falx cerebri
11 Midbrain
12 Pons
13 Cerebellum
14 Medulla oblongata
15 Spinal cord
16 Nasal septum
17 Soft palate
18 Tongue
19 Epiglottis
20 Atlas
21 Axis
22 Intervertebral disc
23 Sphenoid sinus
24 Occipital bone

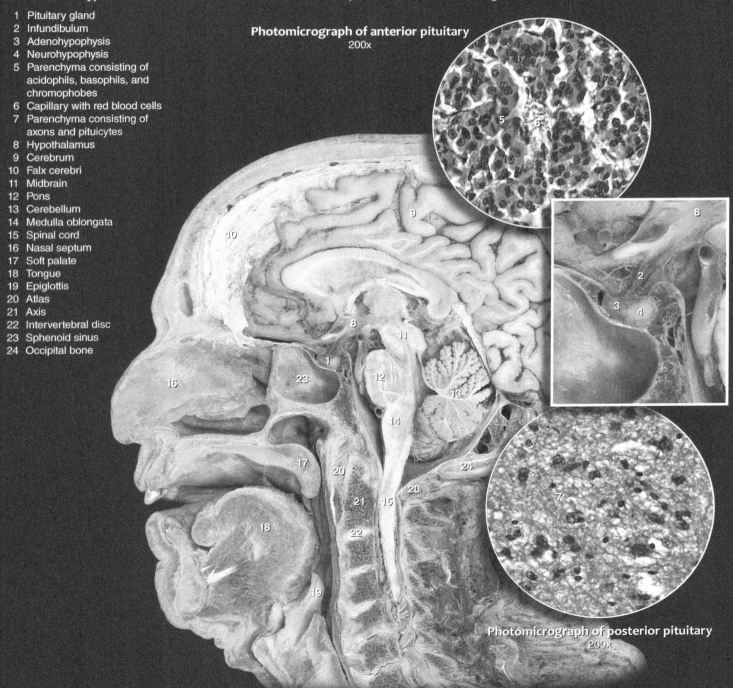

Photomicrograph of anterior pituitary
200x

Photomicrograph of posterior pituitary
200x

Sagittal section of head and neck with enlarged callout of pituitary gland
Medial view

Pineal Gland

The pineal gland, a small reddish-gray body covered with pia mater, is a midline epithelial outgrowth of the embryonic midbrain positioned in a depression between the two superior colliculi on the midbrain's dorsal surface. The distal end of this outgrowth becomes a small mass of secretory cells that resemble the shape of a pine cone. It is from this appearance that it derives its name. The pia mater sends septa into the pineal gland that divide it into cords of secretory cells that are intermingled with numerous blood capillaries. The secretory cells of the pineal gland, called pinealocytes, have arm-like processes that contact both neighboring capillaries and the ependymal cells that line the third ventricle. Hormonal secretions produced in the body of the cell are moved through the arm-like processes where they are released by exocytosis into the capillaries and cerebrospinal fluid. Projecting into these cords of tissue are sympathetic postganglionic neurons from the superior cervical sympathetic ganglion. The gland plays a role in integrating photoperiod and affecting circadian rhythms.

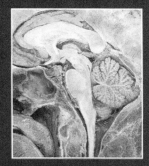

1   Pineal gland
2   Adenohypophysis
3   Neurohypophysis
4   Thalamus
5   Superior colliculi
6   Inferior colliculi
7   Medial geniculate nucleus
8   Cerebral peduncle
9   Medulla oblongata
10 Falx cerebri
11 Corpus callosum
12 Pons
13 Cerebellum
14 Sphenoid sinus
15 Occipital bone
16 Atlas
17 Axis
18 Soft palate
19 Nasopharynx
20 Tongue
21 Middle cerebellar peduncle
22 Fourth ventricle

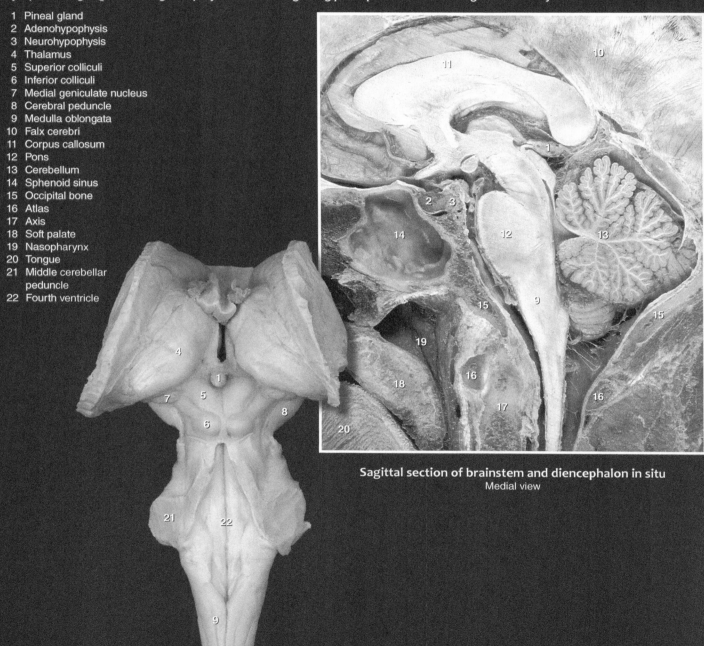

Sagittal section of brainstem and diencephalon in situ
Medial view

Dissection of brainstem and diencephalon
Posterior view

Thymus

The thymus is one of the primary lymphoid organs, but it also has an endocrine component. The thymus provides the specialized environment for the precursor T cells to develop, differentiate, and undergo clonal expansion. This bilobed organ sits just posterior to the superior sternum along the midline. It spans from the top of the sternum, sometimes even projecting into the inferior cervical region, to the level of the fourth costal cartilages and sits anterior to the top of the heart and its great vessels. It has an outer fibrous capsule that sends fibrous septa, connective tissue walls, into the organ forming small lobular subregions. The thymus was once thought to diminish in size with age, but in actuality it does not. Because of its high content of lymphoid tissue and a rich blood supply, it has a reddish appearance in a living body. With age, however, fatty infiltrations replace the lymphoid tissue and it takes on more of the yellowish color of the invading fat. This gives it the false appearance of a reduction in size. The thymus produces hormones that promote the maturation of T cells and may help retard the aging process.

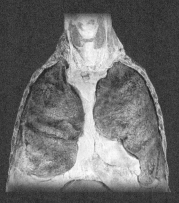

1 Thymus
2 Thymic cortex
3 Thymic medulla
4 Trabeculae
5 Capsule
6 Maturing T cells
7 Epithelioreticular cell
8 Thymic corpuscle
9 Right lung
10 Left lung

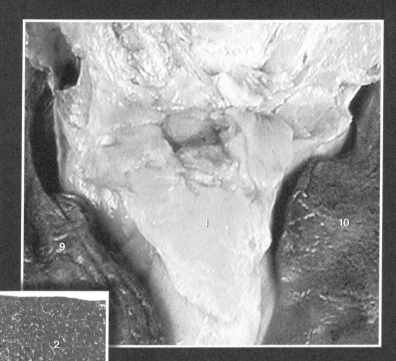

Thymus in situ
Anterior view

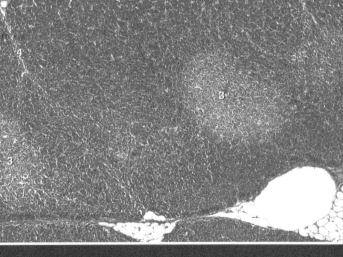

Photomicrograph of thymus
50x

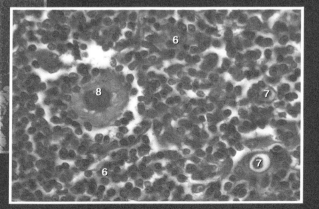

Photomicrograph of thymus
400x

Thyroid Gland

The thyroid gland is a bilobed organ positioned in the anterior neck. This highly vascular organ consists of two lateral lobes of endocrine tissue joined in the middle by a narrow portion of the gland called the isthmus. It is red-brown in color and is enveloped by a thin layer of connective tissue. This connective tissue capsule sends extensions into the gland that divide the vascular and epithelial core into masses of irregular shape and size. The epithelial cells within the compartments of the thyroid gland form the secretory tissues of the organ. The major thyroid secretory cells are arranged into hollow spheres, each of which forms a functional unit called a follicle. In a microscopic section the follicles appear as rings of follicular cells enclosing an inner lumen filled with colloid, a substance that serves as an extracellular storage site for thyroid hormones. Interspersed in the interstitial spaces between the follicles are other secretory cells, the C cells, so called because they secrete the peptide hormone calcitonin.

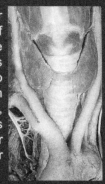

1 Right lobe of thyroid gland
2 Left lobe of thyroid gland
3 Isthmus of thyroid gland
4 Thyroid follicle
5 Follicular cell
6 Thyroglobulin (TGB)
7 Parafollicular (C) cell
8 Trachea
9 Fibromuscular membrane of trachea
10 Esophagus
11 Thyroid cartilage
12 Cricoid cartilage
13 Cricothyroid muscle
14 Brachiocephalic artery
15 Common carotid artery
16 Subclavian artery
17 Aortic arch
18 Vagus nerve
19 Thyrohyoid muscle

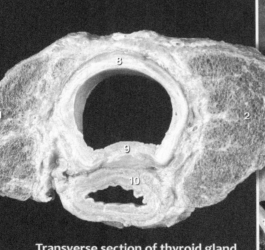

Transverse section of thyroid gland
Inferior view

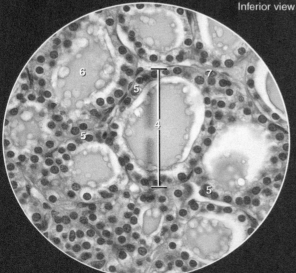

Photomicrograph of thyroid gland
240x

Thyroid gland in situ
Anterior view

Parathyroid Glands

The parathyroid glands are small, oval, light brown glands situated on the posterior border of the two lateral lobes of the thyroid gland. The parathyroid glands sit just beneath the connective tissue capsule of the thyroid gland. There are four parathyroid glands, two superior and two inferior. The endocrine cells of the parathyroid glands are called chief or principal cells. The chief cells form interconnecting columns of cells separated by fenestrated capillaries. The chief cells produce the parathyroid hormone.

1 Superior parathyroid gland
2 Inferior parathyroid gland
3 Left lobe of thyroid gland
4 Right lobe of thyroid gland
5 Isthmus of thyroid gland
6 Pyramidal lobe of thyroid gland

7 Chief cell
8 Oxyphil cell
9 Capillary
10 Arteriole
11 Venule

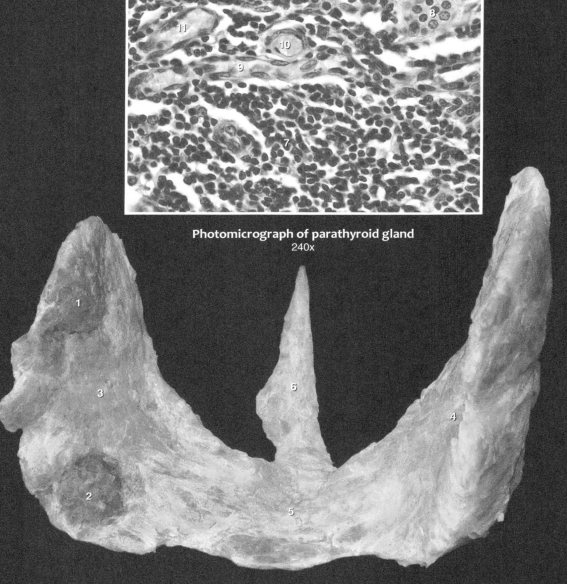

Photomicrograph of parathyroid gland
240x

Thyroid and parathyroid glands (exposed on left)
Posterior view

Suprarenal Glands

There are two yellowish suprarenal or adrenal glands that sit on the superior end of the kidneys. Each gland is surrounded by a thin connective tissue envelope. These highly vascular organs are not symmetrical. The right suprarenal gland is slightly smaller and forms a flat tetrahedron or four-sided polygon. The left suprarenal gland, like the left kidney, is more superior than the right gland and has a semilunar shape that resembles a flattened stocking hat placed on the upper end of the kidney. Each suprarenal gland is actually composed of two endocrine organs, one surrounding the other. The inner portion of the gland, called the suprarenal medulla, forms approximately 20% of the organ. The medulla secretes catecholamines. The more massive outer part of the gland, called the suprarenal cortex, secretes a variety of steroid hormones. The two parts of the gland each have different embryonic origins. The suprarenal medulla forms from the embryonic mesoderm, and the suprarenal cortex forms from embryonic neural crest cells.

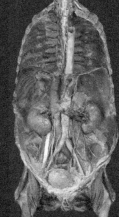

1 Right suprarenal gland
2 Left suprarenal gland
3 Zona glomerulosa of cortex
4 Zona fasciculata of cortex
5 Zona reticularis of cortex
6 Medulla
7 Capsule
8 Kidney
9 Aorta
10 Inferior vena cava
11 Crura of diaphragm
12 Diaphragm
13 Psoas major muscle
14 Bladder
15 Celiac artery
16 Superior mesenteric artery
17 Ureter
18 Common iliac artery
19 Renal vein and artery
20 Autonomic nerve plexus

Deep dissection of abdomen
Anterior view

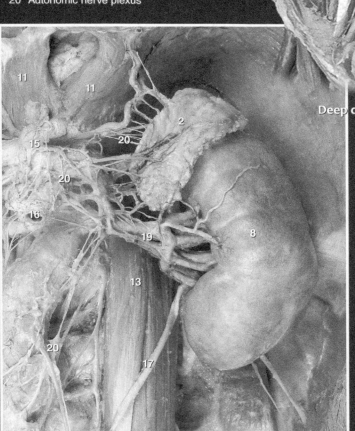

Left suprarenal gland
Anterior view

Photomicrograph of suprarenal gland
100x

Pancreas

The pancreas is a retroperitoneal organ that forms as an outgrowth of the duodenal lining. Situated posterior to the stomach it is pinkish in color and about 15 cm long, running from the loop of the duodenum on the right to the spleen on the left. It has four basic regions: a head, neck, body, and tail. The pancreas has two functional parts, the exocrine pancreas and the endocrine pancreas. The endocrine portion of the pancreas forms as small clusters of cells, the pancreatic islets, distributed among the exocrine acinar cells of the pancreas. They are far less numerous (approximately 5% of the pancreas) than the cells of the exocrine pancreas. There are four distinct cell types within the pancreatic islets: alpha or A cells, beta or B cells, delta or D cells, and F cells. The alpha (20%) and beta (70%) cells constitute the greater part of the pancreatic islets and produce the hormones glucagon and insulin, respectively. The other 10% of the islet cells are delta and F cells, which secrete somatostatin and pancreatic polypeptide, respectively.

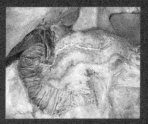

1 Pancreas
2 Pancreatic islet
3 Beta cell
4 Alpha cell
5 Exocrine acinus
6 Pancreatic duct
7 Gallbladder
8 Common bile duct
9 Duodenum
10 Liver

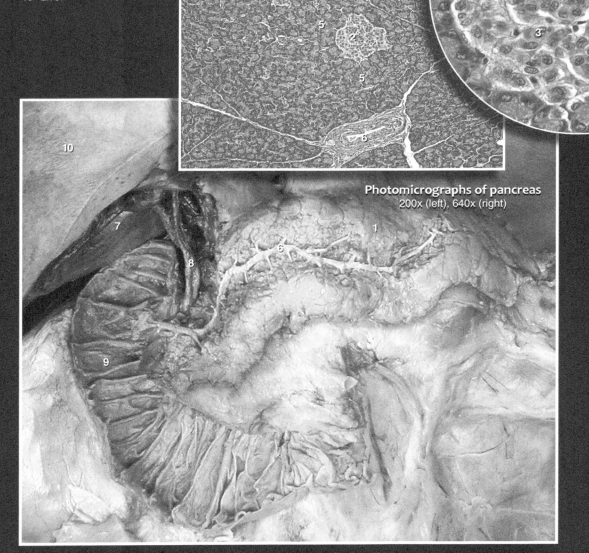

Photomicrographs of pancreas
200x (left), 640x (right)

Pancreas in situ
Anteror view

Ovaries

The ovaries are ovoid organs about the size of an unshelled almond and occupy the boundary zone between the abdominal and pelvic cavities. They consist of a dull white fibrous tissue embedded with oocytes, the "egg" cells of the female. Surrounding the oocytes are numerous follicular cells that undergo changes during the female menstrual cycle. The follicular cells are the endocrine cells of the ovary that produce the female steroidal hormones.

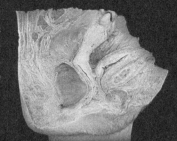

1 Ovary
2 Primordial follicle
3 Primary follicle granulosa cells
4 Secondary follicle granulosa cells
5 Follicular antrum
6 Corpus luteum
7 Primary oocyte
8 Zona pellucidum
9 Corona radiata
10 Uterine tube
11 Uterus
12 Vagina
13 Bladder
14 Urethra
15 Rectum
16 Clitoris
17 Pubic symphysis
18 Parietal peritoneum

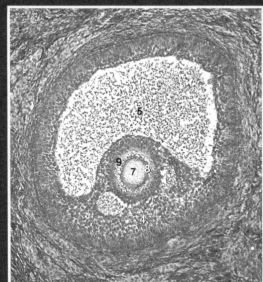

Photomicrograph of mature ovarian follicle
70x

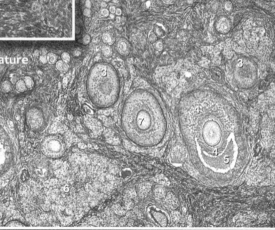

Photomicrograph of ovary
30x

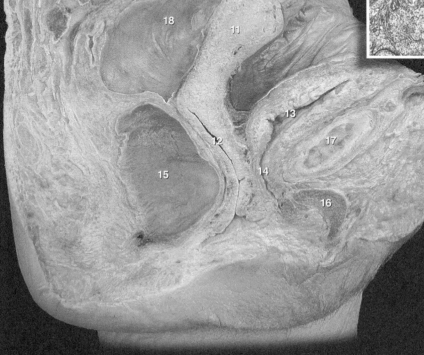

Sagittal section of female pelvis
Medial view

Testes

The testes are oval-shaped organs about 2 inches (5 cm) long and 1 inch (2.5 cm) wide that occupy the scrotal sac of a male. They are covered by a tough fibrous tunic and wrapped in a serous sac that separates them from the external tissues that surround them. Internally, the testes consist of numerous small compartments created by connective tissue bands that project inward from the outer fibrous tunic. Each testicular compartment is occupied by a thin, highly coiled seminiferous tubule. This thin tube is the site of sperm production. Situated between the tubules are the interstitial cells (of Leydig). It is these large interstitial cells that secrete the steroidal hormones in the testis.

1 Testis
2 Interstitial (Leydig) cell
3 Basement membrane
4 Sertoli cell
5 Spermatogonium
6 Primary spermatocyte
7 Secondary spermatocyte
8 Spermatid
9 Seminiferous tubule
10 Tunica albuginea
11 Epididymis
12 Spermatic cord

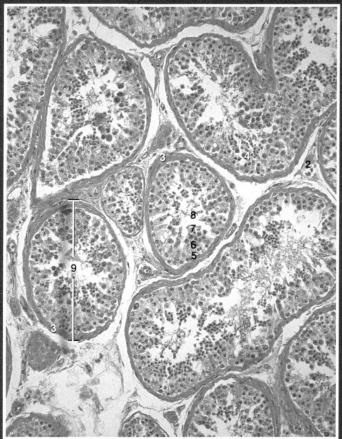

Photomicrograph of testis
40x

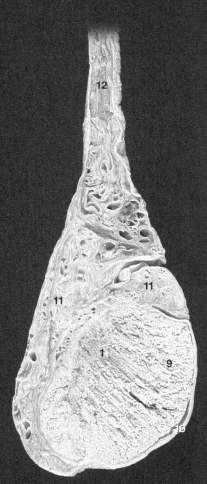

Sagittal section of left testis
Medial view

Other Endocrine Structures

In addition to the endocrine organs discussed on the preceding pages, there are other endocrine tissues in the body. These include tissues in the wall of the gastrointestinal tract that produce hormones such as gastrin and secretin, tissues in the kidney that produce renin and erythropoietin, tissues in the atrium of the heart that produce atrial natriuretic peptide, tissues of the placenta that produce human chorionic gonadotropin, estrogens, and progesterone, and adipose tissue that produces leptin. These hormones have a variety of functions, from stimulating the release of digestive enzymes, to raising blood pressure, to decreasing blood pressure, to regulating reproductive cycles, and suppressing appetite.

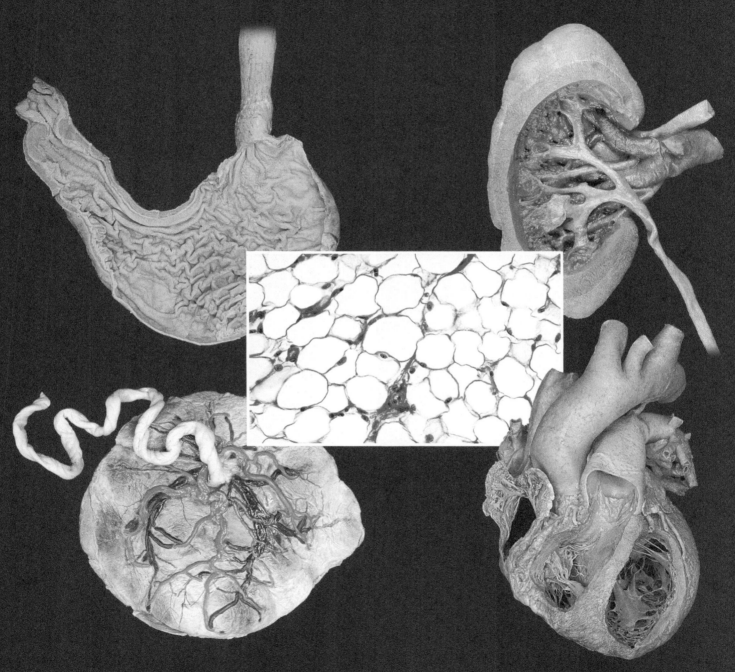

Other organs with endocrine tissues
Stomach (upper left), kidney (upper right), heart (lower right),
placenta (lower left), and adipose tissue (center)

16 | Cardiovascular System

If you have ever planted a garden of significant size, you have probably experienced the importance of an irrigation system. At its simplest, an irrigation system is a network of channels or furrows that deliver needed water from one main source to the roots of all the garden's plants. Like an irrigation system, the body's blood vessels form an extensive network of "irrigation channels" to deliver needed fluid — in this case the homeostatically maintained blood — to all the body's cells. In fact, this delivery system is probably the most phenomenal irrigation network imaginable. Emanating from a muscular pump, the heart, these vessels form an extensive system of tubular roadways that carry nourishing blood away from the heart and toward the tissues. They then make a "U-turn" through small permeable, exchange vessels, the capillaries, which feed all the body's cells. Here, life-supporting molecules, such as water, oxygen, glucose, and amino acids are delivered to the cells, and the by-products of cellular metabolism are picked up from the surrounding tissue fluid. The blood then flows back to the heart through a series of return vessels, the veins, that parallel the delivery vessels. This circular pattern of flow to and from the heart constitutes the vascular (blood vessel) component of the cardiovascular (circulatory) system. This irrigation network is so impressive, that if all the blood vessels of the body were placed end-to-end they would extend 25,000 miles (96,500 km), which is approximately two times the equatorial circumference of the earth.

The irrigation network of blood vessels are of no value without a pump. The heart is the dual, self-regulating pump that generates the pressure to drive the blood through this impressive irrigation network. It pumps the blood through two cycles — a pulmonary cycle to pick up oxygen from the lungs and a systemic cycle to deliver the oxygen to all the cells of the body. Soon after conception, and up until death, the heart pumps blood. It averages approximately 70 beats per minute, or about 3 billion contractions in an average lifetime.

The final aspect of the cardiovascular system is the accessory drainage network — the lymphatics. These small vein-like vessels insure that the cardiac return equals the cardiac output. This chapter will depict the anatomy of this amazing muscular pump and the vascular and lymphatic roadways that distribute the blood throughout the body.

Find more information about the cardiovascular system in
REAL ANATOMY

Blood

In the histology chapter we learned that the fluid material we call blood has been historically classified as a connective tissue. This classification was a result of the fact that, like other connective tissues, blood has more extracellular matrix than cells. More recently, however, blood has been placed in a tissue category of its own — the hematolymphoid complex. The extracellular portion of the blood is a water solution that gives rise to its liquid nature. Blood is closely related to other aqueous fluids within the body, in fact most of the other body fluids, such as interstitial fluid, lymph, cerebrospinal fluid, and aqueous humor, arise from the blood. These extracellular fluids are the water environment that nourish, protect, and exchange with every cell of the body. This water environment is derived from the blood, renewed by the blood, and returned to the blood. Dispersed in the blood plasma are the three groups of blood cells — erythrocytes (red blood cells), leukocytes (white blood cells), and thrombocytes (platelets). The blood smear below depicts the three cell categories.

1　Erythrocyte (red blood cell)
2　Leukocyte - neutrophil (white blood cell)
3　Leukocyte - monocyte (white blood cell)
4　Thrombocyte (platelet)
5　Blood plasma

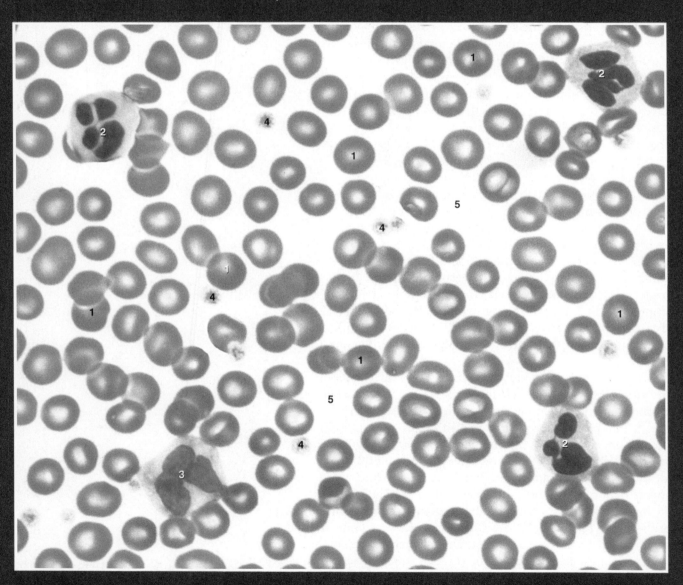

Blood smear
700x

Heart

From its origin in the embryo as a simple pumping tube, the heart develops into a strong fibromuscular organ. During its development the original tubular pump is folded and subdivided into a four chambered organ that has a pyramidal or conical form. It is approximately the size of a closed fist and weighs approximately 300 grams in males and a little less than this in females. For its small size, comprising only one half of one percent of the total body mass, it is an important and functionally amazing organ. The wall of the heart consists of three structural layers that each play significant roles in its function as an efficient pump. While the tissue makeup of this wall is similar at any location in the heart, the thickness can vary considerably. Internally a septum and series of valves divide the heart into four chambers through which the blood moves in a unidirectional flow. The chambers differ in structure and function, which is primarily reflected in the anatomy of their walls. Embedded within the walls of heart is a special electrical conduction system that helps regulate its coordinated pumping action.

1	Right atrium	18	Right coronary artery
2	Left atrium	19	Conus arteriosus branch
3	Right ventricle	20	Marginal branch
4	Left ventricle	21	Anterior interventricular artery
5	Right auricle	22	Lateral branches
6	Left auricle	23	Circumflex branch
7	Aorta	24	Posterior interventricular artery
8	Brachiocephalic artery	25	Anterior cardiac vein
9	Left common carotid artery	26	Great cardiac vein
10	Left subclavian artery	27	Posterior vein of left ventricle
11	Pulmonary trunk	28	Middle cardiac vein
12	Right pulmonary artery	29	Small cardiac vein
13	Left pulmonary artery	30	Right superior pulmonary vein
14	Ligamentum arteriosum	31	Right inferior pulmonary vein
15	Superior vena cava	32	Left superior pulmonary vein
16	Inferior vena cava	33	Left inferior pulmonary vein
17	Coronary sinus		

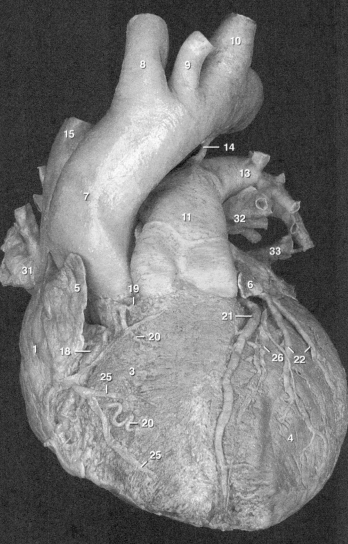

Heart
Anterior view

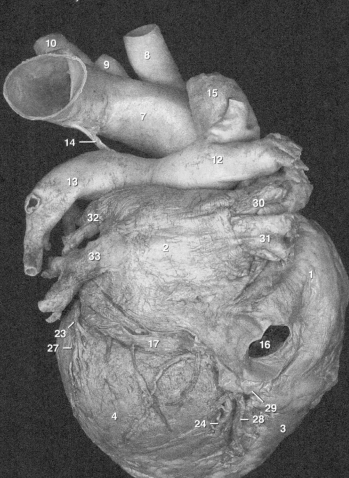

Heart
Posterior view

Heart

1 Parietal pericardium
2 Fibrous pericardium
3 Visceral pericardium
4 Epicardium
5 Myocardium
6 Endocardium
7 Right atrium
8 Right auricle
9 Interatrial septum
10 Fossa ovalis
11 Crista terminalis
12 Valve of inferior vena cava
13 Pectinate muscle
14 Tricuspid valve
15 Chordae tendineae
16 Trabeculae carnae
17 Papillary muscle
18 Right ventricle
19 Pulmonary valve
20 Left atrium
21 Left auricle
22 Bicuspid valve
23 Left ventricle
24 Aortic valve
25 Apex
26 Aorta
27 Brachiocephalic artery
28 Left common carotid artery
29 Left subclavian artery
30 Pulmonary trunk
31 Left pulmonary artery
32 Ligamentum arteriosum
33 Anterior interventricular artery
34 Lateral branches of interventricular artery
35 Superior vena cava
36 Right coronary artery
37 Left coronary artery
38 Right pulmonary veins
39 Left pulmonary veins
40 Diaphragm
41 Lung

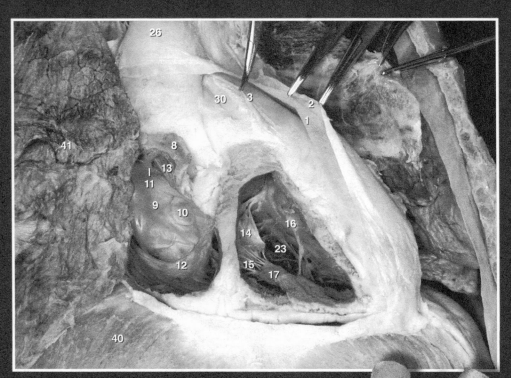

Dissection of heart and pericardial sac
Anterolateral view

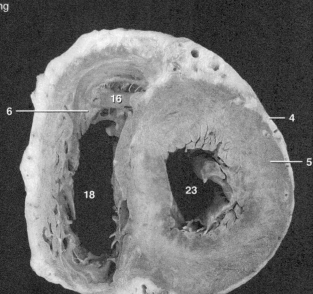

Transverse section of heart comparing ventricle thickness
Inferior view, left ventricle at right

Dissected heart showing interior of chambers
Anterior view

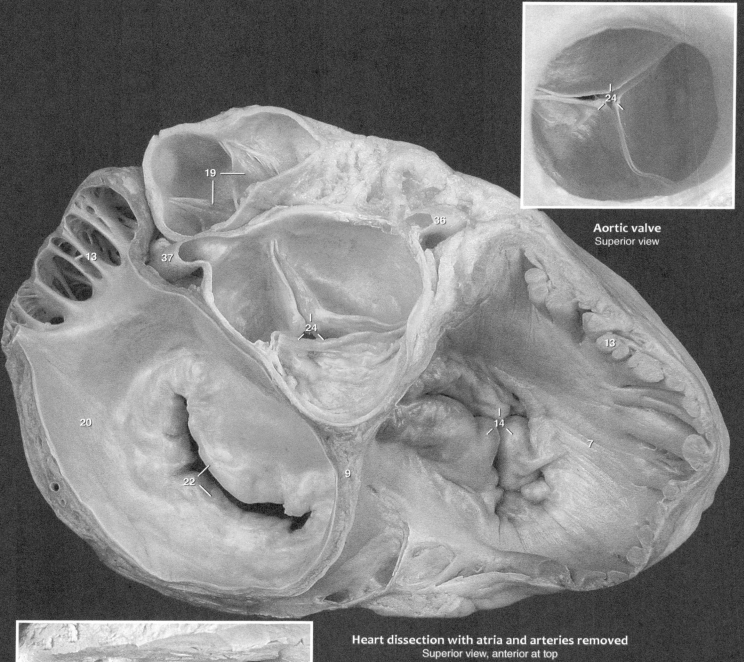

Aortic valve
Superior view

Heart dissection with atria and arteries removed
Superior view, anterior at top

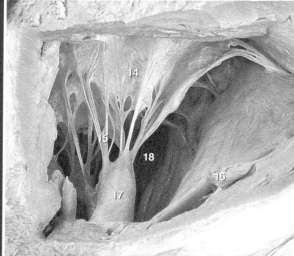

Dissection of heart revealing tricuspid valve
Anterior view

Blood Vessels

Like all tubes in the body, blood vessels have a basic pattern of design that involves three structural tunics, or layers. The inner layer of the vessel is the tunica intima. This consists of the luminal endothelium and a thin network of underlying elastic connective tissue. The middle layer of the vessel is the tunica media, which consists of varying amounts of smooth muscle and elastic connective tissue. Variations in the tunica media define the different types of blood vessels. The outer layer, the tunica externa, is a dense connective tissue outer coat. The designations — elastic arteries, muscular arteries, arterioles, venules, and veins — are based on size differences and the differences in the vessels' tunica media. Elastic arteries have a thick elastic tunica media. Muscular arteries have a tunica media dominated by smooth muscle. Arterioles are tiny arteries with a muscular tunica media. All the venous vessels have a thin, almost non-existent tunica media. The smallest blood vessels, the capillaries, loose all the layers of their wall except the inner endothelium. These microscopic, thin walled tubes become the exchange vessels of the system.

1	Endothelium of tunica intima	6	Red blood cells
2	Internal elastic membrane of tunica intima	7	White blood cells
3	Elastic lamellae of tunica media	8	Venous valves
4	Smooth muscle cells of tunica media	9	Nerve
5	Connective tissue of tunica externa	10	Striated skeletal muscle

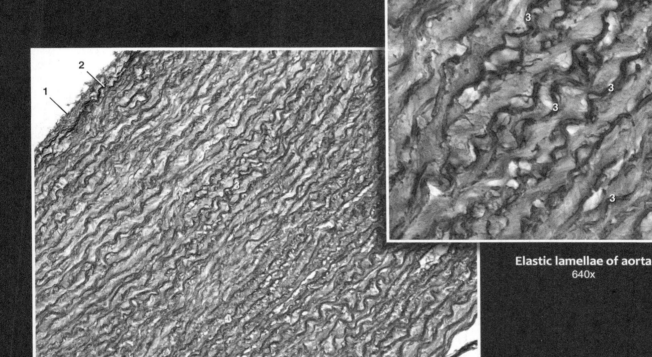

Section of aorta — large elastic artery
100x

Elastic lamellae of aorta
640x

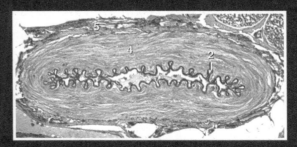

Muscular artery
100x

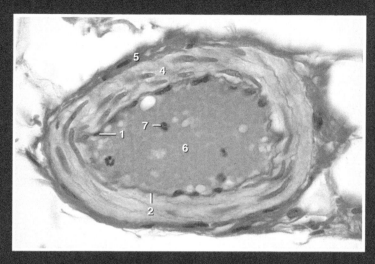

Arteriole
500x

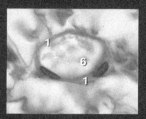

Capillary
1000x

Longitudinal section of vein showing valves
Anterior view

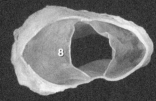

Transverse section of vein showing valves
Superior view

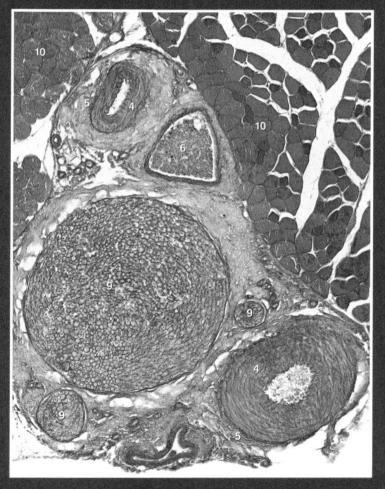

Neurovascular bundle — note thin-walled vein filled with red blood cells (6) compared to thick-walled muscular arteries (4)
100x

Pulmonary Circuit

The vascular system consists of two long circular loops of continuous branched tubing that each begin and end with the heart. Leaving the right ventricle and returning to the left atrium is the smaller pulmonary circulation. This circular loop courses through the lung tissues where its smallest vessels form an extensive interface with the small air sacs of the lungs. This important interface is the site of exchange of O_2 and CO_2 between the blood and air.

1 Heart
2 Pulmonary trunk
3 Right pulmonary artery
4 Left pulmonary artery
5 Right superior pulmonary vein
6 Right inferior pulmonary vein
7 Left superior pulmonary vein
8 Left inferior pulmonary vein
9 Aorta
10 Right coronary artery
11 Left coronary artery
12 Right common carotid artery
13 Right subclavian artery
14 Left common carotid artery
15 Left subclavian artery
16 Superior vena cava
17 Inferior vena cava
18 Trachea
19 Right principal bronchus
20 Left principal bronchus
21 Esophagus
22 Thyroid gland
23 Vagus nerve
24 Pulmonary plexus
25 Posterior vagal trunk
26 Esophageal plexus
27 Anterior vagal trunk
28 Anterior scalene muscle
29 Cricothyroid muscle

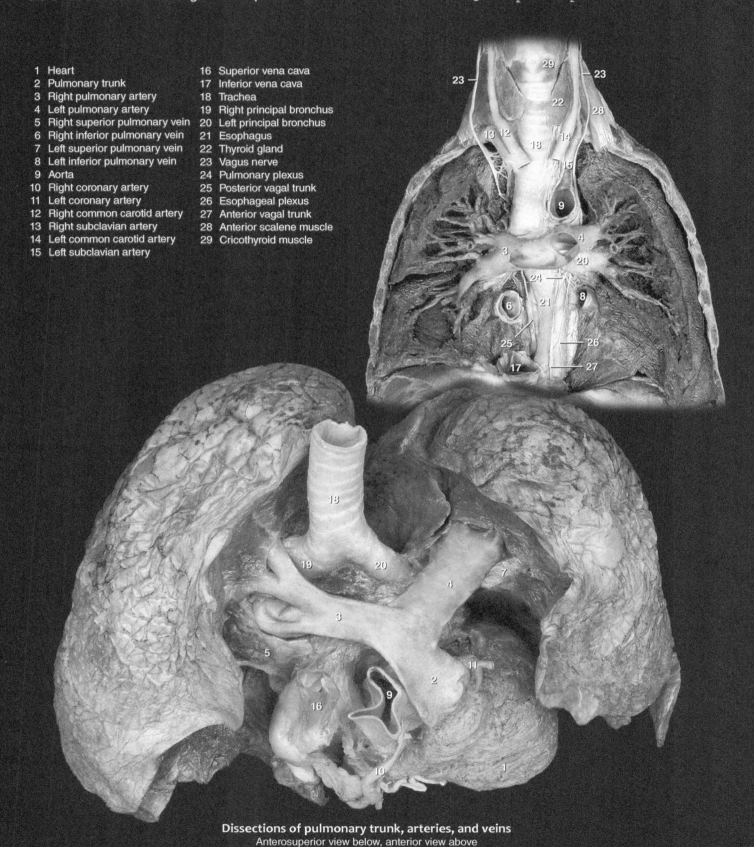

Dissections of pulmonary trunk, arteries, and veins
Anterosuperior view below, anterior view above

Systemic Circuit

The left ventricle pumps blood into the much larger systemic circulation, which is distributed throughout all the body's tissues. Unlike the smaller pulmonary circuit, the extensive systemic circuit serves a multitude of functions before returning to the right atrium: (1) it distributes the necessary nutrients and other supplies to all the body cells while removing their metabolic wastes; (2) it acquires metabolic fuel through the lining of the digestive system to distribute throughout the body; (3) it expels wastes and excess water and adjusts the body's electrolyte composition through its association with the tubes of the kidney; (4) it distributes generated heat throughout the body and plays an important role in adjusting heat loss to the external environment as it courses through the skin; and (5) it distributes hormones, regulatory chemical-messenger molecules secreted by endocrine glands, to various sites of action throughout the body.

1 Aorta
2 Brachiocephalic artery
3 Right common carotid artery
4 Right subclavian artery
5 Right internal thoracic artery
6 Left common carotid artery
7 Left subclavian artery
8 Left axillary artery
9 Left brachial artery
10 Left ulnar artery
11 Left radial artery
12 Left radial recurrent artery
13 Coeliac trunk
14 Common hepatic artery
15 Left gastric artery
16 Splenic artery
17 Superior mesenteric artery
18 Right renal artery
19 Left renal artery
20 Inferior mesenteric artery
21 Common iliac arteries
22 Internal iliac arteries
23 External iliac artery
24 Femoral artery
25 Deep femoral artery
26 Popliteal artery
27 Azygos vein
28 Thyroid gland
29 Trachea
30 Ligamentum arteriosum
31 Vagus nerve
32 Phrenic nerve
33 Anterior scalene muscle
34 Brachialis muscle
35 Brachioradialis muscle
36 Innermost intercostal muscles
37 Quadratus lumborum muscle
38 Psoas major muscle
39 Clavicle
40 First rib

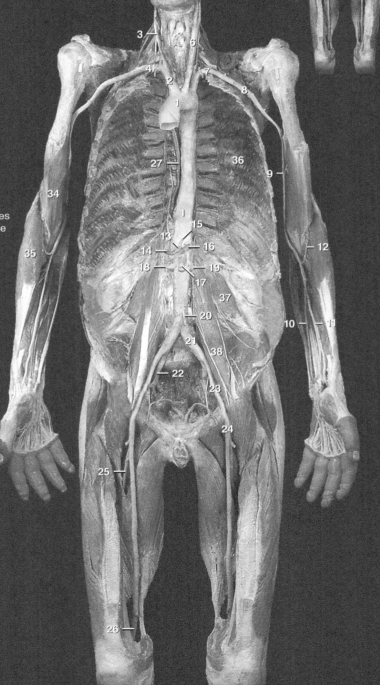

Dissection of major arterial pathways
Anterior view

Dissection of aortic arch and its branches
Anterior view

Heart Vessels

The coronary arteries are the first branches of the aorta. These important vessels provide the constantly needed blood supply to the heart. The left coronary artery is, on average, larger than the right coronary artery and supplies a greater percentage of the heart tissue. Accompanying the branches of the coronary arteries, a series of cardiac veins emerge from the capillaries of the heart to return blood to the right atrial chamber, either by entering directly or by joining the large coronary sinus, which enters the right atrium from the posterior side.

1 Coronary sinus
2 Right coronary artery
3 Conus arteriosus branch
4 Marginal branch
5 Anterior interventricular artery
6 Lateral branches
7 Circumflex branch of left coronary
8 Posterior interventricular artery
9 Anterior cardiac vein
10 Great cardiac vein
11 Posterior vein of left ventricle
12 Middle cardiac vein
13 Oblique vein
14 Aorta
15 Pulmonary trunk
16 Superior vena cava
17 Left atrium
18 Right atrium
19 Right ventricle
20 Left ventricle
21 Pulmonary veins
22 Pulmonary artery
23 Inferior vena cava
24 Ligamentum arteriosum
25 Brachiocephalic artery
26 Left common carotid artery
27 Left subclavian artery

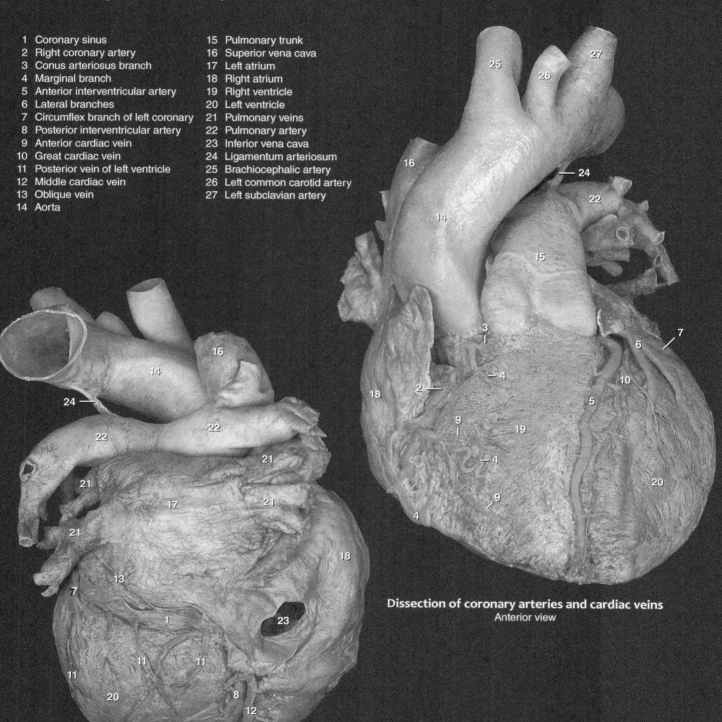

Dissection of coronary arteries and cardiac veins
Anterior view

Dissection of coronary arteries, coronary sinus, and cardiac veins
Posterior view

Head Vessels

Like the heart, which needs a constant, uninterrupted blood supply, the brain tissue also must be guaranteed of a continuous perfusion in order to maintain its crucial functions. The common carotid arteries, arising from the aortic arch, bifurcate into external and internal carotids. The external carotid supplies all tissues of the head except the brain, while the function of the internal carotid is to supply the brain. Because of the brain's critical vascular needs the internal carotid artery has a partner, the vertebral artery, which courses cranially from the subclavian artery to assist with the essential blood supply to the brain.

1 Internal carotid artery
2 Basilar artery
3 Vertebral artery
4 Posterior cerebral artery
5 Posterior communicating artery
6 Middle cerebral artery
7 Posterior inferior cerebellar artery
8 Posterior superior cerebellar artery
9 Common carotid artery
10 External carotid artery
11 Superior thyroid artery
12 Ascending pharyngeal artery
13 Lingual artery
14 Facial artery
15 Occipital artery
16 Posterior auricular artery
17 Superficial temporal artery
18 Transverse facial artery
19 Maxillary artery
20 Optic chiasm
21 Thyroid gland
22 Trigeminal nerve
23 Lateral pterygoid muscle
24 Temporal lobe of cerebrum
25 Zygomatic arch

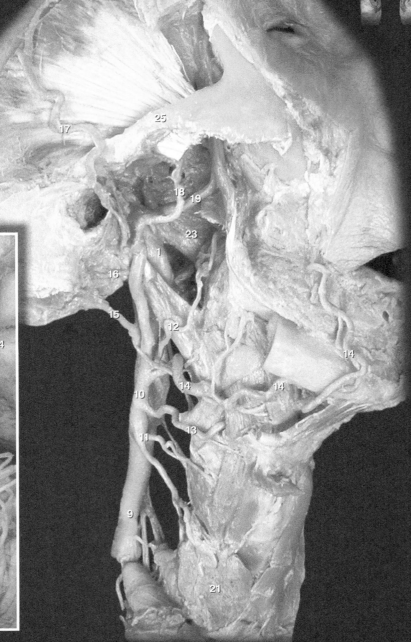

Dissection of branches of external carotid artery
Lateral view

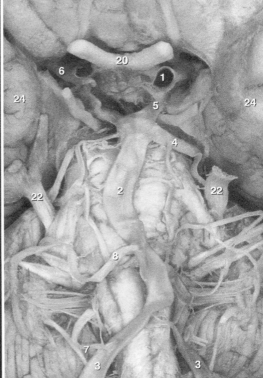

Dissection of basilar artery
Inferior view

Head Vessels

1 Internal carotid artery
2 Vertebral artery
3 Basilar artery
4 Middle cerebral artery
5 Anterior cerebral artery
6 Anterior communicating artery
7 Posterior communicating artery

8 Cerebral veins
9 Cerebellar veins
10 Superior sagittal sinus
11 Transverse sinus
12 Inferior sagittal sinus
13 Sigmoid sinus
14 Opening of straight sinus

15 Confluence of the sinuses
16 Dura mater
17 Pia-arachnoid mater
18 Spinal cord
19 Vertebral body
20 Cervical transverse process
21 Temporal lobe of cerebrum

22 Pituitary gland
23 External acoustic meatus
24 Pons
25 Frontal lobe of cerebrum
26 Vagus nerve
27 Cervical sympathetic trunk
28 Superior cervical ganglion

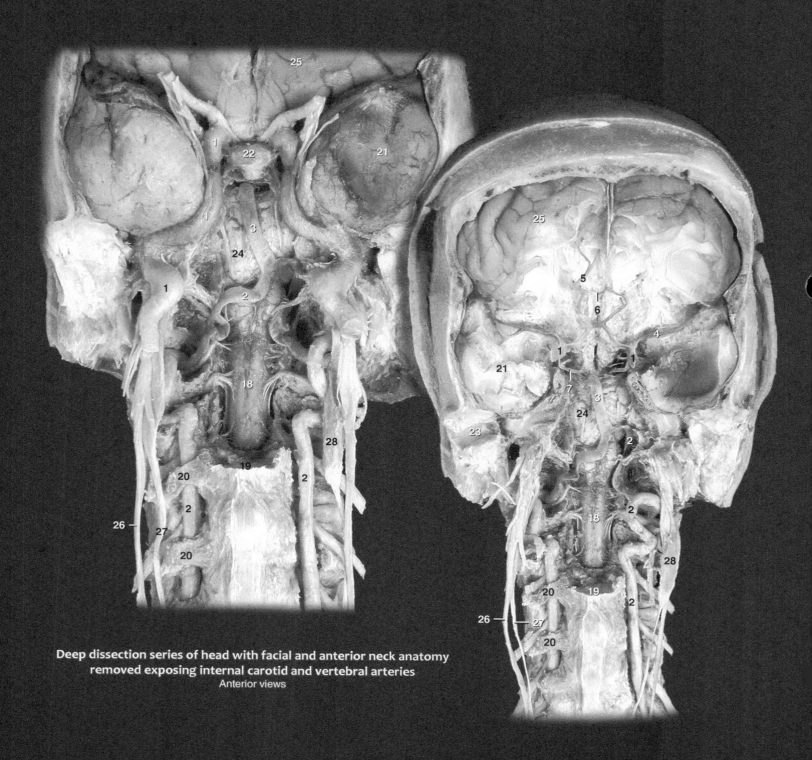

Deep dissection series of head with facial and anterior neck anatomy removed exposing internal carotid and vertebral arteries
Anterior views

Unlike the internal and external carotid arteries, the internal and external jugular veins form a wide array of collateral circuitry. The major structural difference of the venous pathways in the head is the existence of dural venous sinuses within the skull. The dural venous sinuses are non-collapsible, endothelial lined spaces within the tough meningeal dura mater. All the smaller veins draining capillaries within the brain tissue enter into the dural venous sinuses. These dural sinuses converge with one another throughout the skull to exit the cranial vault via the internal jugular vein.

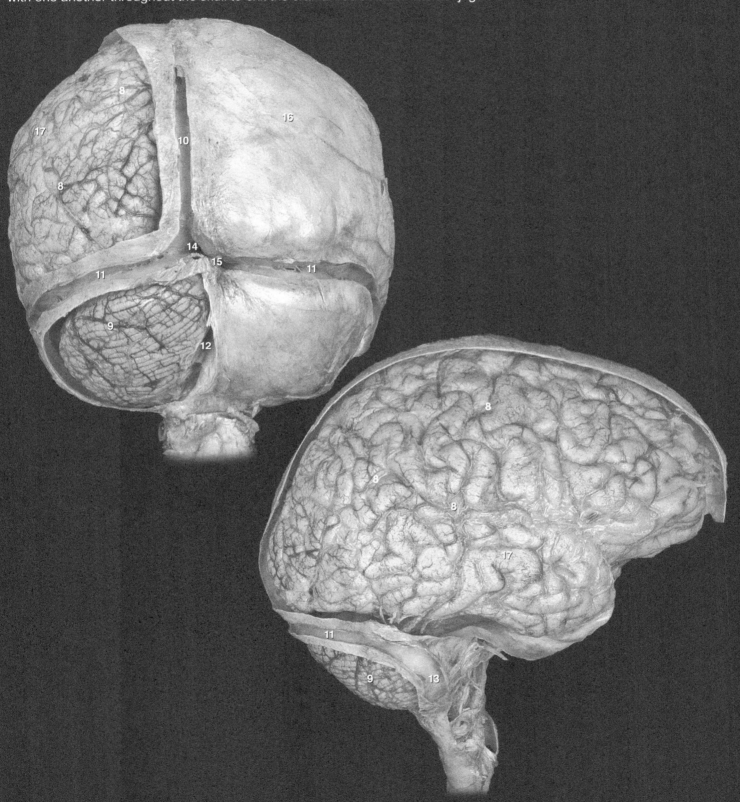

Dissections of dural venous sinuses and cerebral veins
Posterior view (top), lateral view (bottom)

Superior Limb Vessels

The arterial pathway into the upper limb consists of a single, major arterial roadway that gradually tapers as it gives rise to the various branches that supply the tissues of the limb. This large arterial roadway begins as the subclavian artery, takes on regional names — the axillary artery and brachial artery — as it tapers distally, then branches into the radial and ulnar arteries, which course through the antebrachium, paralleling the bones of the same names. The radial and ulnar arteries terminate as the collateral arches in the hand. This central pathway through the limb is the sole blood supply to this region, supplying the integument, muscles, bones, joints, and connective tissues of the upper limb. The deep venous pathways follow the arteries and have similar names. However, superficial veins that have no arterial counterparts aid the deep veins in returning blood to the heart.

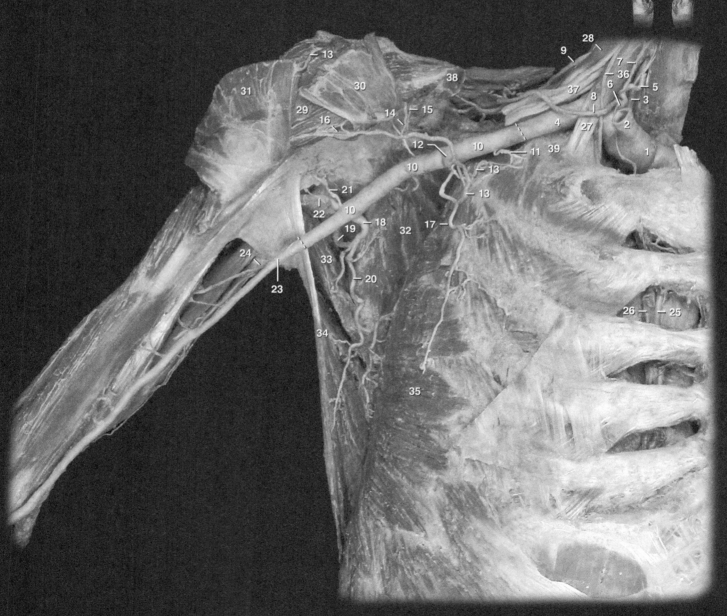

Dissection of subclavian and axillary arteries
Anterior view

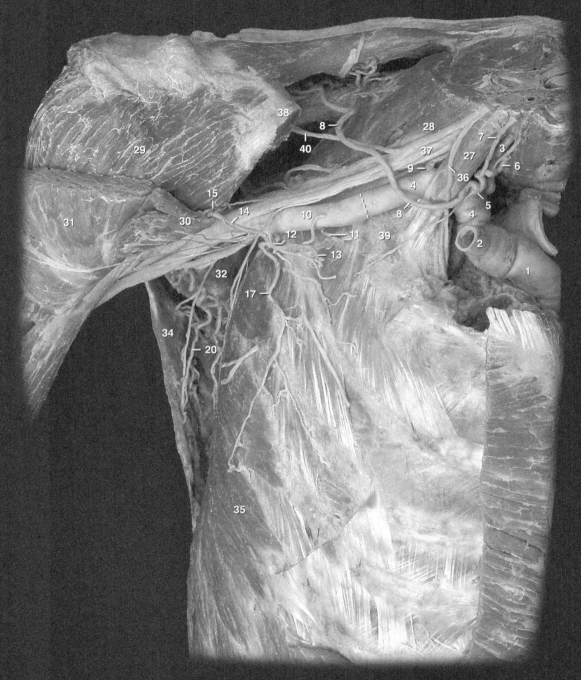

1 Brachiocephalic artery	11 Superior thoracic artery	21 Posterior circumflex humeral artery	31 Pectoralis major muscle
2 Common carotid artery	12 Thoracoacromial trunk	22 Anterior circumflex humeral artery	32 Subscapularis muscle
3 Vertebral artery	13 Pectoral artery	23 Brachial artery	33 Teres major muscle
4 Subclavian artery	14 Acromial artery	24 Deep artery of arm	34 Latissimus dorsi muscle
5 Thyrocervical trunk	15 Clavicular artery	25 Internal thoracic artery	35 Serratus anterior muscle
6 Inferior thyroid artery	16 Deltoid artery	26 Internal thoracic vein	36 Phrenic nerve
7 Ascending cervical artery	17 Lateral thoracic artery	27 Anterior scalene muscle	37 Brachial plexus
8 Suprascapular artery	18 Subscapular artery	28 Middle scalene muscle	38 Clavicle
9 Dorsal scapular artery	19 Circumflex scapular artery	29 Deltoid muscle	39 First rib
10 Axillary artery	20 Thoracodorsal artery	30 Pectoralis minor muscle	40 Suprascapular nerve

Dissection of subclavian and axillary arteries
Anterosuperior view

Superior Limb Vessels

1 Brachial artery
2 Ulnar artery
3 Radial artery
4 Anterior interosseous artery
5 Superficial palmar arch
6 Common digital artery
7 Proper digital artery

8 Deep palmar arch
9 Cephalic vein
10 Median cubital vein
11 Basilic vein
12 Median antebrachial vein
13 Accessory cephalic vein
14 Brachial vein

15 Interosseous membrane
16 Transverse carpal ligament
17 Supinator muscle
18 Pronator quadratus muscle
19 Flexor digitorum superficialis tendons
20 Flexor digitorum profundus tendons
21 Biceps brachii muscle

22 Triceps brachii muscle
23 Pectoralis major muscle
24 Deltoid muscle
25 Deltopectoral groove
26 Serratus anterior muscle
27 Brachioradialis muscle
28 Coracobrachialis muscle

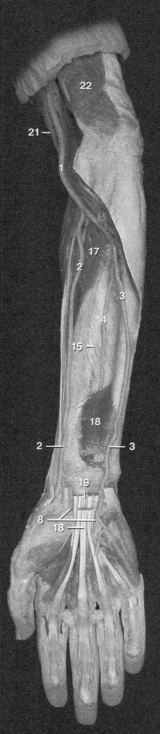

Dissection of antebrachial arteries
Anterior view

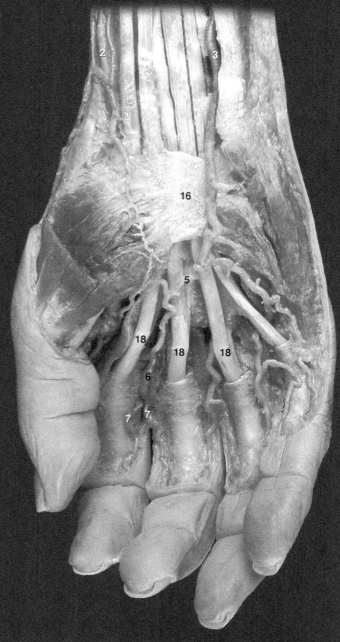

Dissection of palmar arterial arch and branches to digits
Anterior view

Within the upper limb there are two sets of veins: deep veins that accompany the arteries, and superficial veins that course through the hypodermis without arterial counterparts. The deep veins, running with the arteries of the upper limb, have the same names as their arterial counterparts. These veins are significantly smaller than the arteries they accompany and form vena comitans with anastomotic channels around the arteries. The superficial veins of the upper limb are large and numerous. There are three major superficial veins into which all the other superficial veins flow; they are the basilic vein, cephalic vein, and median cubital vein. The median cubital vein is a connecting vein between the cephalic vein and the basilic vein. The cephalic and basilic veins eventually pass deep to join the axillary vein at the proximal end of the limb. Most of the venous return from the upper limb passes through the superficial veins.

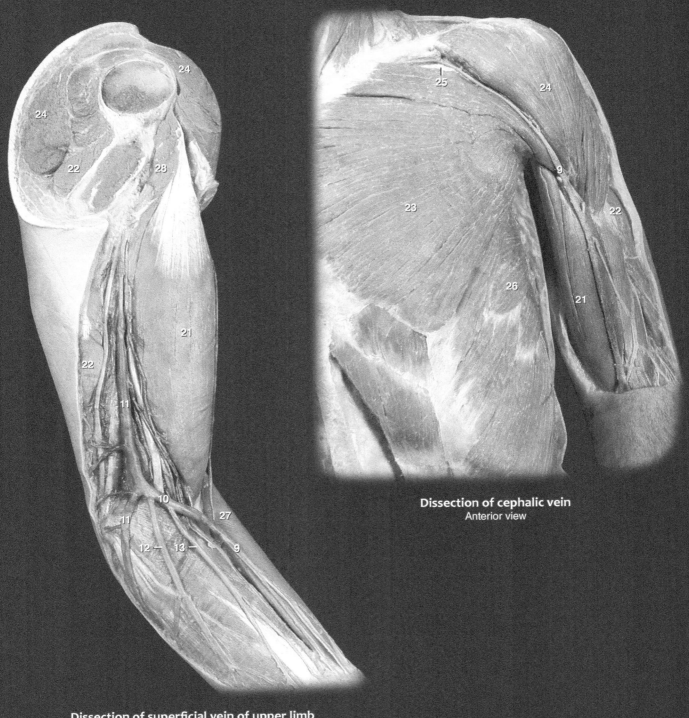

Dissection of cephalic vein
Anterior view

Dissection of superficial vein of upper limb
Medial view of left upper limb

Thoracic Vessels

The branches of the aorta that supply the thoracic region can be divided into two principal groups — those that supply the thoracic body wall and those that supply thoracic viscera. Two arterial supply routes carry blood into the thoracic body wall. Posteriorly the aorta courses vertically down the vertebral column, while anteriorly the internal thoracic arteries arise from the subclavian arteries and course vertically down the inside of the sternum. Between these anterior and posterior supply arteries are interconnecting collateral arteries. These collateral vessels are the anterior intercostal arteries and the posterior intercostal arteries, which supply the tissues of the intercostal spaces and form collateral circuits between the anterior and posterior arterial pathways. All thoracic viscera receive their blood supply from branches of the aorta. The thoracic viscera include the heart, lungs with their associated bronchial tubes, and the esophagus.

1 Aorta
2 Posterior intercostal artery
3 Posterior intercostal vein
4 Azygos vein
5 Hemi-azygos vein
6 Accessory hemi-azygos vein
7 Superior vena cava
8 Brachiocephalic vein
9 Subclavian vein
10 Internal jugular vein
11 Inferior vena cava
12 Right atrium (cut)
13 Left subclavian artery
14 Left common carotid artery
15 Right common carotid artery
16 Hepatic vein
17 Trachea
18 Diaphragm
19 Esophageal hiatus
20 Subcostal muscle
21 Innermost intercostal muscle
22 Esophagus
23 Sympathetic trunk nerve
24 Thoracic lymphatic duct

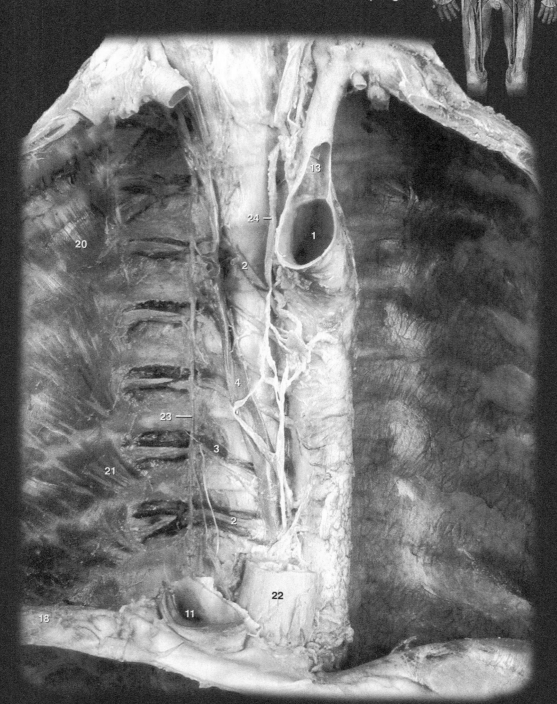

Dissection of vessels of posterior thoracic wall
Anterior view

Like the arterial supply to the thoracic wall, the venous drainage returns via both anterior-wall and posterior-wall drainage veins. The veins of the anterior wall have the same names as their arterial counterparts, while the veins of the posterior wall differ in name and structure. Unlike the aorta, which is the posterior-wall supply artery, the superior vena cava and inferior vena cava diverge from the posterior thoracic wall to enter the thoracic cavity and return their contents to the heart. In the absence of vena cavae in the posterior thoracic wall, an azygos system of veins is formed to drain the body wall and the thoracic viscera. These azygos veins communicate with the superior vena cava to return their contents to the heart. With the exception of the azygos veins, the veins are similar to the arteries in name and distribution.

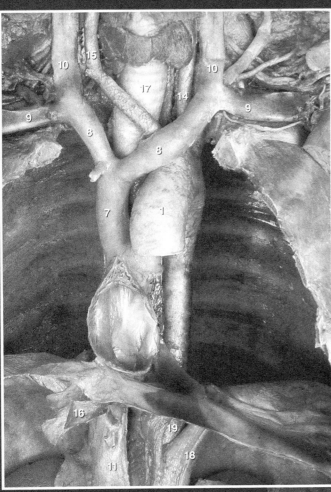

Dissection of vena cavae and tributaries
Anterior view

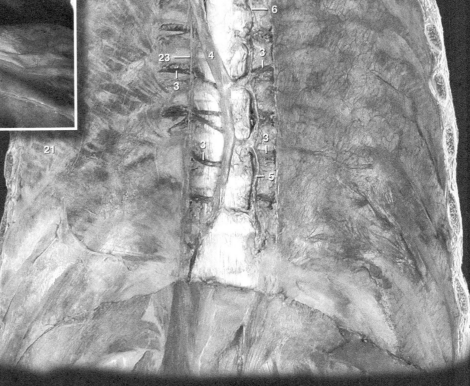

Dissection of azygos veins
Anterior view

Abdominal Vessels

Like the thorax, the abdomen has somatic arteries that supply the abdominal muscle wall and visceral arteries that supply the viscera of the abdominal cavity. These vessels follow the same pattern observed in the thoracic region; that is, the abdominal body wall has both anterior (epigastric arteries) and posterior (aorta) supply pathways that form interconnecting collateral arteries, while the viscera receive branches from the aorta — celiac artery to the foregut, superior mesenteric artery to the midgut, inferior mesenteric artery to the hindgut, and renal arteries to the kidneys.

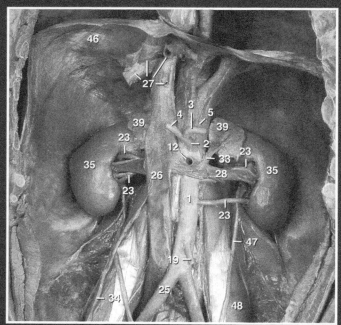

Deep dissection of abdomen showing renal vessels
Anterior view

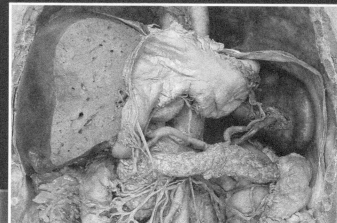

Branches of celiac artery

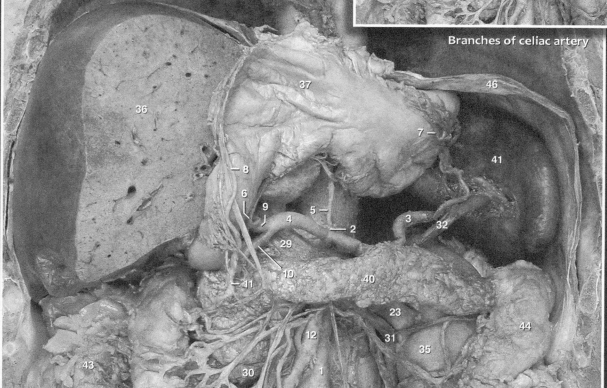

Dissection of abdomen showing celiac branches and supply of foregut viscera
Anterior view, stomach reflected upward

280

1	Aorta	13	Middle colic artery	25	Common iliac artery	37	Stomach
2	Celiac artery	14	Marginal artery	26	Inferior vena cava	38	Transverse colon
3	Splenic artery	15	Right colic artery	27	Hepatic vein	39	Suprarenal gland
4	Common hepatic artery	16	Ileocolic artery	28	Renal vein	40	Pancreas
5	Left gastric artery	17	Jejunal arteries	29	Hepatic portal vein	41	Spleen
6	Right gastric artery	18	Ileal arteries	30	Superior mesenteric vein	42	Duodenum
7	Left gastro-omental artery	19	Inferior mesenteric artery	31	Inferior mesenteric vein	43	Ascending colon
8	Right gastro-omental artery	20	Left colic artery	32	Splenic vein	44	Descending colon
9	Proper hepatic artery	21	Sigmoid artery	33	Suprarenal vein	45	Ileum
10	Gastroduodenal artery	22	Superior rectal artery	34	Testicular vein	46	Diaphragm
11	Superior pancreaticoduodenal artery	23	Renal artery	35	Kidney	47	Ureter
12	Superior mesenteric artery	24	Segmental arteries	36	Liver	48	Psoas major muscle

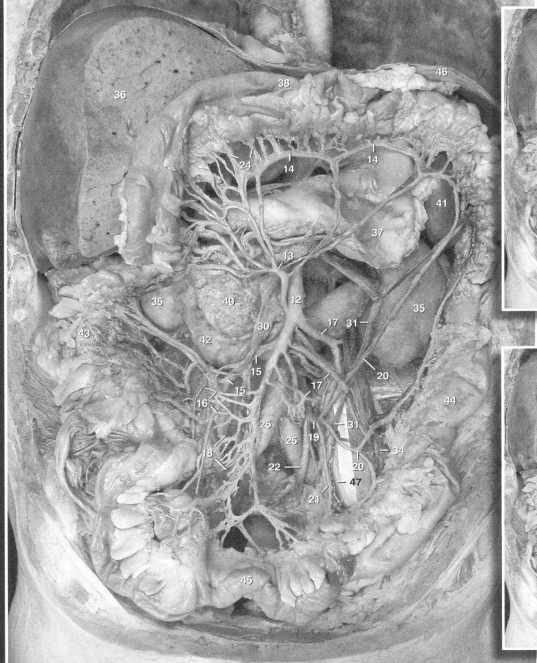

Dissection of abdomen showing arterial supply of midgut and hindgut viscera
Anterior view

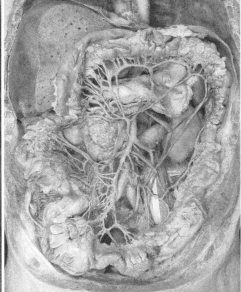

Superior mesenteric artery

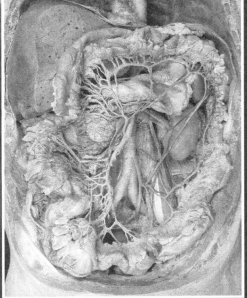

Inferior mesenteric artery

Abdominal Vessels

The major difference between the arteries and veins of the abdomen is the fact that all the visceral venous return from the capillaries of the digestive system and spleen pass via the hepatic portal system to the capillaries of the liver before returning to the heart. Within the liver, both the hepatic artery and hepatic portal vein branch to form a complex network of specialized capillaries called the hepatic sinusoids. The hepatic sinusoids then drain into the hepatic veins to return the blood to the inferior vena cava.

1 Inferior vena cava
2 Hepatic portal vein
3 Superior mesenteric vein
4 Right colic vein
5 Inferior mesenteric vein
6 Renal vein
7 Superior mesenteric artery
8 Inferior mesenteric artery
9 Middle colic artery
10 Marginal artery
11 Left colic artery
12 Common iliac artery
13 External iliac artery
14 Internal iliac artery
15 Superior gluteal artery
16 Inferior gluteal artery
17 Obturator artery
18 Internal pudendal artery
19 Lateral sacral artery
20 Superior vesical artery
21 Vaginal artery
22 Obliterated umbilical artery
23 Uterus
24 Bladder
25 Prostate
26 Rectum
27 Stomach
28 Kidney
29 Upper bands of sacral plexus
30 Sympathetic trunk
31 Inferior vesical artery
32 Middle rectal artery
33 Obturator nerve
34 Uterine artery

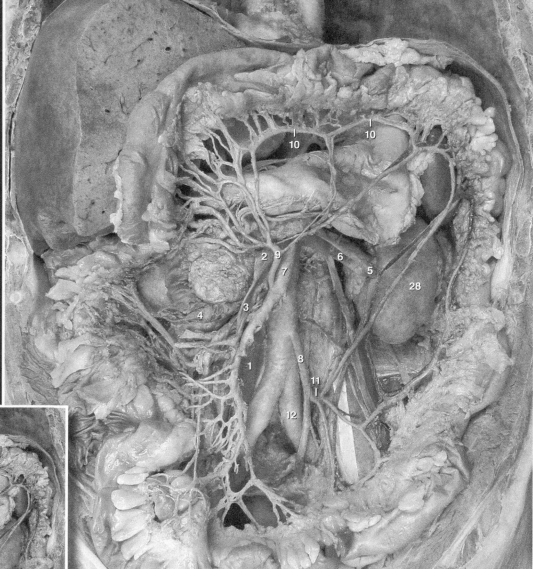

Dissection of abdomen showing arteries and veins of the intestines
Anterior view

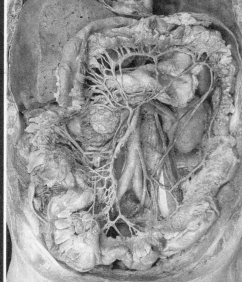

Abdominal veins

Pelvic Vessels

The common iliac arteries, the terminal branches of the aorta, carry all of the blood supply to the lower limbs and pelvis. All pelvic viscera, along with the body wall anatomy of the pelvis and perineal regions, receive their blood supply from the internal iliac artery. Numerous branches arise from the internal iliac artery to supply the pelvic wall, the perineum, and the gluteal region. Other branches course into the pelvic cavity to supply the viscera. The veins are similar in name and course with the corresponding arteries.

Dissection of pelvic arteries of female
Medial view, anterior at left

Dissection of pelvic arteries of male
Medial view, anterior at right

Inferior Limb Vessels

As in the upper limb, the main arterial pathway into the lower limb consists of a single, major arterial roadway that gradually tapers as it gives rise to numerous branches on its pathway through the limb. This large arterial roadway begins as the external iliac artery in the pelvis, passes beneath the inguinal ligament to enter the thigh as the femoral artery, passes to the back of the knee to become the popliteal artery, and in the proximal aspect of the leg bifurcates into the anterior tibial and posterior tibial arteries, which course through the leg and into the foot.

1 Superior gluteal artery
2 Inferior gluteal artery
3 Internal pudendal artery
4 Femoral artery
5 Deep artery of thigh
6 Muscular branches of femoral
7 Femoral vein
8 Great saphenous vein
9 External iliac artery
10 Internal iliac artery
11 External iliac vein
12 Common iliac artery
13 Aorta
14 Gluteus maximus muscle
15 Sacrotuberous ligament
16 Piriformis muscle
17 Spermatic cord (cut)
18 Penis (cut)
19 Adductor longus muscle
20 Rectus femoris muscle
21 Vastus intermedius muscle
22 Gracilis muscle
23 Vastus lateralis muscle
24 Vastus medialis muscle
25 Fascia lata
26 Sartorius muscle
27 Iliacus muscle

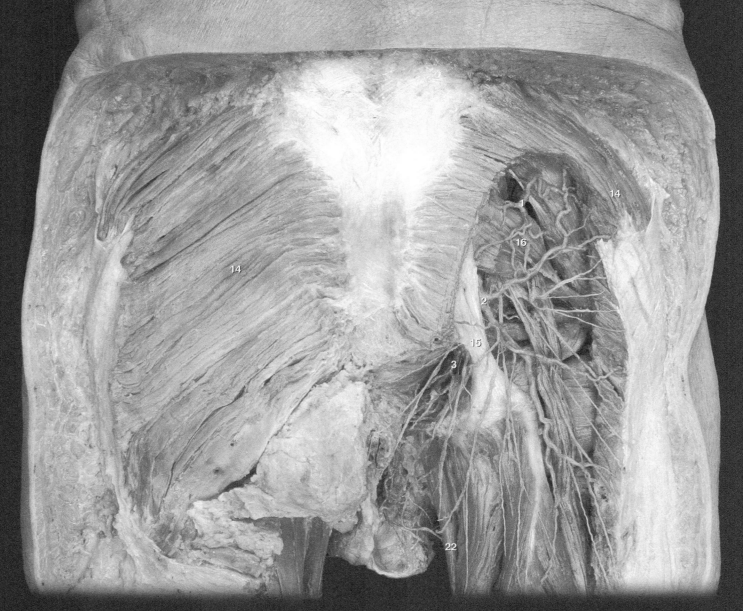

Dissection of gluteal region showing gluteal arteries and nerves
Posterior view

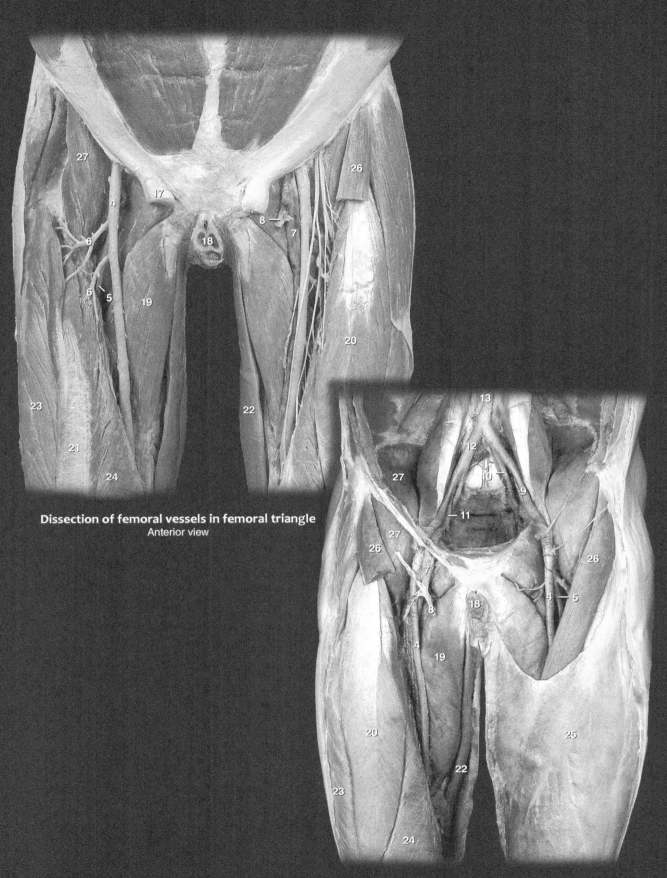

Dissection of femoral vessels in femoral triangle
Anterior view

Dissection of vessels of inferior limb
Anterior view

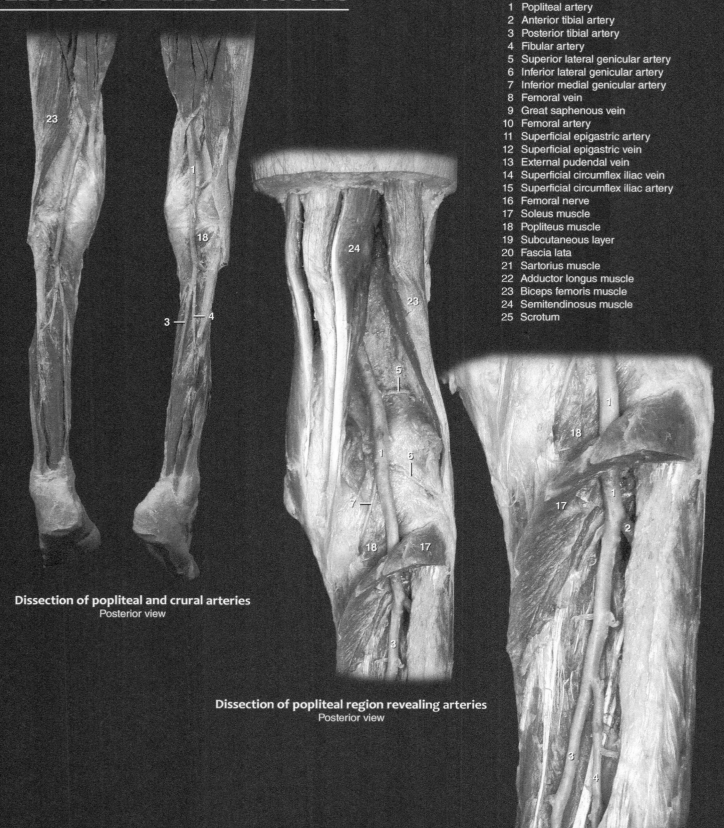

1 Popliteal artery
2 Anterior tibial artery
3 Posterior tibial artery
4 Fibular artery
5 Superior lateral genicular artery
6 Inferior lateral genicular artery
7 Inferior medial genicular artery
8 Femoral vein
9 Great saphenous vein
10 Femoral artery
11 Superficial epigastric artery
12 Superficial epigastric vein
13 External pudendal vein
14 Superficial circumflex iliac vein
15 Superficial circumflex iliac artery
16 Femoral nerve
17 Soleus muscle
18 Popliteus muscle
19 Subcutaneous layer
20 Fascia lata
21 Sartorius muscle
22 Adductor longus muscle
23 Biceps femoris muscle
24 Semitendinosus muscle
25 Scrotum

Dissection of popliteal and crural arteries
Posterior view

Dissection of popliteal region revealing arteries
Posterior view

Dissection of proximal crus revealing arteries
Posterior view

Similar to the veins of the upper limb, the venous pathways in the lower limb consist of both deep veins that accompany the arteries, and superficial veins that course through the hypodermis. In the foot and leg, the deep veins form vena comitans with their arterial counterparts; however, the more proximal popliteal and femoral veins are large single vessels accompanying their associated arteries. Two major superficial venous channels receive numerous tributaries from smaller superficial veins throughout the lower limb. These major superficial veins are the small saphenous vein and the great saphenous vein. Unlike the upper limb, the majority of venous blood flow through the lower limb passes via the deep veins. Anastomosing veins between the saphenous veins and the deep veins have one-way valves. The valves direct blood flow to the deep veins where contractions of surrounding skeletal muscles facilitate movement of the blood toward the heart.

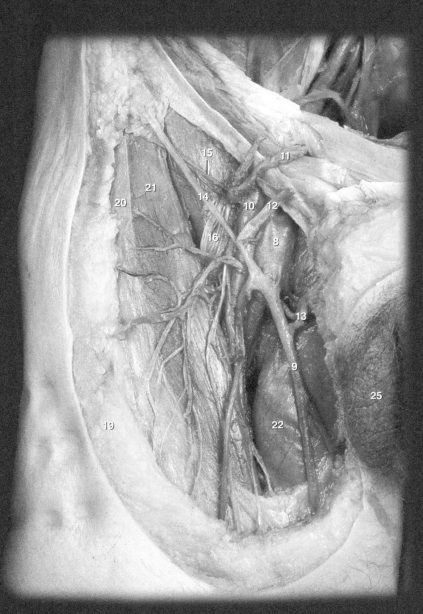

Dissection of femoral vein and tributaries in femoral triangle
Anterior view

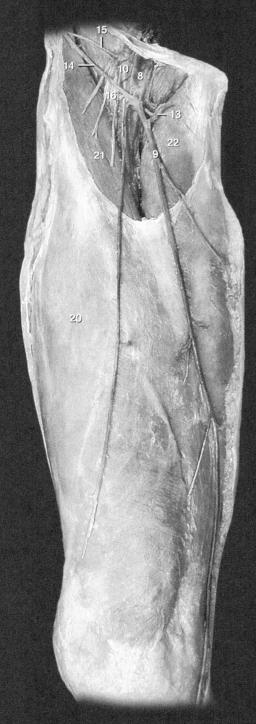

Dissection of great saphenous vein
Anteromedial view

Lymphatics

Even under normal circumstances, slightly more fluid is filtered out of the capillaries into the interstitial fluid than is reabsorbed from the interstitial fluid back into the plasma. On average, the net filtration pressure starts at 11 mm Hg at the beginning of the capillary, whereas the net reabsorption pressure only reaches 9 mm Hg by the vessel's end. Because of this pressure differential more fluid is filtered out of the first half of the capillary than is reabsorbed in its last half. If this extra filtered fluid were not drained away, the consequence of this unbalanced exchange would be accumulation of excess interstitial fluid, or edema. To circumvent this potentially disastrous problem, a system of accessory drainage vessels, the lymphatic vessels, evolved in vertebrate animals. This lymphatic system of vessels consists of an extensive network of one-way tubes that provide an accessory route through which fluid is returned from the interstitial fluid to the blood to keep the cardiac output and return equal.

1 Superficial inguinal lymph node
2 Afferent lymphatic vessels
3 Efferent lymphatic vessels
4 Great saphenous vein
5 Femoral vein
6 Femoral artery
7 Spermatic cord
8 Penis
9 Sartorius muscle
10 Rectus femoris muscle
11 Femoral nerve

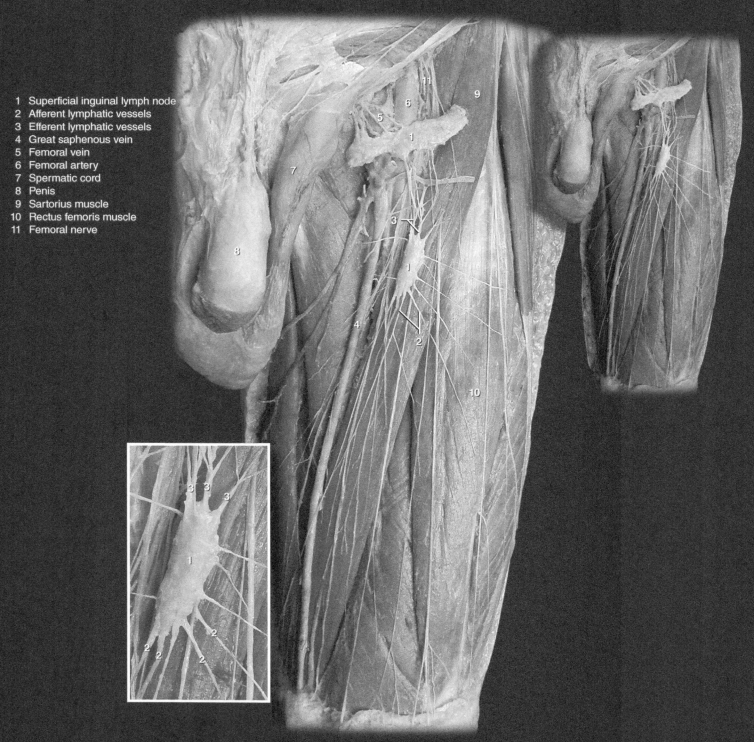

Dissection of lymphatic vessels and nodes in the thigh
Anterior view

17 | Respiratory System

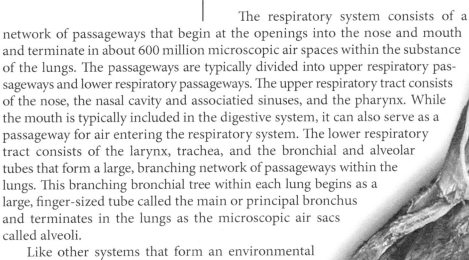

The respiratory system consists of a network of passageways that begin at the openings into the nose and mouth and terminate in about 600 million microscopic air spaces within the substance of the lungs. The passageways are typically divided into upper respiratory passageways and lower respiratory passageways. The upper respiratory tract consists of the nose, the nasal cavity and associatied sinuses, and the pharynx. While the mouth is typically included in the digestive system, it can also serve as a passageway for air entering the respiratory system. The lower respiratory tract consists of the larynx, trachea, and the bronchial and alveolar tubes that form a large, branching network of passageways within the lungs. This branching bronchial tree within each lung begins as a large, finger-sized tube called the main or principal bronchus and terminates in the lungs as the microscopic air sacs called alveoli.

Like other systems that form an environmental exchange surface with the cardiovascular system, the respiratory system forms an extensive surface area in contact with the capillaries. It is estimated that the surface area of the small dead-end air sacs in the lungs is about the size of a tennis court. This extensive interface is essential for the exchange of oxygen and carbon dioxide between the inhaled air and the blood. If body cells are deprived of oxygen, they cannot function and they die as a result. So the acquisition of oxygen through the respiratory passageways and its subsequent exchange with the capillary blood is an important function of the respiratory system.

In addition to gas exchange, the portion of the respiratory passageways referred to as the larynx is responsible for generating the sound waves that we manipulate into voice. Internal folds in the lining of the larynx, the vocal folds, vibrate as air passes upward from the lungs to produce the vibrations. For this reason the larynx is often referred to as the voice box.

Find more information about the respiratory system in REAL ANATOMY

Upper Respiratory Tract

The upper respiratory tract consists of the initial series of passageways that carry the inspired air through the head. The various sections of the head seen on this and the facing page show the passageways of the upper respiratory tract, which include the nose and nasal vestibule, the nasal cavity, the paranasal sinuses, nasopharynx, oropharynx, laryngopharynx, and even the oral cavity. The nasal cavity functions in filtering, warming, and humidifying the inspired air, while also detecting chemical odorants.

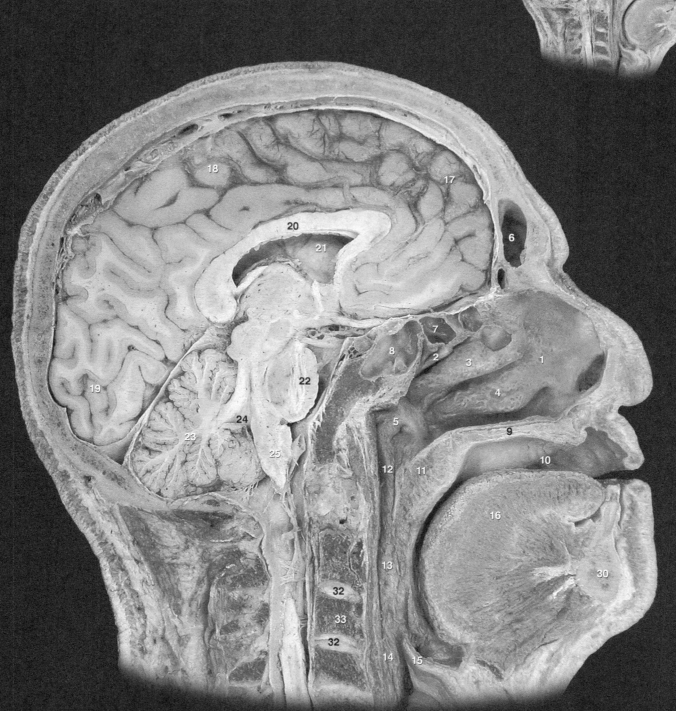

Sagittal section of head
Medial view

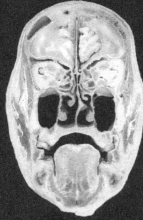

1 Nasal cavity
2 Superior nasal concha
3 Middle nasal concha
4 Inferior nasal concha
5 Torus tuberius
6 Frontal sinus
7 Ethmoid air cell
8 Sphenoidal sinus
9 Hard palate
10 Oral cavity
11 Soft palate
12 Nasopharynx
13 Oropharynx
14 Laryngopharynx
15 Epiglottis
16 Tongue
17 Frontal lobe
18 Parietal lobe
19 Occipital lobe
20 Corpus callosum
21 Lateral ventricle
22 Pons
23 Cerebellum
24 Fourth ventricle
25 Medulla oblongata
26 Nasal septum
27 Maxillary sinus
28 Temporalis
29 Masseter
30 Mandible
31 Orbit
32 Intervertebral disc
33 Vertebral body

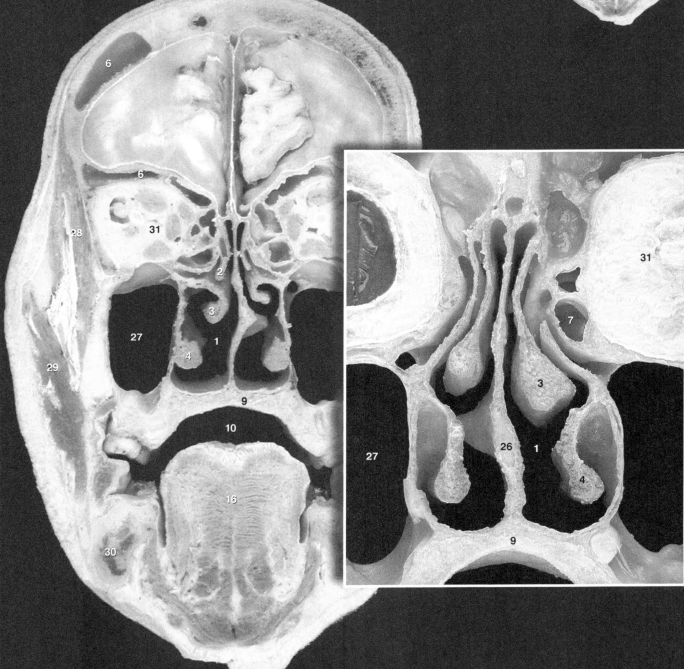

Frontal section of head
Anterior view

Lower Respiratory Tract

The lower respiratory tract arises as an outgrowth of the tubular gut during embryonic development. This anterior outgrowth of the gut tube begins at the larynx (voice box), which is the upper expanded portion of the lower respiratory tract. It continues from the neck into the thorax as the trachea (windpipe), and forms a large branching network of tubes that enter the lungs, the bronchial tree. The pages that follow show the tubular organs and histology of the lower respiratory tract.

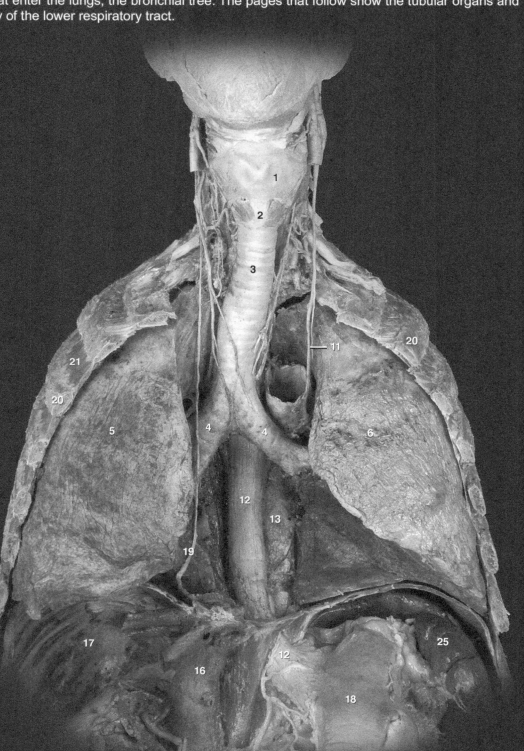

Lower respiratory tract and lungs in situ
Anterior view

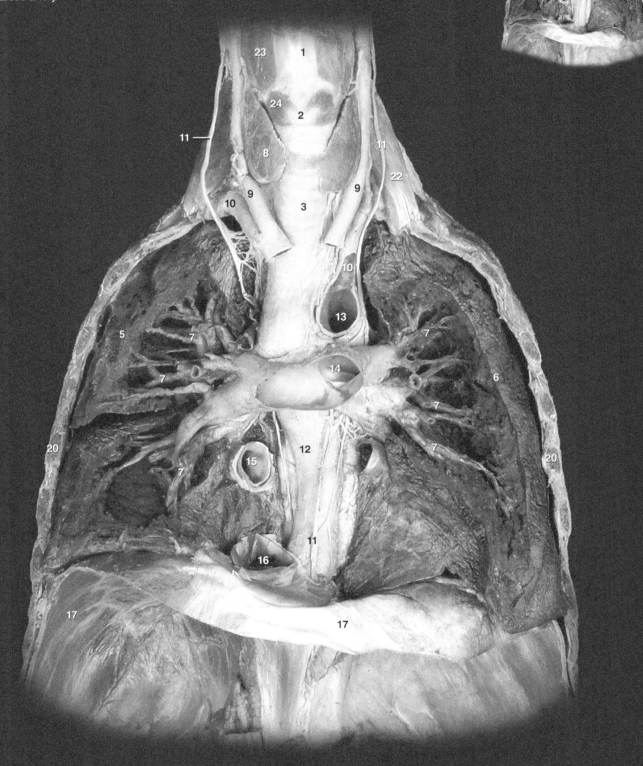

1 Thyroid cartilage of larynx
2 Cricoid cartilage of larynx
3 Trachea
4 Main (primary) bronchus
5 Right lung
6 Left lung
7 Bronchial tree
8 Thyroid gland
9 Common carotid artery
10 Subclavian artery
11 Vagus nerve
12 Esophagus
13 Aorta
14 Pulmonary artery
15 Pulmonary vein
16 Inferior vena cava
17 Diaphragm
18 Stomach
19 Phrenic nerve
20 Rib
21 Intercostal muscle
22 Anterior scalene muscle
23 Thyrohyoid muscle
24 Cricothyroid muscle
25 Spleen

Dissection of lower respiratory tract and lungs in situ
Anterior view

Larynx

The entrance to the trachea is an expanded region called the larynx, or voice box. A series of large cartilages form the walls of this region. The soft tissue lining of the laryngeal cartilages folds into the larynx to form the vocal folds, flaps of tissue that lie across the opening of the larynx. Within the edges of the vocal folds are the vocal cords, two bands of elastic tissue that can be stretched and positioned in different shapes by laryngeal cartilages and muscles. As air is moved past the taut vocal cords, they vibrate to produce the many different sounds of speech. During swallowing, the vocal cords assume a function not related to speech; they are brought into tight apposition to each other to close off the rima glottidis, the entrance to the lower larynx and trachea.

1   Epiglottis
2   Thyroid cartilage
3   Thyroid tubercle (Adam's apple)
4   Superior cornu
5   Inferior cornu
6   Cricothyroid membrane
7   Cricoid cartilage
8   Arytenoid cartilage
9   Corniculate cartilage
10 Trachea
11 Vocal fold
12 Vocal ligament
13 Rima glottidis

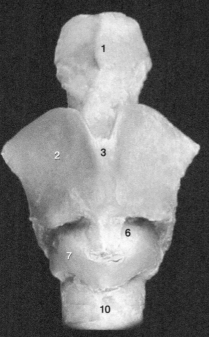

Laryngeal cartilages
Anterior view

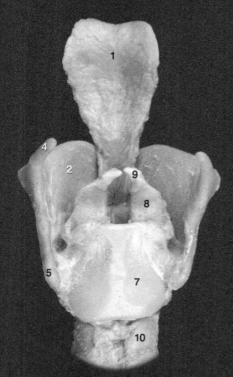

Laryngeal cartilages
Posterior view

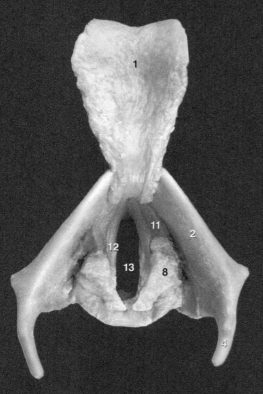

Laryngeal cartilages
Superior view

Trachea and Bronchial Tree

The trachea, "windpipe," is the conduction tube that transports the air to and from the lungs. It is reinforced by U-shaped cartilages.The trachea branches into two tubes called bronchi that enter the lungs. Each bronchus serves as the trunk of a highly branched, tree-like network of bronchial tubes that become progressively narrower, shorter, and more numerous as they spread throughout the tissues of the lung. These small tubes eventually terminate as the small, dead-end air sacs called alveoli, the principal site of gas exchange between air and blood.

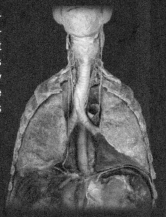

1 Epiglottis
2 Thyroid cartilage
3 Cricoid cartilage
4 Trachea
5 Right main (primary) bronchus
6 Left main (primary) bronchus
7 Lobar (secondary) bronchus
8 Segmental (tertiary) bronchus
9 Bronchiole
10 Fibromuscular membrane
11 Tracheal ring
12 Hyaline cartilage of tracheal ring
13 Tunica mucosa (pseudostratified)
14 Tela submucosa (areolar ct)
15 Tunica adventitia (dense ct)
16 Bronchiole cartilage (hyaline)
17 Alveolar spaces
18 Vein with red blood cells (rbc)
19 Pulmonary vein with rbcs

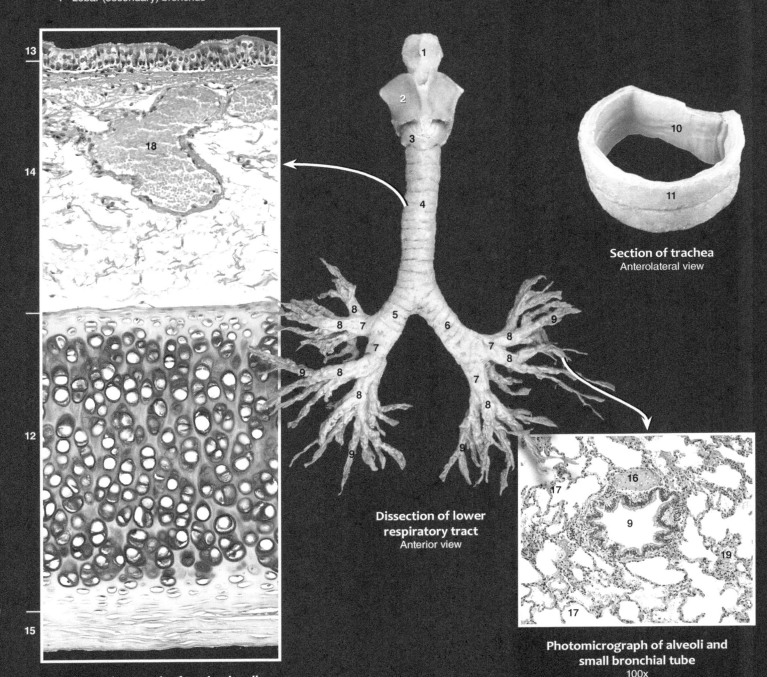

Section of trachea
Anterolateral view

Dissection of lower respiratory tract
Anterior view

Photomicrograph of tracheal wall
100x

Photomicrograph of alveoli and small bronchial tube
100x

Lungs

The lungs are the spongy, pyramidal-shaped organs that house the bronchial tree and the extensive pulmonary vascular network. Each lung is surrounded by a thin mesothelial covering, the visceral pleura, and sits on either side of the heart within the thoracic cavity. The vascular and respiratory passageways enter each lung on its medial aspect at the hilum. The wide base of the lung sits on the diaphragm inferiorly and tapers to a narrow apex superiorly. The right lung has three lobes and the left lung two lobes.

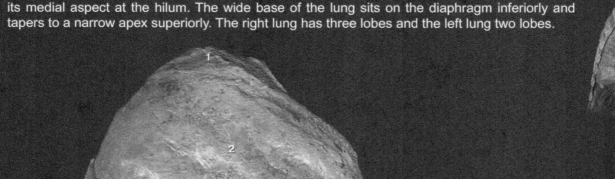

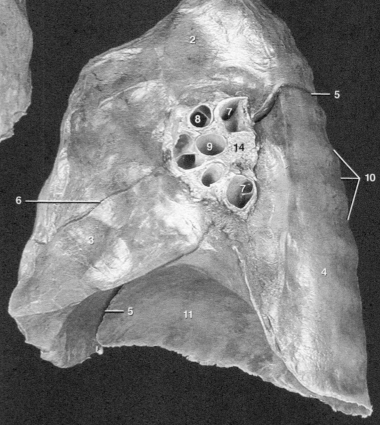

Right lung
Lateral view, anterior to the right

Right lung
Medial view, anterior to the left

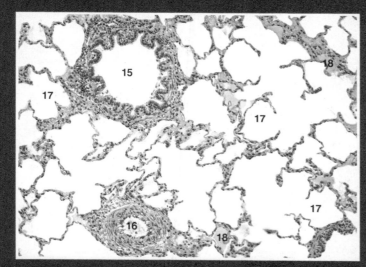

Photomicrograph of lung tissue
100x

1 Apex
2 Superior lobe
3 Middle lobe
4 Inferior lobe
5 Oblique fissure
6 Transverse fissure
7 Segmental (tertiary) bronchus
8 Pulmonary artery
9 Pulmonary vein
10 Costal impression
11 Diaphragmatic surface
12 Aortic impression
13 Cardiac notch
14 Hilum
15 Bronchiole
16 Small artery
17 Alveolar spaces
18 Blood vessels with rbcs
19 Lingula

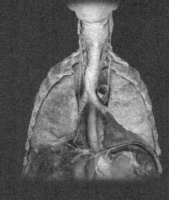

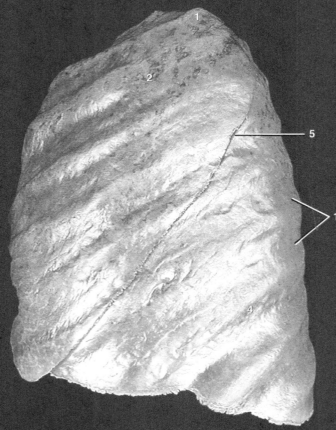

Left lung
Lateral view, anterior to the left

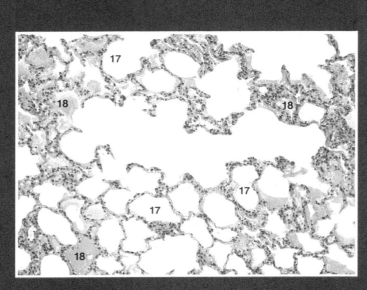

Photomicrograph of lung tissue
100x

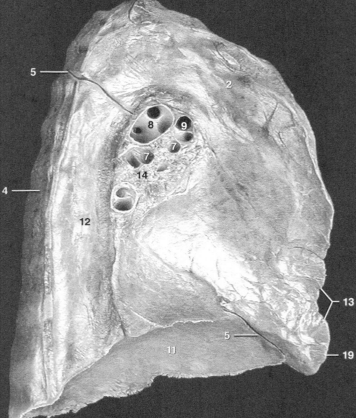

Left lung
Medial view, anterior to the right

Cast of Trachea and Bronchial Tree

The cast below is from a large dog's lungs and is approximately the same size as human lungs. The casts were created by forcing liquid latex into the respiratory passageways of the lungs and then letting the latex harden. The lungs were then placed in a weak acid until the organic tissue of the lungs was digested away. These views of the cast allow you to visualize the extensive nature of the bronchial tree as it branches out to the larger alveolar passageways within the lungs. The smaller alveolar spaces did not get incorported into the casts.

1 Trachea
2 Right main (primary) bronchus
3 Left main (primary) bronchus
4 Lobar (secondary) bronchus
5 Segmental (tertiary) bronchus
6 Branching bronchiole network

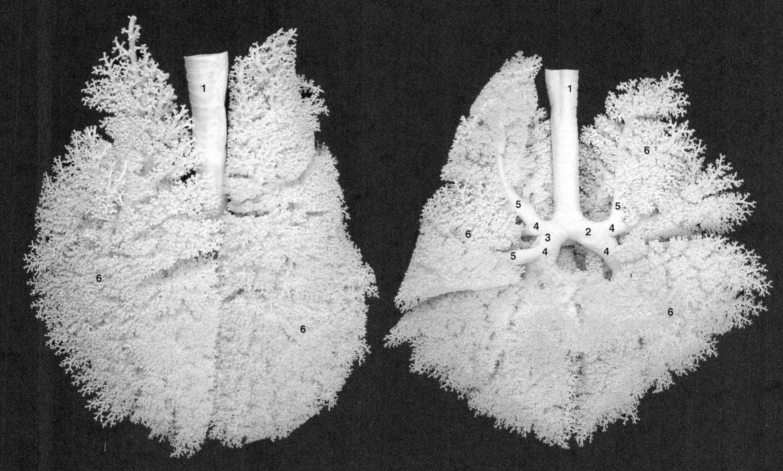

Latex cast of respiratory passageways of trachea and lungs of a dog
Anterior view at left, posterior view at right

18 | Digestive System

The digestive system is the extensive environmental interface that makes it possible to transfer nutrients, water, and electrolytes from the food we eat into the body's internal environment. This is made possible by a complex lining, which through a series of folds and a variety of small to microscopic projections greatly increases the surface interface between the digested contents within the gastrointestinal organs and the numerous small capillaries beneath this lining. To better appreciate the degree of this surface increase, realize that the average total surface area of the skin of an adult human is about 20 square feet, while the surface area of the digestive system is approximately 2,500 square feet, or about the size of a tennis court. To make the transfer across this extensive surface area possible, the food we eat must be broken down into small molecules that can be absorbed from the digestive tract into the circulatory system, which then distributes the molecular metabolites to the cells. Therefore, the digestive organs also function in the mechanical and chemical breakdown of the food.

Developmentally the digestive system begins as a simple tube called the gut tube or gut. As this simple tube develops into the highly convoluted organs of the adult anatomy, it undergoes structural changes that account for its various functions. Though these structural changes lead to differences in the tube from one region to the next, there is a basic pattern of design throughout the length of the tube. This structural pattern is responsible for the general function of the digestive system. Modifications of this pattern allow for the variation in structure and function along its length. This chapter will highlight the structural variation and underlying design of the digestive system.

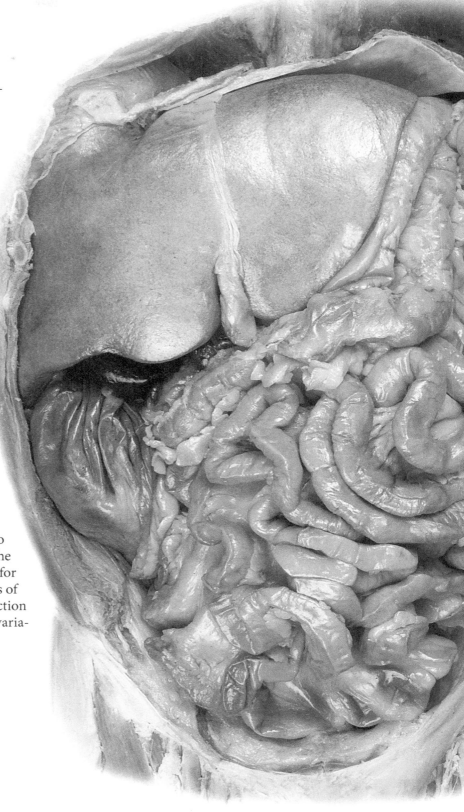

Find more information about the digestive system in
REAL ANATOMY

Digestive System Organs

The digestive system begins at the mouth, where food and drink enter this tubular organ system to be processed by the teeth and tongue. From the mouth the broken-down food moves through the transport tube called the esophagus to the storage and mixing organ called the stomach. The stomach thoroughly mixes digestive juices and mucous with the food as it tosses it around to produce a softened substance called chyme. The chyme is slowly moved into the small intestine where powerful digestive chemicals are added from the pancreas. As the chyme slowly moves through the long small intestine, the digestive enzymes break it into small metabolic fuel molecules that the intestine absorbs. The material that cannot be digested and absorbed is passed into the large intestine where the nondigested remains are held until they can be removed through the anus as feces. The photos on this and the facing page depict the digestive organs and their related mesenteries.

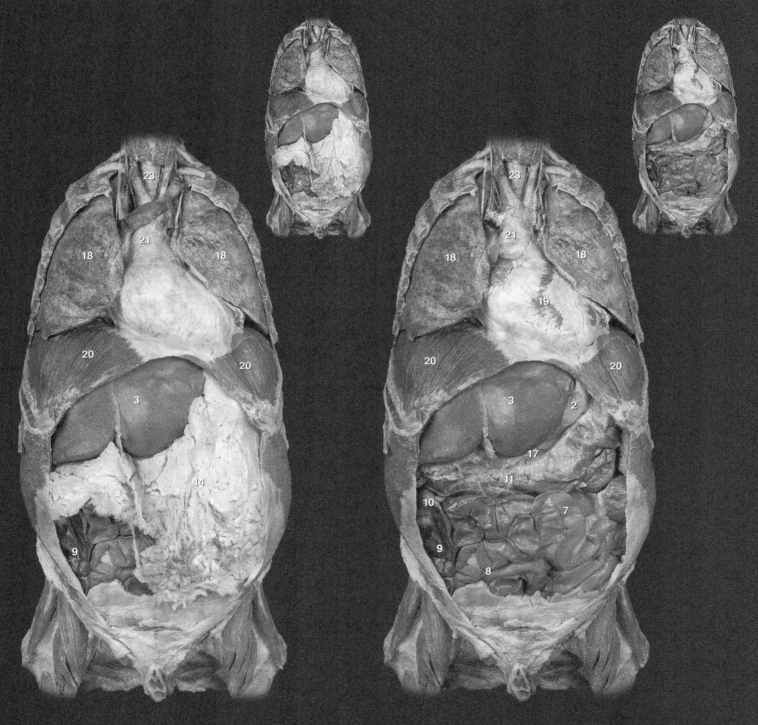

Superficial dissection of abdominal viscera
Anterior view

Intermediate dissection of abdominal viscera
Anterior view

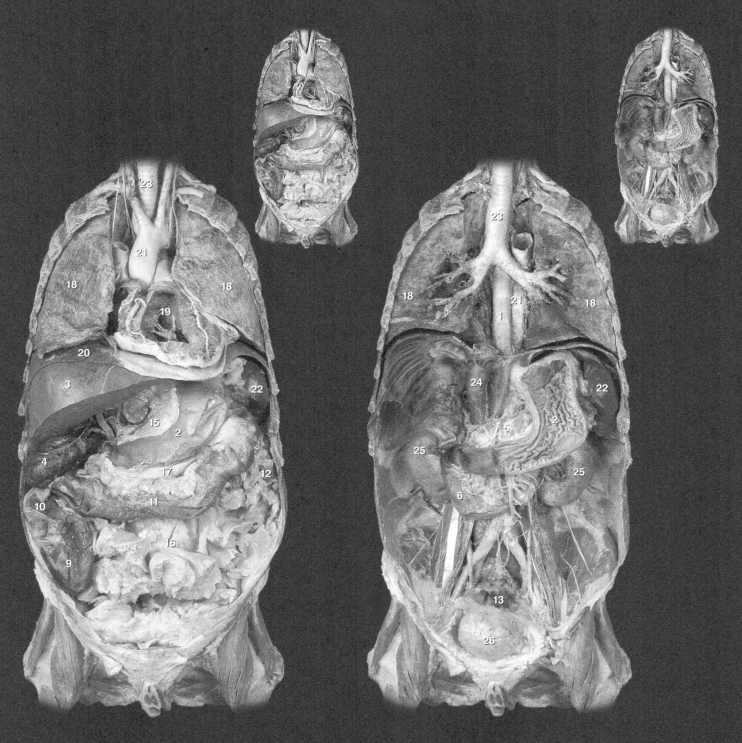

1 Esophagus
2 Stomach
3 Liver
4 Gallbladder
5 Pancreas
6 Duodenum
7 Jejunum
8 Ileum
9 Cecum

10 Ascending colon
11 Transverse colon
12 Descending colon
13 Rectum
14 Greater omentum
15 Lesser omentum
16 Mesentery
17 Transverse mesocolon
18 Lungs

19 Heart
20 Diaphragm
21 Aorta
22 Spleen
23 Trachea
24 Inferior vena cava
25 Kidney
26 Bladder

Intermediate dissection of abdominal viscera
Anterior view

Deep dissection of abdominal viscera
Anterior view

Design of the Gut Wall

The wall of the digestive tract has a basic pattern of design that is found throughout its length. This pattern consists of three tunics or layers of anatomy. The tunica mucosa and its subdivisions, including the tela submucosa, form the inner layer of the wall and consist of an extensive epithelial lining with an underlying vascullar connective tissue. The middle layer, or tunica muscularis, consists of smooth muscle that provides for the varied types of movements that occur within the digestive organs. The majority of the organs have an outer layer, the tunica serosa, comprised of a lubricated meosthelial membrane that reduces friction as the organs move against one another. The image below, from the small intestine, illustrates the basic layers of the digestive tract wall.

1 Simple columnar epithelium
2 Lamina propria
3 Muscularis mucosae
4 Submucosal (Brunner's) glands
5 Villi

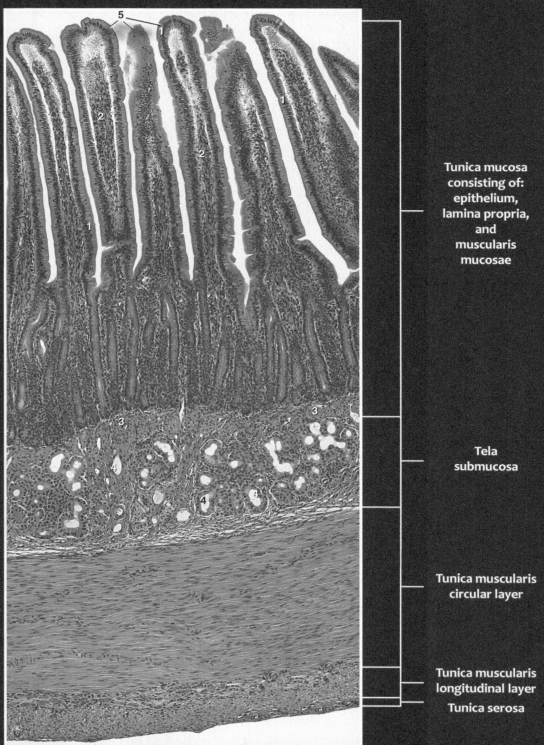

Tunica mucosa consisting of: epithelium, lamina propria, and muscularis mucosae

Tela submucosa

Tunica muscularis circular layer

Tunica muscularis longitudinal layer

Tunica serosa

Photomicrograph of small intestine wall
40x

Mouth and Pharynx

The mouth, or oral cavity, is the entryway into the digestive system. In addition to serving as the portal to the tubular gut, the mouth contains structures, such as the tongue, teeth, and salivary glands, that help initiate the digestive process. The boundaries of this region are defined by the lips and cheeks, which form the anterior and lateral walls, the palate, which forms the roof, and numerous muscles, the most conspicuous being the muscles of the tongue, which form the floor of the mouth. The pharynx is the first portion of the gut tube and is divided into three regions. Each region communicates with a different cavity — the nasopharynx with the nasal cavity, the oropharynx with the oral cavity, and the laryngopharynx with the cavity of the larynx.

1 Lips
2 Teeth
3 Tongue
4 Hard palate
5 Soft palate
6 Nasopharynx
7 Oropharynx
8 Laryngopharynx
9 Parotid gland
10 Submandibular gland
11 Parotid duct
12 Serous acini
13 Mucous acini
14 Vein
15 Trabecula
16 Masseter
17 Sternocleidomastoid
18 Sphenoid sinus
19 Epiglottis
20 Vertebral column
21 Cerebrum
22 Spinal cord

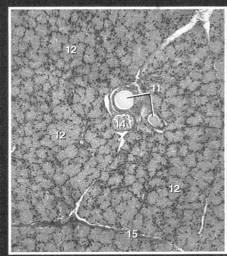

Photomicrograph of parotid gland
100x

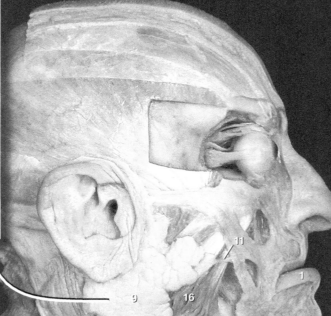

Dissection of head showing salivary glands
Lateral view

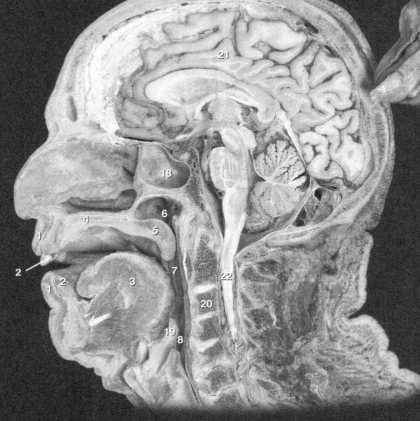

Sagittal section of head and neck
Medial view

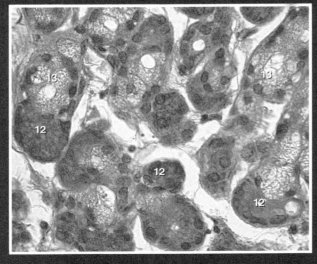

Photomicrograph of submandibular gland
240x

Esophagus

Below the laryngopharynx the gut tube branches into an anterior respiratory tube, the larynx and a posterior digestive tube, the esophagus. The esophagus is a narrow, collapsed muscular tube coursing from the laryngopharynx to the stomach. It is approximately 25 cm in length and begins near the level of the sixth cervical vertebra, where it runs inferiorly against the anterior surface of the thoracic vertebral column. At the level of the tenth thoracic vertebra it deviates slightly to the left passing through the esophageal hiatus of the diaphragm to enter the stomach. It functions as a muscular tube of transmission.

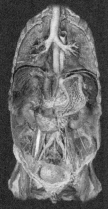

1	Esophagus	7	Lamina propria
2	Tunica mucosa	8	Muscularis mucosae
3	Tela submucosa	9	Tunica adventitia
4	Tunica muscularis circular layer	10	Stomach
5	Tunica muscularis longitudinal layer	11	Pharynx - dorsal wall
6	Stratified squamous epithelium	12	Vagus nerve

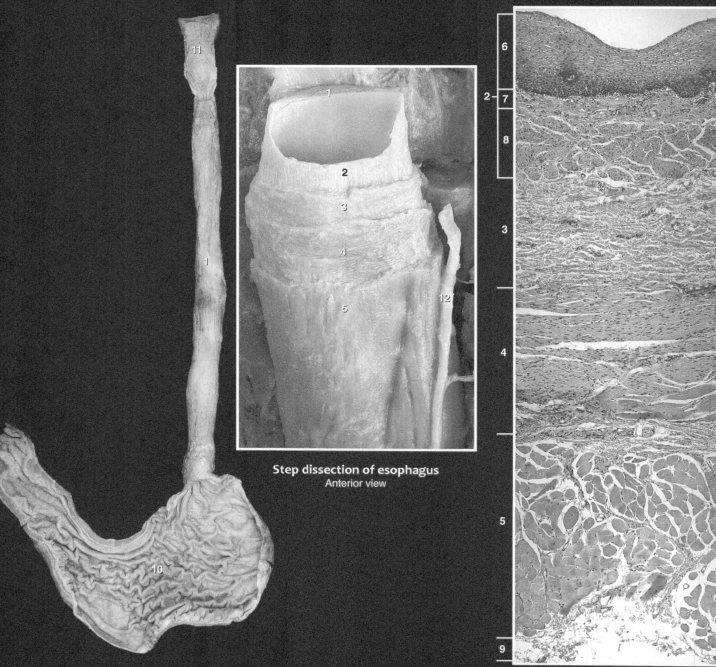

Step dissection of esophagus
Anterior view

Pharynx, esophagus, and stomach
Anterior view

Photomicrograph of esophageal wall
40x

Stomach

The stomach is a J-shaped organ of variable size and shape and has the greatest diameter of any part of the gut tube. It occupies the upper left quadrant of the abdominal cavity, where it is anchored to the posterior abdominal wall by a mesentery. The stomach performs several functions, the most important of which is to store ingested food until it can be emptied into the small intestine at a rate that allows for optimal digestion and absorption.

1 Stomach
2 Cardia of stomach
3 Fundus of stomach
4 Body of stomach
5 Pyloric antrum
6 Pyloric canal
7 Pylorus
8 Pyloric sphincter
9 Gastric rugae
10 Greater curvature
11 Lesser curvature
12 Gastric pit
13 Surface mucous cell
14 Lamina propria
15 Mucous neck cell
16 Gastric glands
17 Liver
18 Gallbladder
19 Spleen
20 Greater omentum

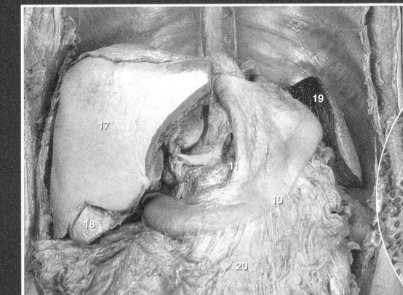

Abdominal dissection revealing stomach
Anterior view

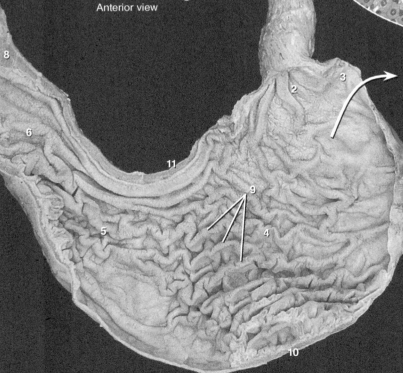

Frontal section of stomach
Anterior view

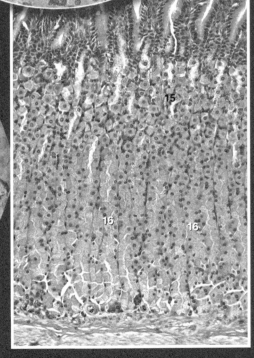

**Photomicrograph of stomach mucosa
with callout above**
40x and 100x

305

Small Intestine

The small intestine is a highly coiled tube with a fairly consistent diameter from beginning to end. It is approximately 6 to 7 meters long in the cadaver but, because of its muscle tone only around 4 to 5 meters in the living. The small intestine occupies the greater part of the mid- to lower abdominal cavity and consists of three regions. The retroperitoneal first part is called the duodenum and is about 30 cm in length. This C-shaped region receives the secretions from the pancreas and liver. The remaining parts of the small intestine are the jejunum and ileum, which make up the bulk of the organ and are attached to the posterior wall of the abdomen by the mesentery. The small intestine is the principal site of digestion and absorption.

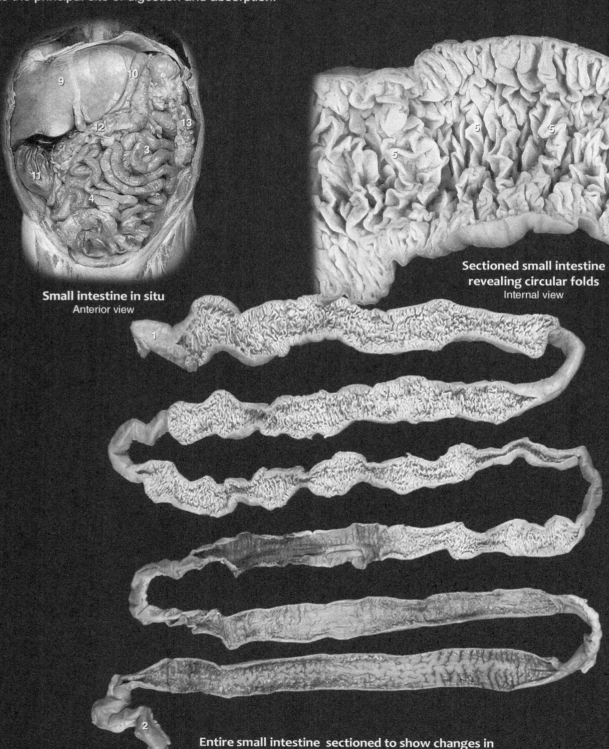

Small intestine in situ
Anterior view

Sectioned small intestine revealing circular folds
Internal view

Entire small intestine sectioned to show changes in internal surface from the duodenal end to the ileal end
Internal view

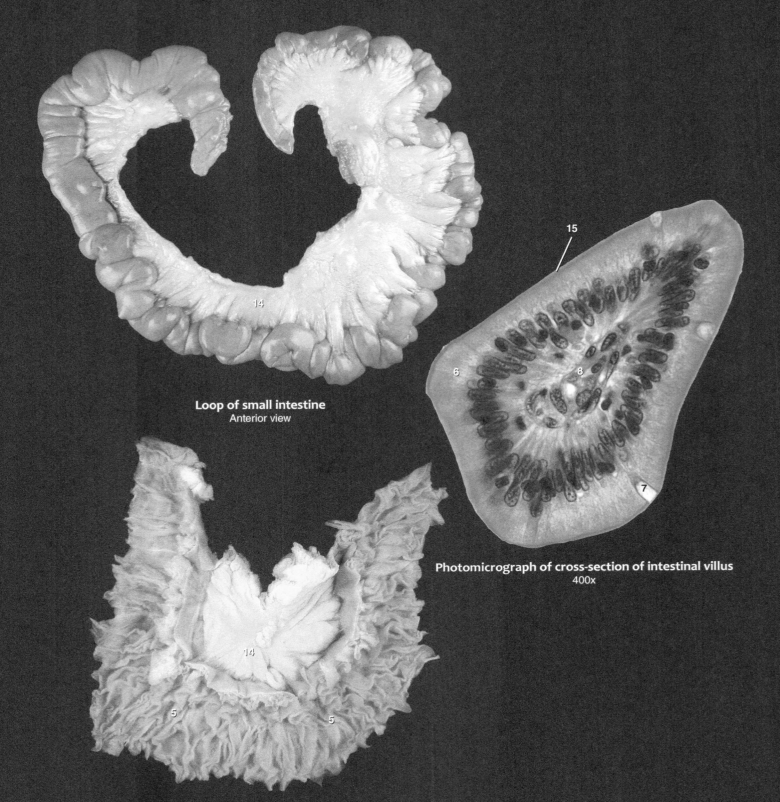

1 Duodenal end
2 Ileal end
3 Jejunum
4 Ileum
5 Circular folds
6 Simple columnar epithelium
7 Goblet cell
8 Lamina propria
9 Liver
10 Stomach
11 Cecum
12 Transverse colon
13 Descending colon
14 Mesentery
15 Microvillus brush border

Loop of small intestine
Anterior view

Photomicrograph of cross-section of intestinal villus
400x

Loop of small intestine from unembalmed cadaver, opened to show circular folds
Anterior view

Pancreas

The pancreas is a pinkish glandular structure situated posterior to the stomach in the retroperitoneal space of the abdominal cavity. It arises as an outgrowth of the duodenum during development and retains this connection via the pancreatic duct. It is a dual glandular organ consisting of both exocrine and endocrine glandular tissue. It has four basic regions: a head, neck, body, and tail. The exocrine glands and ducts produce and deliver the powerful digestive enzymes to the small intestine.

1 Tail of pancreas
2 Body of pancreas
3 Head of pancreas
4 Uncinate process of pancreas
5 Pancreatic duct (of Wirsung)
6 Major duodenal papilla
7 Exocrine acinus
8 Pancreatic ductule
9 Pancreatic islet (endocrine cells)
10 Trabecula
11 Duodenum
12 Liver
13 Gallbladder
14 Common bile duct
15 Spleen
16 Diaphragm

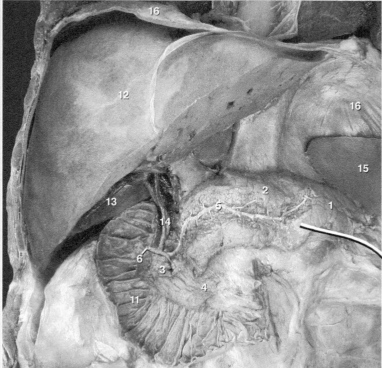

Abdominal dissection with part of liver and peritoneal organs removed
Anterior view

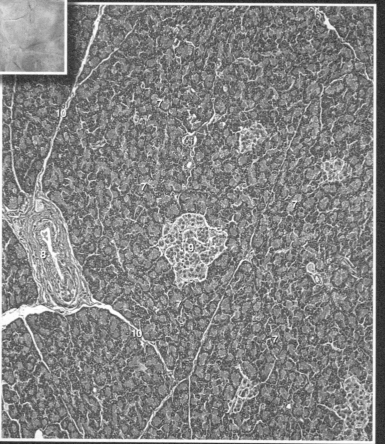

Photomicrograph of pancreas
100x

Liver and Gallbladder

Besides pancreatic juice, the other secretory product emptied into the duodenum is bile. The biliary system, which also develops as an embryonic outgrowth of the duodenum, includes the liver, the gallbladder, and associated ducts. The rounded, wedge-shaped liver, the largest organ of the abdomen, occupies a major portion of the upper right peritoneal cavity. The gallbladder is a pear-shaped, saccular organ situated in a depression on the inferior surface of the right lobe of the liver where it is a storage organ of the bile that is produced in the liver. Connecting the gallbladder to the common hepatic bile duct is the cystic bile duct. The junction of these ducts forms the main bile duct that drains into the duodenum. The liver is the largest and most important metabolic organ in the body, which in addition to producing the important bile salts associated with digestion, performs a myriad of metabolic functions.

1 Right lobe of liver
2 Left lobe of liver
3 Caudate lobe of liver
4 Quadrate lobe of liver
5 Gallbladder
6 Cystic bile duct
7 Hepatic artery
8 Hepatic portal vein
9 Round ligament
10 Inferior vena cava
11 Hepatocytes
12 Central vein
13 Hepatic sinusoid
14 Branch of hepatic artery
15 Bile duct
16 Branch of hepatic portal vein

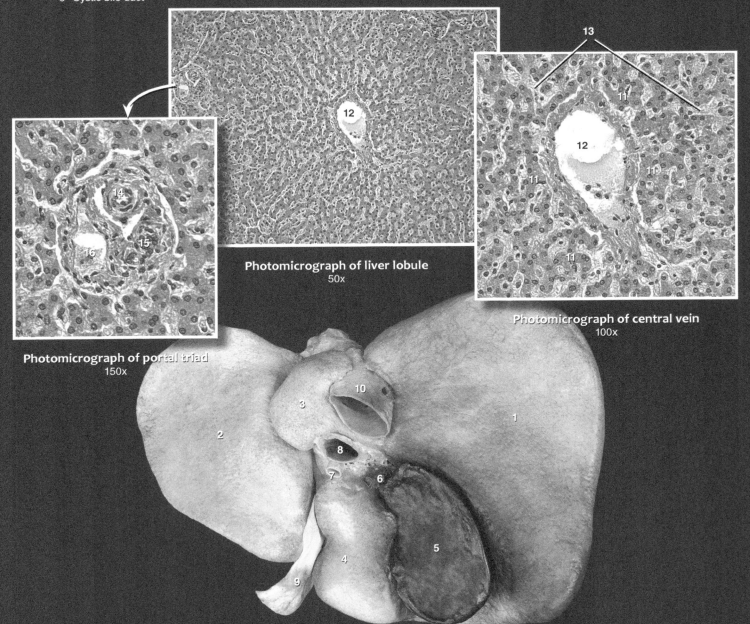

Photomicrograph of liver lobule
50x

Photomicrograph of central vein
100x

Photomicrograph of portal triad
150x

Liver and gall bladder
Inferior view, posterior at top

Large Intestine

The large intestine is much shorter than the small intestine, averaging about 1.5 meters in length, but typically has a greater diameter, therefore the name. The large intestine consists of the cecum, appendix, colon, and rectum. The cecum receives indigestible material from the small intestine and then moves it through the subdivisions of the colon — the ascending colon, transverse colon, descending colon, and sigmoid colon — before it enters the terminal portion of the gut tube, the rectum. The **large intestine** is primarily a drying and storage organ of indigestible plant fibers. Minimal absorption of fluids occurs in the large intestine as the fecal contents are stored prior to evacuation.

1 Cecum
2 Vermiform appendix
3 Ascending colon
4 Right colic (hepatic) flexure
5 Transverse colon
6 Left colic (splenic) flexure
7 Descending colon
8 Sigmoid colon

9 Rectum
10 Omental or fatty appendices
11 Haustra
12 Taeniae coli
13 Absorptive cells
14 Goblet cells
15 Intestinal glands
16 Muscularis mucosae

17 Lamina propria
18 Tela submucosa
19 Ileum (cut)
20 Duodenal-jejunal junction (cut)
21 Stomach
22 Root of the mesentery (cut)

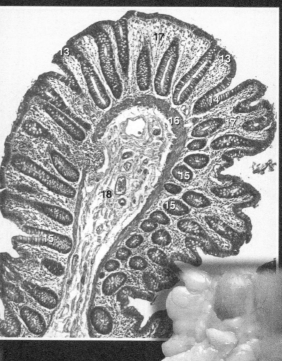

Photomicrograph of of large intestine mucosa
100x

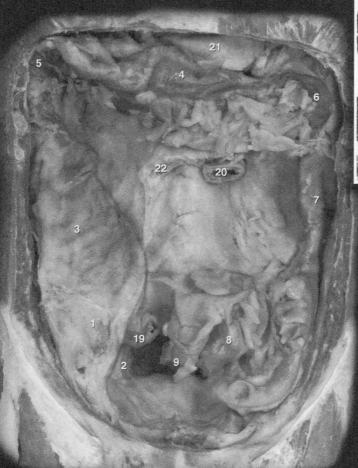

Dissection of abdominal cavity with jejunum and ileum removed
Anterior view

Portion of descending colon
Anterior view

Mesenteries

Mesenteries are reflections of the serous peritoneal membrane from the parietal layer lining the posterior abdominal wall to the visceral layer covering the peritoneal abdominal organs. The mesenteries not only support the digestive organs and help anchor them in the abdominal cavity, but also are the pathways for the vessels and nerves that supply the peritoneal organs.

1 Transverse mesocolon
2 The mesentery partially dissected to reveal vessels
3 Greater omentum
4 Superior mesenteric vein and tributaries
5 Branches of superior mesenteric artery
6 Cecum
7 Ascending colon
8 Transverse colon
9 Gallbladder
10 Cystic bile duct
11 Common hepatic bile duct
12 Common bile duct
13 Omental or fatty appendices
14 Stomach
15 Small intestine
16 Aorta

17 Heart
18 Vertebral column
19 Trachea
20 Aortic arch
21 Pulmonary trunk
22 Brain
23 Tongue
24 Sternum
25 Rectum
26 Bladder
27 Prostate
28 Testis
29 Penis
30 Pubic symphysis
31 Diaphragm
32 Esophagus

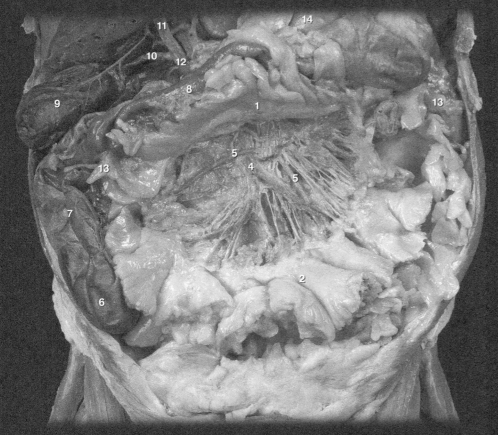

Dissection of the mesentery with jejunum and ileum removed
Anterior view

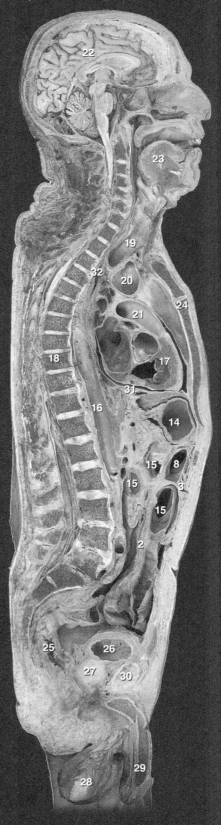

Sagittal section of head and trunk
Medial view

Omenta

Omenta are mesenteric structures that unite two digestive organs. These reflections of the peritoneal membrane course from one abdominal digestive organ to another abdominal digestive organ, rather than from organ to body wall. There are two omenta in the abdominal cavity. The greater omentum is a peritoneal reflection between the greater curvature of the stomach and the transverse colon. The lesser omentum is a peritoneal reflection between the lesser curvature of the stomach and the liver.

1 Greater omentum
2 Lesser omentum
3 Hepatogastric ligament of lesser omentum
4 Hepatoduodenal ligament of lesser omentum
5 Hepatorenal part of coronary ligament
6 Falciform ligament
7 Transverse mesocolon
8 Liver
9 Stomach
10 Duodenum
11 Transverse colon

12 Fossa for removed gallbladder
13 Gallbladder
14 Common hepatic bile duct
15 Common bile duct
16 Caudate lobe of liver
17 Lung
18 Heart
19 Breast
20 Diaphragm
21 Epiploic foramen
22 Spleen

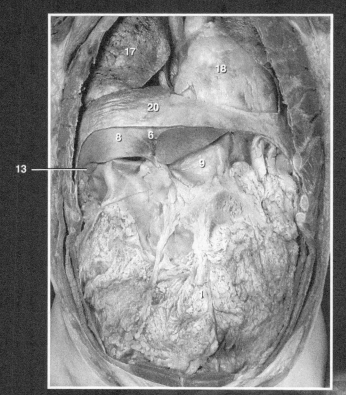

Anterior body wall removed exposing body cavity
Anterior view

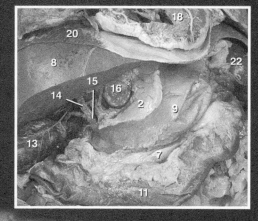

Dissection of abdominal cavity with anterior aspect of liver removed
Antero-inferior view

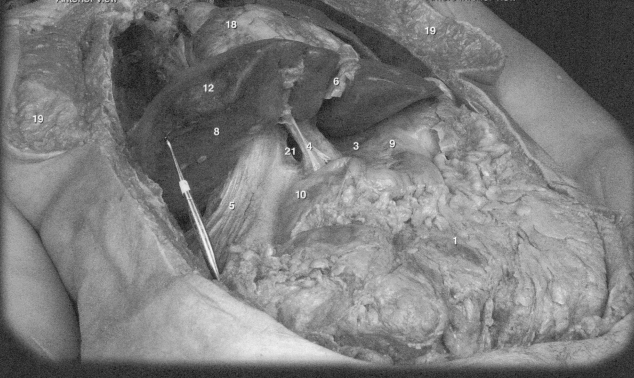

Superficial dissection of abdominal cavity with liver elevated
Antero-inferior view

312

19 | Urinary System

Like the respiratory and digestive systems, the urinary system is an environmental exchange system. Like all the exchange systems of the body, the urinary system forms an immense interface with the cardiovascular system for the single purpose of regulating the homeostatic balance of the water environment (extracellular matrix) that surrounds every cell in the body. To make this exchange possible a large network of microscopic urinary tubes form an intimate interface with an equally large network of cardiovascular capillaries. The urinary system consists of two blood processing centers called the kidneys, two transport tubes called the ureters, a storage organ called the bladder, and a drain called the urethra. The kidneys continually produce urine, which is then moved via the ureters to the storage organ, the bladder. When it is convenient to remove the urine from the body, contractions in the wall of the bladder expell the urine through the urethra.

In order to survive, every body cell requires a water environment that is similar to the composition of the oceans in which cellular life first arose. The kidneys help maintain this intercellular water environment by filtering the blood and regulating its contents so the blood can help maintain the correct composition of the extracellular fluid that bathes every cell. By adjusting the amount of water in the plasma and the various plasma constituents, which are either conserved for the body or eliminated in the urine, the kidneys are able to maintain water and electrolyte balance within the very narrow range compatible with life, despite wide variations in intake and losses of these constituents through other avenues.

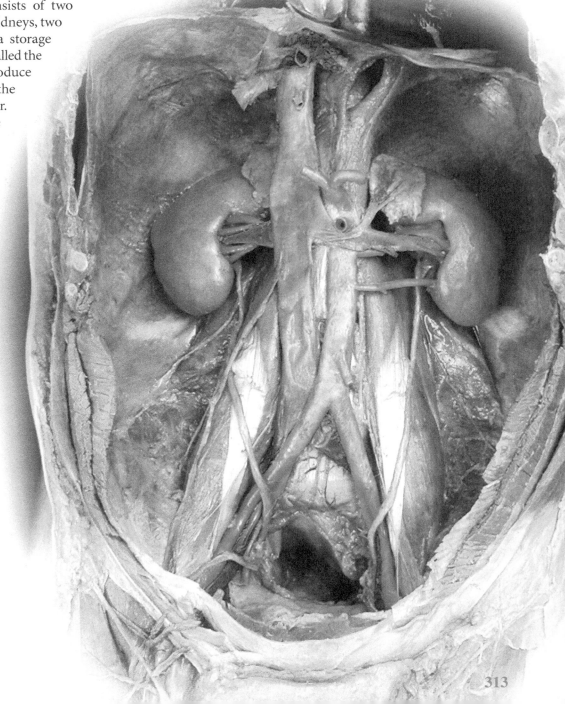

Find more infomation about the urinary system in

REALANATOMY

Urinary Organs

The organs of the urinary system include the paired kidneys, paired ureters, bladder, and urethra. The urinary organs occupy the retroperitoneal and subperitoneal spaces in the abdominopelvic cavity, where they are surrounded by a large amount of adipose tissue and some areolar connective tissue. The dissection images on this and the facing page depict the organs of the urinary system and their relations to other organs in the abdominopelvic cavity.

1 Kidney	7 Adrenal gland	13 Liver
2 Renal pelvis	8 Aorta	14 Lumbar vertebra
3 Ureter	9 Inferior vena cava	15 Hilum
4 Bladder	10 Diaphragm	16 Perirenal fat
5 Renal vein	11 Common iliac artery	17 Intestines
6 Renal artery	12 Psoas major muscle	18 Mesenteric fat

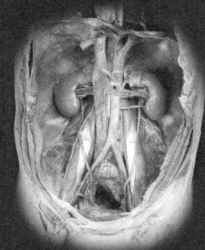

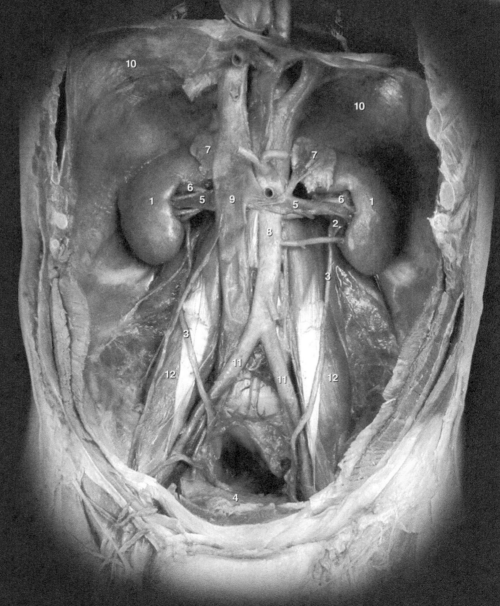

Dissection of the retroperitoneal space of the abdominal cavity
Anterior view

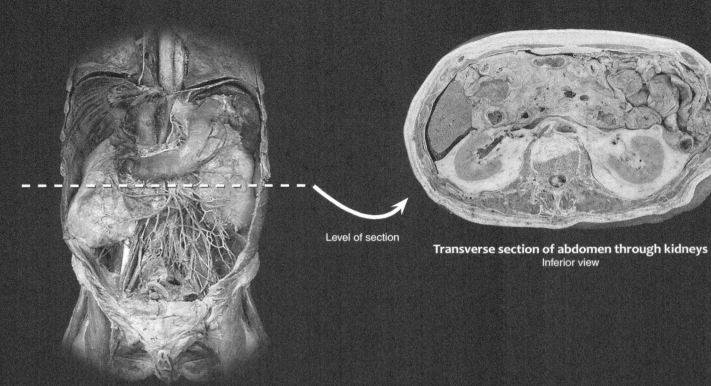

Dissection of abdomen showing perirenal fat
Anterior view

Level of section

Transverse section of abdomen through kidneys
Inferior view

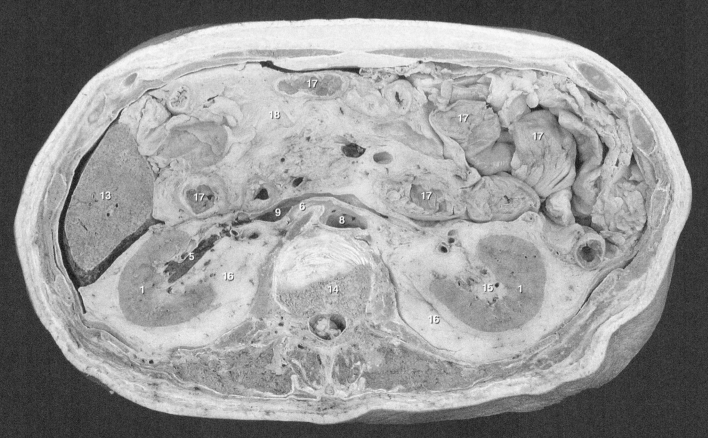

Transverse section of abdomen at level of first lumbar vertebra
Inferior view

Kidneys and Ureters

The paired kidneys are the processing organs of the urinary system that filter the blood for the purpose of regulating the water and electrolyte balance of the tissue fluid, while removing unwanted waste products from the body. They occupy the retroperitoneal space of the abdominal cavity immediately anterior to the 12th ribs. The ureters descend from the kidneys lateral to the lumbar vertebrae, cross anterior to the psoas musculature and the common iliac vessels, and enter the pelvis to join the bladder.

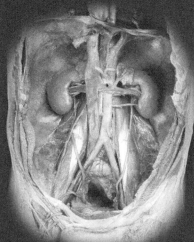

1	Hilum	6	Renal artery
2	Renal pelvis	7	Segmental artery
3	Ureter	8	Segmental vein
4	Renal capsule	9	Major calyx
5	Renal vein	10	Minor calyx

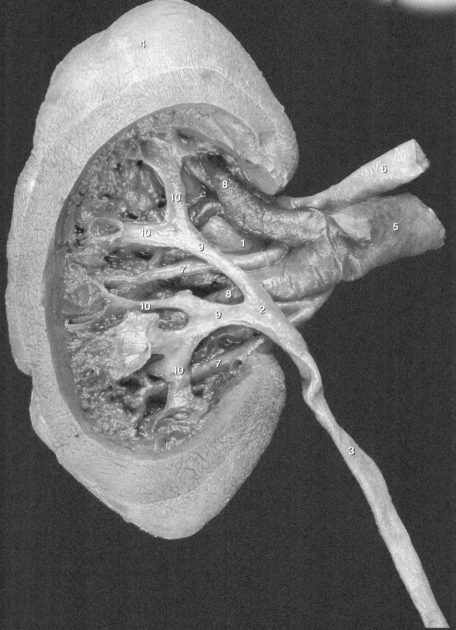

Dissection into medulla of left kidney
Posterior view

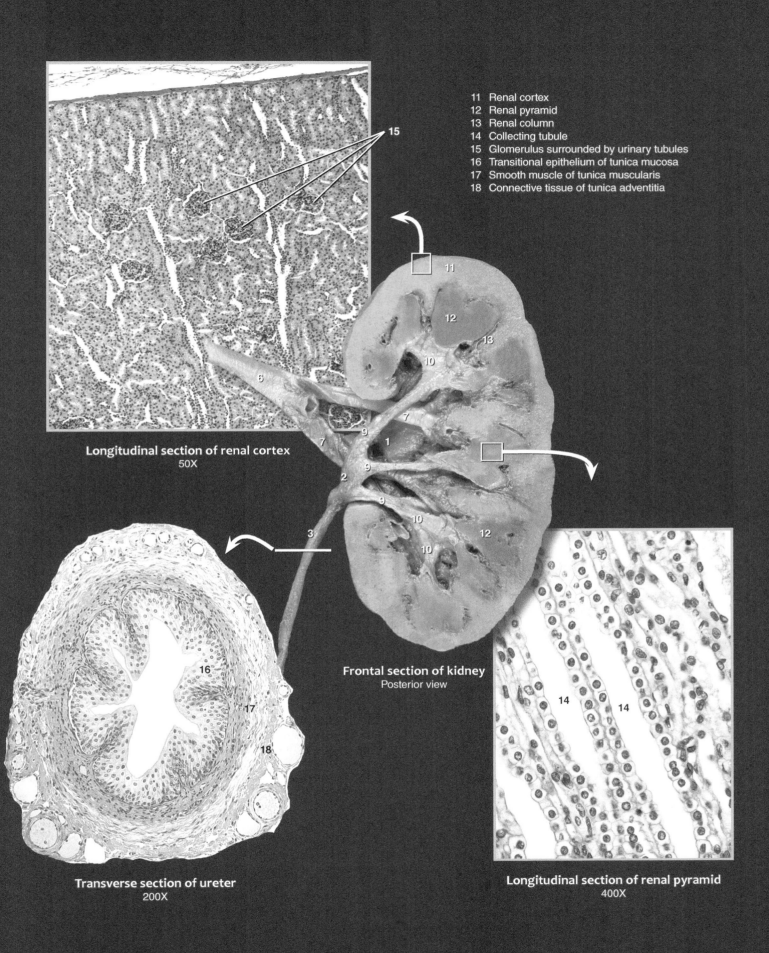

11 Renal cortex
12 Renal pyramid
13 Renal column
14 Collecting tubule
15 Glomerulus surrounded by urinary tubules
16 Transitional epithelium of tunica mucosa
17 Smooth muscle of tunica muscularis
18 Connective tissue of tunica adventitia

Longitudinal section of renal cortex
50X

Frontal section of kidney
Posterior view

Transverse section of ureter
200X

Longitudinal section of renal pyramid
400X

Bladder and Urethra

The bladder is the convenience organ of the urinary system that stores the urine, which is continually being produced by the kidneys, until it is convenient to remove it from the body. Arising from the inferior surface of the bladder is the drain for the bladder called the urethra. It is a short tube in females and a much longer tube in males. The male urethra not only transports urine, but also is the passageway for sperm as it exits during ejaculation.

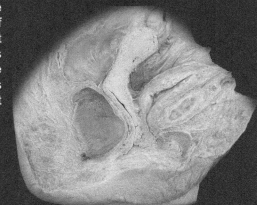

Female	Male (opposite page)
1 Bladder	1 Bladder
2 Urethra	2 Prostatic urethra
3 Clitoris	3 Spongy urethra
4 Vagina	4 Prostate
5 Uterus	5 Penis
6 Rectum	6 Testis
7 Pubis	7 Scrotum
8 Anus	8 Rectum
9 Labia majora	9 Anus
	10 Pubis
	11 Transitional epithelium of tunica mucosa

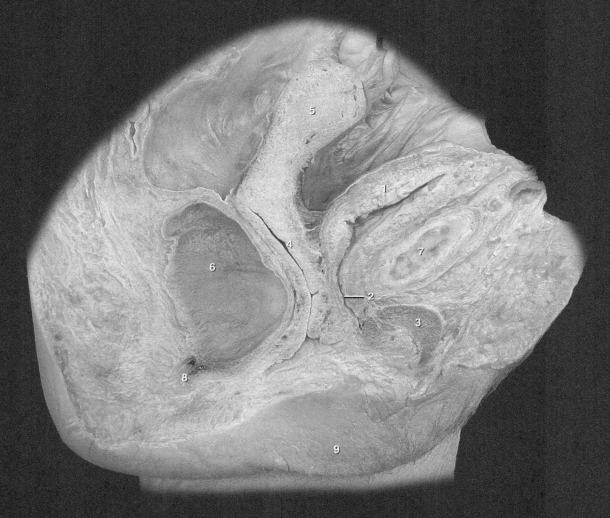

Sagittal section of female pelvis
Medial view

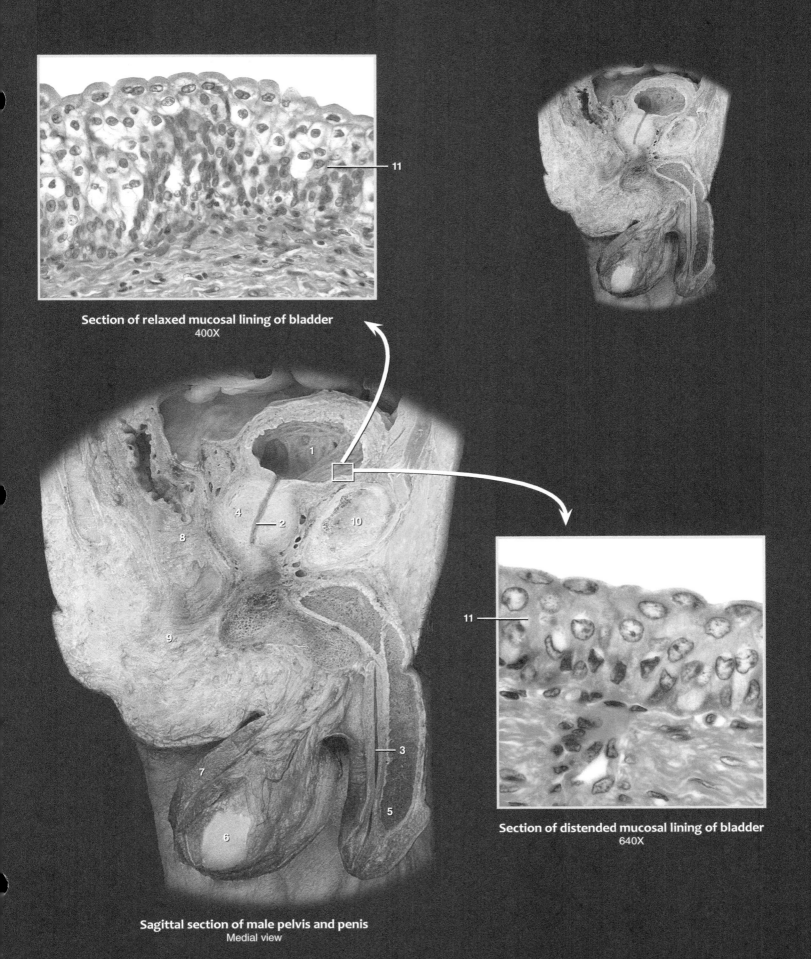

Section of relaxed mucosal lining of bladder
400X

11

Section of distended mucosal lining of bladder
640X

11

Sagittal section of male pelvis and penis
Medial view

1

4

2

8

10

9

3

7

5

6

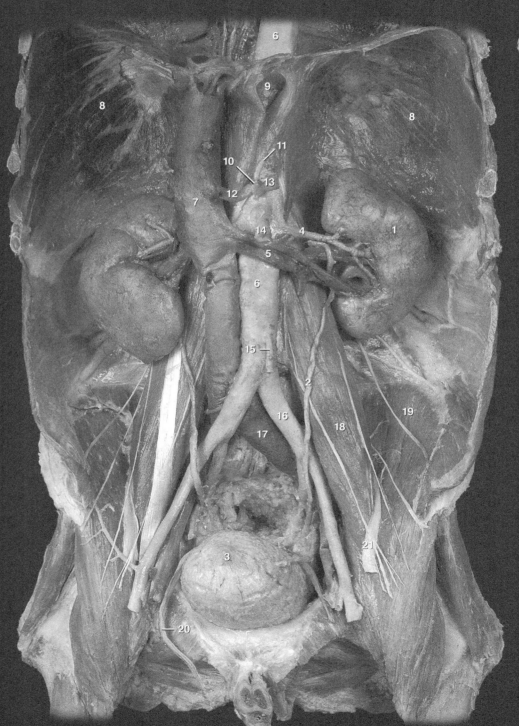
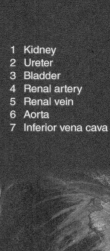

1 Kidney	8 Diaphragm	15 Inferior mesenteric artery
2 Ureter	9 Esophageal hiatus	16 Common iliac artery
3 Bladder	10 Celiac artery	17 Common iliac vein
4 Renal artery	11 Left gastric artery	18 Posas major muscle
5 Renal vein	12 Splenic artery	19 Iliacus muscle
6 Aorta	13 Common hepatic artery	20 Ductus deferens
7 Inferior vena cava	14 Superior mesenteric artery	21 Femoral nerve

Dissection of urinary system
Anterior view

20 | Reproductive Systems

The organs of the male and female reproductive (genital) systems have, as their primary role, the responsibility of producing the specialized cells called gametes and making it possible for these cells to unite to form a new individual. The male gametes, the sperm, arise in the testes from meiotic divisions in the walls of the numerous seminiferous tubules. From here hundreds of millions of sperm make their way during ejaculation through a series of tubes — rete testis, efferent ductules, epididymis, ductus deferens, ejaculatory duct, prostatic urethra, intermediate urethra, spongy urethra — that move the sperm out of the male genital system and introduce them into the female genital system. During this passage secretions are added to the sperm by the prostate, seminal, and bulbourethral glands to help protect and nurture the sperm in their journey to unite with the female gamete.

The sperm are introduced by the male intromittent organ, the penis, into the female vagina, which serves the dual function of being a penile receptacle and the birth canal. Sperm deposited in the fornices of the vagina then enter the os of the uterine cervix and propel themselves to the top of the uterine cavity. Here the sperm enter the openings into the uterine tubes where they continue their journey toward the ovulated female gamete.

After rupturing the surface of the ovary in an event called ovulation, the female gamete, the primary oocyte, is swept into the ostium of the uterine tube by the fingerlike fimbriae. Ciliary action of the uterine tube mucosa carry the the oocyte down the uterine tube where the sperm and oocyte make contact. If a sperm penetrates the oocyte's surrounding cells and membranes, then fertilization occurs and the DNA of the two cells unite to form a new individual called a zygote. Cell divisions give rise to the embryo, and ciliary actions and muscular contractions in the wall of the tube move the embryo into the uterus, the mammalian equivalent of a nest, where the remainder of development will occur.

Find more information about the reproductive system in

REAL ANATOMY

Female Reproductive Organs

The female genital organs consist of the internal genitalia and the external genitalia. The ovary, uterine tube, uterus, and vagina form the internal genitalia. These organs are responsible for production of the female gamete, the oocyte, and for nourishing, protecting, and delivering the new life that results from fertilization of the oocyte by the sperm. The external genitalia consist of the erectile tissues, glands, and folds of skin that proctect the entry into the female internal genitalia. These organs are the clitoris, vestibular glands, and labia majora and minora.

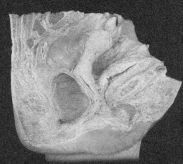

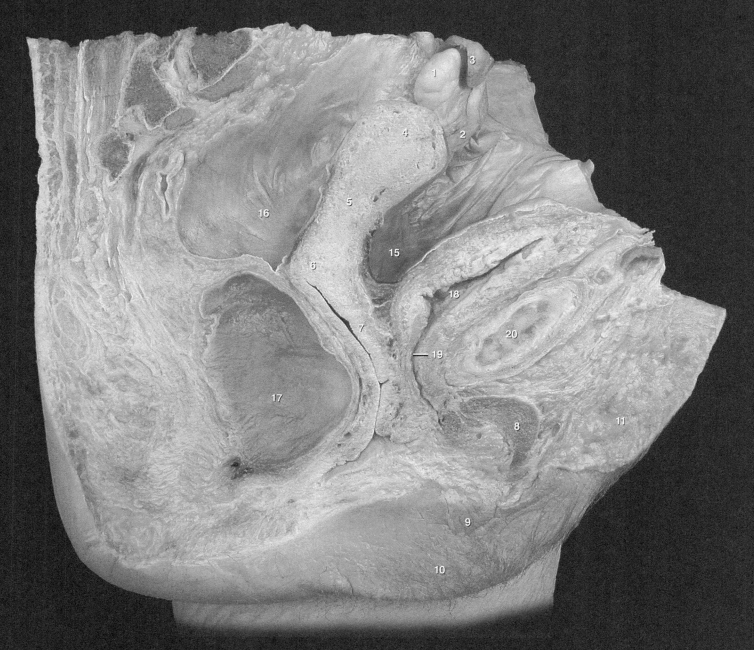

Sagittal section of female pelvis
Medial view

1 Ovary
2 Uterine tube
3 Fimbriae
4 Fundus of uterus
5 Body of uterus
6 Cervix of uterus
7 Vagina
8 Clitoris
9 Labia minora
10 Labia majora
11 Mons pubis
12 Broad ligament
13 Round ligament of uterus
14 Ovarian ligament
15 Vesicouterine pouch
16 Rectouterine pouch
17 Rectum
18 Bladder
19 Urethra
20 Pubic symphysis
21 Cecum
22 Sigmoid colon
23 Ileum
24 Mesentery

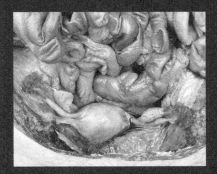

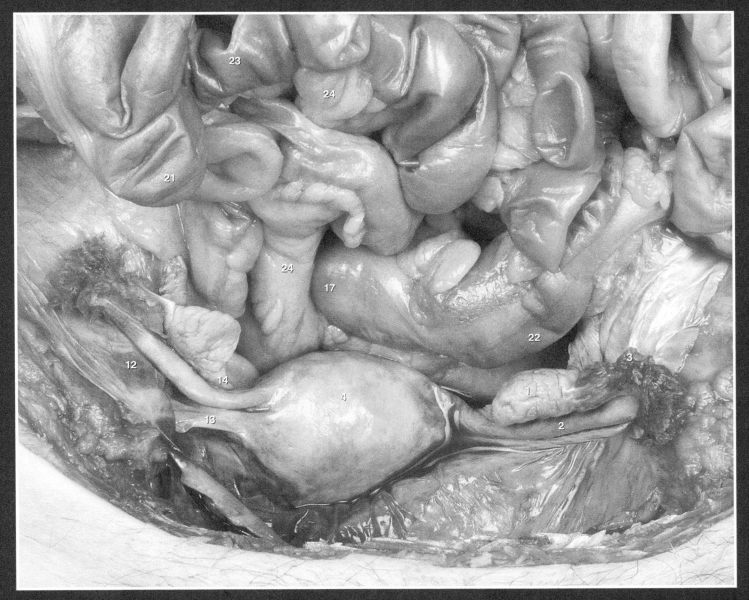

Dissection of female abdominoplevic cavity
Superoanterior view

Ovary

The ovaries are the site of oocyte, "egg," production in the female. These solid organs are approximately the size of an unshelled almond and project into the lower abdominal cavity at the boundary of the pelvis where they are covered and supported by folds of the peritoneum. During embryonic life, millions of oogonia, potential oocytes, surrounded by nursing follicular cells begin their development. Of these millions of cells only about 500 are ever ovulated during the female's reproductive life. The follicular cells not only nurse the ooytes, but also are the endocrine cells of the ovary that produce the estrogens and progesterone.

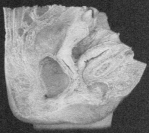

1 Ovary
2 Tunica albuginea
3 Primordial follicle
4 Granulosa cells
5 Theca folliculi
6 Zona pellucida
7 Primary oocyte
8 Seconary follicle
9 Follicular antrum
10 Corona radiata
11 Corpus luteum
12 Infundibulum of uterine tube
13 Ampulla of uterine tube
14 Isthmus of uterine tube
15 Fimbriae of uterine tube
16 Round ligament of uterus
17 Ovarian ligament
18 Uterus

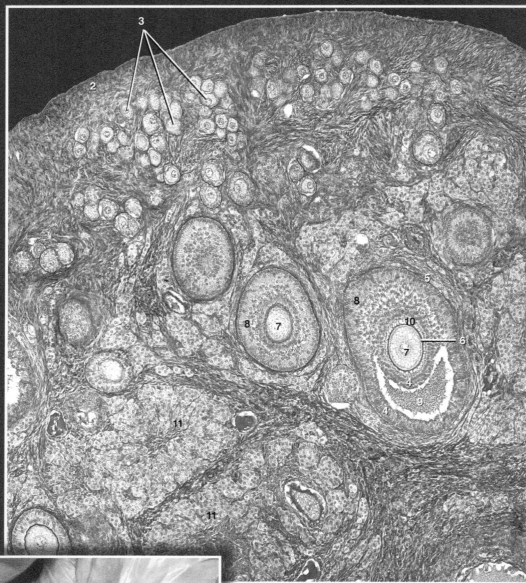

Photomicrograph of ovary
50x

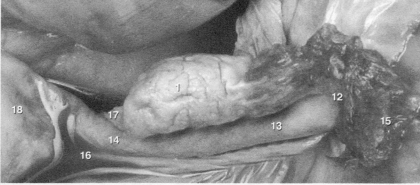

Ovary in situ
Anterior view

Uterus and Uterine Tubes

The uterine tubes, also called the oviducts or fallopian tubes, are suspended in the peritoneal fold, the broad ligament, along with the ovaries. In addition to transporting the oocyte toward the uterus, they are the site of fertilization of the oocyte by the sperm. The uterus is the thick smooth muscle organ that functions as the internal nest of mammalian animals. Note the vascular and glandular changes exhibited by the uterine endometrium as it progresses through the menstrual cycle.

1 Uterine tube
2 Fimbriae
3 Mesosalpinx
4 Fundus of uterus
5 Body of uterus
6 Cervix of uterus
7 Vagina
8 Mucosa of uterine tube
9 Muscularis of uterine tube

10 Peg cells
11 Ciliated columnar cells
12 Lamina propria
13 Perimetrium

Endometrium:
14 Stratum functionalis
15 Stratum basalis

Myometrium:
16 Inner longitudinal muscle
17 Middle circular muscle
18 Outer longitudinal muscle

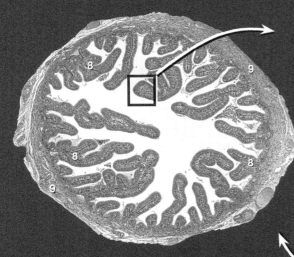

Photomicrograph of uterine tube
25x

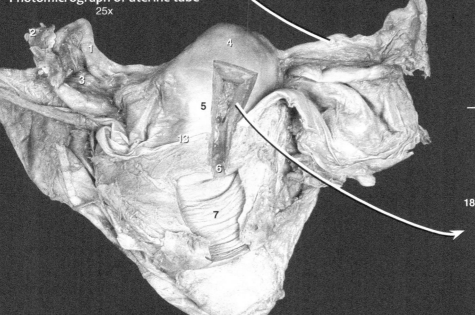

Photomicrograph of tunica mucosa of uterine tube
400x

Female internal genitalia
Anterior view

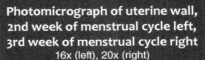

Photomicrograph of uterine wall, 2nd week of menstrual cycle left, 3rd week of menstrual cycle right
16x (left), 20x (right)

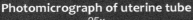

Vagina

The vagina, from the Latin word meaning sheath, is the receptacle for the penis during sexual intercourse, the birth canal, and the outlet for the menstrual flow. This muscular tube has a protective mucosal lining of stratified squamous epithelium. Approximately 10 cm (4 inches) in length, it expands at its superior end to form a cufflike wrapping around the cervix of the uterus. The caverns of the cufflike superior end are called the fonices, and it is in this region that the sperm are deposited during intercourse.

1 Vagina
2 Nonkeratinized stratified squamous epithelium of the mucosa
3 Lamina propria of the mucosa
4 Inner circular layer of tunica muscularis
5 Outer longitudinal layer of tunica muscularis
6 Adventitia
7 Fundus of uterus
8 Body of uterus
9 Cervix of uterus
10 Bladder
11 Urethra
12 Rectum
13 Rectouterine pouch
14 Vesicouterine pouch
15 Pubic symphysis
16 Clitoris

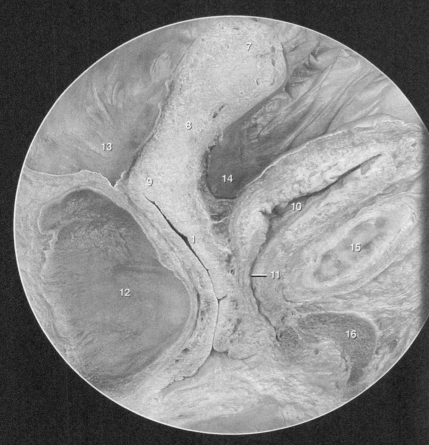

Sagittal section showing vagina in situ
Medial view

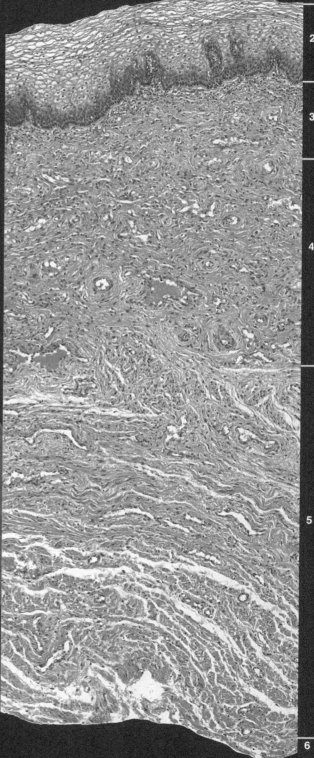

Photomicrograph of vaginal wall
25x

Female External Genitalia

Surrounding the openings of the vagina and ure-

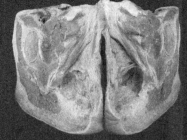

thra in the perineum of the female are the external genital structures. Bounding the openings on either side are the folds of skin called the labia majora and labia minora. Between these folds is the common entry way to both urethra and vagina, the vestibule. Deep to the labial skin are the erectile tissues of the female, the clitoris and bulb of the vestibule. The greater vestibular glands empty their lubricating secretions into the vestibule and opening of the vagina.

1 Body of clitoris	7 Ischiocavernosus muscle	13 Gluteus maximus muscle
2 Crura of clitoris	8 Bulbospongiosus muscle	14 Gluteus medius muscle
3 Bulb of vestibule	9 Ischioanal fossa	15 Ischium
4 Greater vestibular gland	10 Perineal membrane	16 Gracilis muscle
5 Vestibule	11 Deep perineal fascia	17 Adductor muscles
6 Transverse perenei superficialis	12 Head of femur	18 Femoral artery

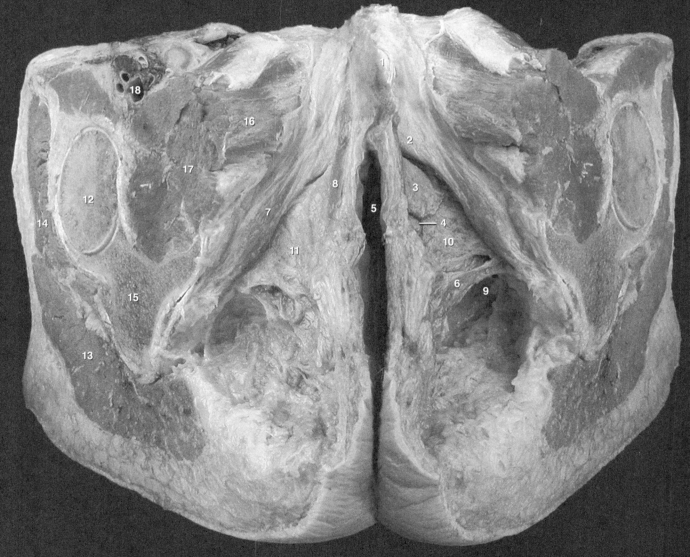

Perineal dissection revealing details of external genitalia
Inferior view

Male Reproductive Organs

Like the female, there are both internal and external genital organs in the male.

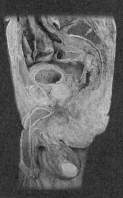

The major difference between the sexes is the enlargement of the erectile tissue organs of the male and the descent of the gonads, the testes, from an internal position to a suspended position outside the body cavity. The male genital organs include the testes suspended in the scrotum. The testes consist of an extensive tubular system that gives rise to the sperm, which then pass through the tubular ducts of egress — the rete testis, epididymis, ductus deferens, ejaculatory duct, and urethra — to exit from the male body. Accessory glands of the male join the ducts of egress and add secretions to the sperm, and the erectile intromittant organ, the penis, introduces the sperm into the female system.

1 Scrotum
2 Testis
3 Glans penis
4 Corpus cavernosum penis
5 Corpus spongiosum penis
6 Bulb of penis
7 Spongy urethra
8 Crus of penis
9 Bulbourethral gland
10 Prostate gland
11 Seminal vesicle
12 Bladder
13 Pubic symphysis
14 Rectus abdominis
15 Rectum
16 Sigmoid colon
17 Small intestine
18 Sacrum

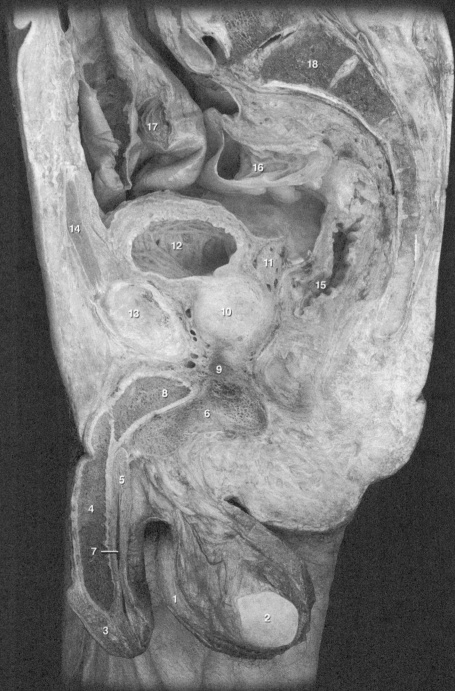

Parasagittal section of male pelvis
Medial view

Testis and Epididymis

The testes are the site of sperm production in the male. Unlike the solid, cellular ovaries, the testes are collections of small, highly coiled tubes, the seminiferous tubules. Beginning at puberty the spermatogonia, sperm stem cells, in the walls of the seminiferous tubules begin meiosis and produce hundreds of millions of sperm cells daily. From the testis the sperm are moved into the epididymis where they are stored and reach maturity prior to passing into the ductus deferens.

1 Coelom of testis
2 External spermatic fascia
3 Cremaster muscle
4 Tunica albuginea of testis
5 Epididymis
6 Seminiferous tubules
7 Rete testis
8 Spermatic cord
9 Spermatogonium
10 Primary spermatocyte
11 Secondary spermatocyte
12 Spermatid
13 Sertoli cell
14 Basement membrane
15 Interstitial cells (of Leydig)
16 Sperm in lumen of epididymis
17 Mucosa of epididymis
18 Stereocilia

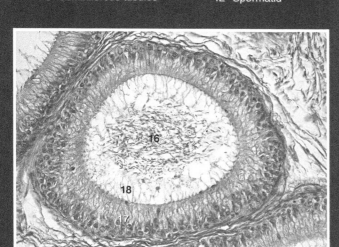

Photomicrograph of epididymis
200x

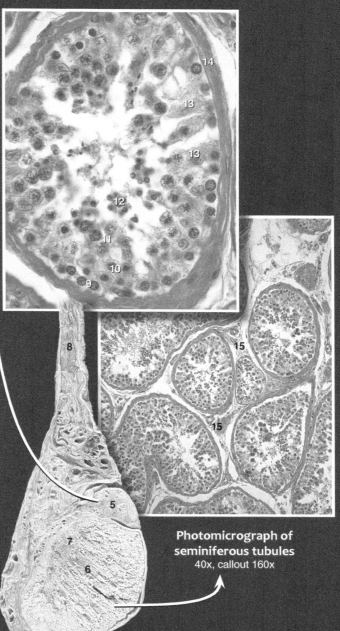

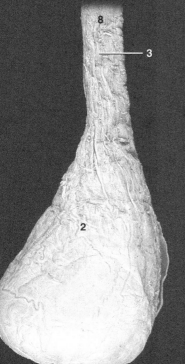

**Testis and spermatic cord
with fascial coverings**
Medial view

**Testis and spermatic cord
with fascia removed**
Medial view

**Sagittal section of testis and
spermatic cord**
Medial view

**Photomicrograph of
seminiferous tubules**
40x, callout 160x

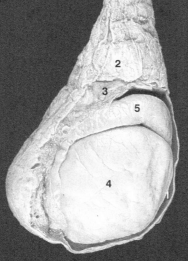

Ductus Deferens and Spermatic Cord

The ductus (vas) deferens is the muscular tube that transports sperm from the epididymis to the ejaculatory duct within the prostate gland. Peristaltic muscle contractions in the tube move the sperm. The ductus deferens accompanies the testicular vessels and nerves within a wrapping of fascia and muscle, called the spermatic cord. The cord extends from the testis to the superficial inguinal ring in the abdominal wall.

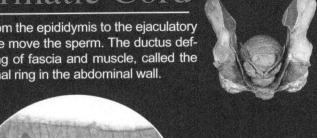

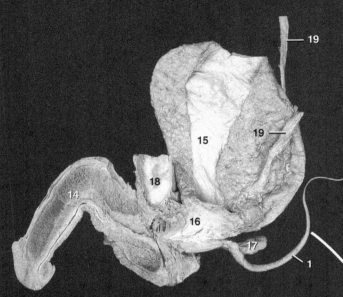

Dissection of male genital structures
Medial view

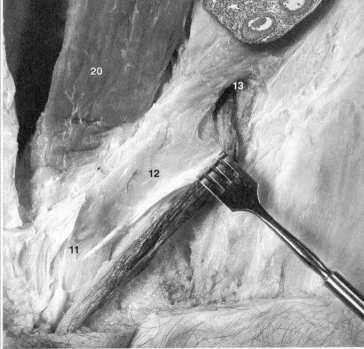

Photomicrograph of ductus deferens
30x, callout 400x

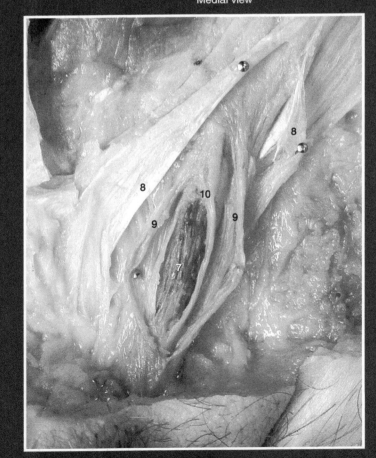

Dissection of spermatic cord exiting superficial inguinal ring
Anterior view

Dissection of inguinal canal and spermatic cord
Anterior view

1	Ductus deferens	10	Internal spermatic fascia	19	Ureter	28	Psoas major muscle
2	Pseudostratified columnar epithelium	11	Superficial inguinal ring	20	Rectus abdominis	29	Iliacus muscle
3	Lamina propria	12	Inguinal canal	21	Superior ramus of pubis (cut)	30	Sacrum
4	Inner longitudinal muscle layer	13	Deep inguinal ring	22	Inferior ramus of pubis (cut)	31	Levator ani muscle
5	Middle circular muscle layer	14	Penis	23	Body of pubis (cut)	32	Sciatic nerve
6	Outer longitudinal muscle layer	15	Bladder	24	Pudendal nerve and vessels	33	Testis
7	Testicular blood vessels	16	Prostate gland	25	Rectum (enlarged)	34	Obturator internus muscle
8	External spermatic fascia	17	Seminal vesicle	26	Internal iliac artery	35	Tendinous arch of levator ani
9	Cremaster fascia	18	Pubic symphysis	27	External iliac artery (cut)	36	Ampulla of ductus deferens

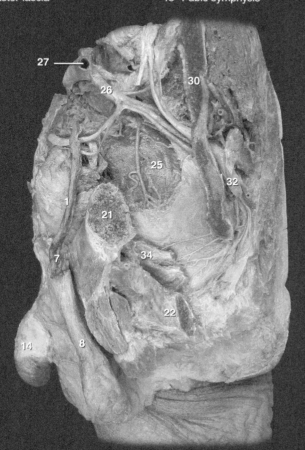

Lateral dissection of male pelvis
Lateral view

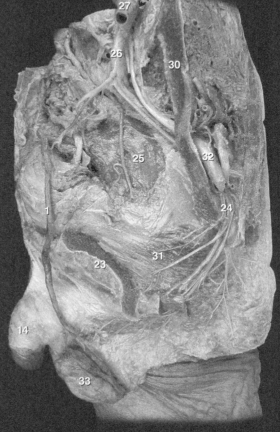

Lateral disscetion of male pelvis
Lateral view

Dissection of male pelvic cavity
Superior view, bladder removed

Male Accessory Glands

Associated with the male ducts of egress are three glands, often referred to as the accessory sex glands of the male. The three named glands are the paired seminal glands (vesicles), the unpaired prostate gland, and the paired bulbourethral glands. They arise as epithelial outgrowths of terminal end of the male ducts of egress at the base of the bladder. They produce secretions that protect and nourish the sperm.

1 Seminal vesicle
2 Prostate gland
3 Bulbourethral gland
4 Secretory epithelium
5 Trabecula
6 Blood vessel

7 Bladder
8 Ductus deferens
9 Ampulla of ductus deferens
10 Rectum
11 Pubic symphysis
12 Bulb of penis

13 Crus of penis
14 Ilium
15 Ischial tuberosity
16 Obturator internus muscle
17 Levator ani muscle
18 Deep transverse perineal muscle

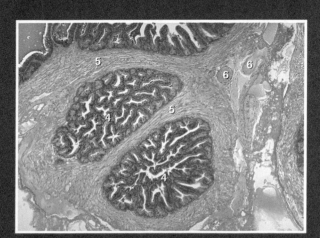

Photomicrograph of seminal vesicle
50x

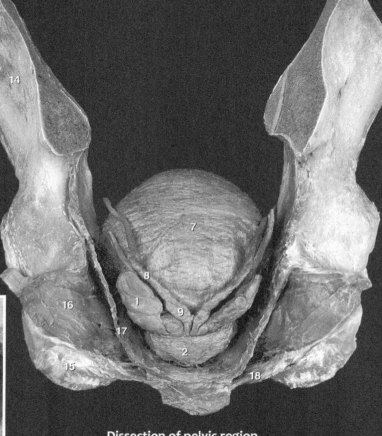

Dissection of pelvic region
Posterior view

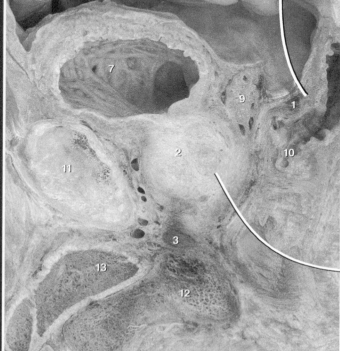

Parasagittal section revealing prostate and bulbourethral glands
Medial view

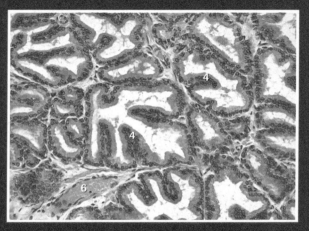

Photomicrograph of prostate gland
200x

Penis

The penis is the intromittent organ of the male external genitalia through which the long urethra, in comparison to the female, courses as it transports both urine and semen from the male body. Along with the urethra, the penis consists of three masses of erectile tissue. On the dorsal aspect of the body of the penis are the paired corpora cavernosae. These erectile tissue bodies are the principal tissues of penile erection. At the base of the penis each corpus cavernosum extends laterally to form the crura of the penis. Each crus attaches to the inferior pubic ramus. On the ventral aspect of the penis is the slender unpaired corpus spongiosum, which surrounds the spongy urethra. The corpus spongiosum expands distally as the glans penis, which forms the expanded tip of the penis. It expands proximally to form the bulb of the penis in the perineum beneath the prostate gland. The glans is covered by a hood of skin, the prepuce, which can be removed via circumcision.

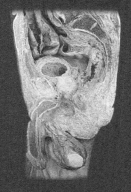

1 Glans penis
2 Corpus cavernosum penis
3 Corpus spongiosum penis
4 Crus of penis
5 Bulb of penis
6 Spongy urethra
7 Deep dorsal vein
8 Tunica albuginea of corpus spongiosum
9 Tunica albuginea of corpus cavernosum
10 Deep (cavernous) artery of penis
11 Intermediate (membranous) urethra
12 Prostatic urethra
13 Ampula of ductus deferens
14 Pubic symphysis
15 Testis
16 Ejaculatory duct
17 Bladder
18 Suspensory ligament of penis

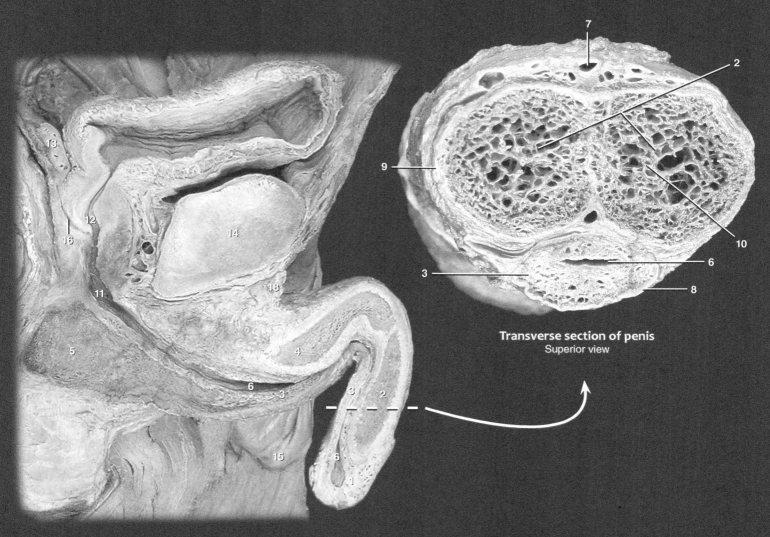

Transverse section of penis
Superior view

Sagittal section of penis in situ
Medial view

Index

fossa ovalis 264
fourth ventricle 224, 227, 236, 241,242, 245, 248, 250, 252, 291
fovea for ligament of head 109, 135
friction ridges 21
frontal angle 52
frontal belly of occipitofrontalis 145, 146, 148, 155, 227
frontal bone 36, 38, 42, 47, 49
frontal border 52
frontal crest 50
frontal lobe of cerebrum 155, 224, 234, 236, 242, 245, 250, 272, 291
frontal notch or foramen 50
frontal process of maxilla 60
frontal process of zygomatic bone 65
frontal tuber 50
fundus of stomach 305
fundus of uterus 323, 325, 326

G
galea aponeurotica 227
gallbladder 257, 301, 305, 308, 309, 311, 312
gastric glands 305
gastric pit 305
gastric rugae 305
gastrocnemius 136, 142, 197, 202, 203, 205, 206, 207
gastroduodenal artery 281
geniculate ganglion 224
genioglossus 151, 155
geniohyoid 151, 155, 168, 229
genital branch of genitofemoral nerve 218
genitofemoral nerve 218
gingiva 124
glabella 50
glandular lumen 8
glans penis 328, 333
glenoid cavity 88, 133
glenoid labrum 133
globus pallidus 242
glomerulus surrounded by urinary tubules 317
glossopharyngeal nerve 224, 241
gluteal muscles 178
gluteal tuberosity 109
gluteus maximus 142, 159, 173, 174, 197, 199, 203, 218, 284, 327
gluteus medius 135, 170, 173, 199, 202, 203, 218, 220, 327
gluteus medius 199
gluteus minimis 135, 170, 173, 199, 201, 218
goblet cell 307, 310
gracilis 197, 199, 201, 202, 203, 284, 327
granular foveolae 49
granulosa cells 324
gray communicating ramus 222
great auricular nerve 226
great auricular nerve 216
great cardiac vein 263, 270
great saphenous vein 284, 286, 288
greater cornu of hyoid bone 153
greater curvature of stomach 305
greater horn of hyoid bone 72
greater occipital nerve 215, 221, 226
greater omentum 301, 305, 311, 312
greater palatine foramen 44
greater palatine groove 60, 66
greater palatine nerve 226
greater sciatic notch 105
greater splanchnic nerve 222
greater trochanter 109, 135
greater tubercle 90, 180, 182, 186, 187
greater tympanic spine 56
greater vestibular gland 327
greater wing 58
groove for extensor muscle tendons 94
groove for fibularis longus 117, 118
groove for flexor hallucis longus 116, 117
groove for occipital sinus 54
groove for popliteus 109
groove for radial nerve 90

groove for sigmoid sinus 47, 52, 56
groove for superior sagittal sinus 50, 52, 54
groove for transverse sinus 54
groove for ulnar nerve 90
groove for vertebral artery 74
grooves for middle meningeal artery 52
growth plate 27
gustatory hair 230
gustatory receptor cell 230
gyrus 234

H
hair 22, 23
hair follicle 23
hamate 97, 99
handle of malleus 71
hard palate 155, 227, 291, 303
haustra 310
head of rib 81
head of ulna 92
head of femur 135, 327
head of malleus 71
head of mandible 62
head of metacarpal 101
head of pancreas 308
head of phalanx 101
head of radius 134
head of stapes 71
heart 268, 301, 311, 312
helicis major 147, 149
helicis minor 147, 149
hemi-azygos vein 278
hepatic artery 309
hepatic portal vein 281, 282, 309
hepatic sinusoid 309
hepatic vein 278, 281
hepatocytes 309
hepatoduodenal ligament of lesser omentum 312
hepatogastric ligament of lesser omentum 312
hepatorenal part of coronary ligament 312
hiatus for greater petrosal nerve 56
hiatus for lesser petrosal nerve 56
hilum of lung 297
hilum of kidney 314, 316
hook of hamate or hamulus 99
horizontal plate 66
humerus 30, 32, 84, 133, 184, 187
hyaline cartilage of tracheal ring 295
hyaline ground substance 12
hyoglossus 149, 151, 155
hyoid bone 34, 151, 155, 156
hypoglossal canal 54
hypoglossal nerve 216, 224, 228, 241
hypoglossal tubercle 241
hyponychium 22
hypophysial fossa 58
hypothalamus 236, 245, 250, 251
hypothenar muscles 177, 178

I
ileal arteries 281
ileocolic artery 281
ileum 281, 301, 307, 310, 323
iliac crest 106, 164
iliac fossa 106
iliac tuberosity 106
iliocostalis cervicis muscle 160
iliocostalis lumborum muscle - lumbar part 160 168
iliocostalis lumborum muscle - thoracic part 160
iliocostalis muscle 159, 164, 220
iliohypogastric nerve 218
iliohypogastric nerve 220
ilioinguinal nerve 218
iliopsoas muscle 170
iliotibal tract 178, 197, 199, 201, 203
ilium 106, 199, 203, 332

impression for costoclavicular ligament 86
incisive canal 47, 60
incisive fossa 44
incus 71
inferior alveolar nerve 228
inferior angle of scapula 88, 182, 186
inferior articular process/facet 74, 76, 77
inferior articular surface 111
inferior border of nasal bone 68
inferior cardiac plexus 222
inferior cerebellar peduncle 241
inferior cluneal nerve 218, 221
inferior colliculus 236, 241, 245, 252
inferior compartment of articular cavity 132
inferior costal facet 76
inferior epigastric vessels 170, 201
inferior frontal gyrus 234, 238
inferior gemellus 173, 199, 202, 203, 218
inferior gluteal artery 282, 284
inferior gluteal line 106
inferior gluteal nerve 218
inferior lateral brachial cutaneous nerves 221
inferior lateral genicular artery 286
inferior lobe 297
inferior longitudinal muscle 151, 155
inferior medial genicular artery 286
inferior mesenteric artery 269, 281, 282, 320
inferior mesenteric vein 281, 282
inferior nasal concha 36, 40, 47, 155, 227, 291
inferior nasal meatus 36, 40, 47
inferior nuchal line 54
inferior oblique 149
inferior orbital fissure 40
inferior parathyroid gland 255
inferior pharyngeal constrictor 147, 149, 151, 153, 155, 229
inferior pubic ramus 107, 331
inferior rectus 149, 155
inferior sagittal sinus 272
inferior temporal gyrus 238
inferior temporal line 52
inferior thyroid artery 275
inferior trunk of brachial plexus 217
inferior vena cava 126,173, 223, 256, 263, 268, 270, 278, 281, 282, 293, 301, 309, 314, 320
inferior vermis 240
inferior vertebral notch 74, 76, 77
inferior vesical artery 282
infra-orbital foramen 60
infra-orbital groove 60
infraglenoid tubercle 88
infraorbital nerve 226, 228
infrapatellar bursa 136, 138
infrapatellar fat pad 136
infraspinatus 160, 163, 173, 178, 180, 182, 184, 187
infraspinous fossa 88
infratemporal crest 58
infundibulum of pituitary gland 224, 241, 251
infundibulum of uterine tube 324
inguinal canal 331
inguinal ligament 170, 177, 201
inner circular layer of tunica muscularis 326
inner lip of crest of ilium 106
inner longitudinal muscle 325, 331
innermost intercostal muscle 167, 170, 173, 220, 222, 269, 278
insular lobe 224, 242
interalveolar septum 60
interatrial septum 264
intercalated disc 15
interchondral (synchondrosis) 126
interchondral (synovial) 126
intercondylar eminence 111
intercondylar fossa 109
intercostal muscle 293
intercostal nerve 222

intermediate (membranous) urethra 333
intermediate cell layer 8
intermediate cuneiform 114
intermediate sacral crest 78
intermediate zone of crest of ilium 106
intermetacarpal muscle 178
internal oblique muscle 170, 173
internal acoustic meatus 56
internal capsule 242
internal carotid artery 216, 224, 242, 245, 271, 272
internal elastic membrane of tunica intima 266
internal iliac artery 269, 282, 284, 331
internal intercostal muscle 159, 160, 163, 164, 167, 168, 170, 173, 180, 184
internal jugular vein 213, 229, 242, 278
internal lamina of rectus sheath 170
internal oblique muscle 167, 215, 222
internal occipital protuberance 54
internal pudendal artery 282, 284
internal spermatic fascia 331
internal table of calvaria 47
internal thoracic artery 275
internal thoracic vein 275
interosseous border of tibia 111, 112
interosseous border of ulna 92
interosseous border of radius 94
interosseous membrane 19, 124, 188, 205, 206, 276
interspinales lumborum muscle 164
interspinales thoracis muscle 164
interspinous ligament (vertebral syndesmosis) 126
interstitial (leydig) cell 259, 329
interthalamic adhesion 236
intertransversarii laterales lumborum dorsal part 164
intertransversarii laterales lmborum ventral part 164
intertransversarii muscle 159
intertransversarii thoracic muscle 215
intertrochanteric crest 109
intertrochanteric line 109
intertubercular sulcus or groove 90
interventricular foramen 245
intervertebral disc 126, 155, 251, 291
intervertebral foramen 73
intestinal glands 310
intestine 135, 314
ischial ramus 107
ischial spine 107
ischial tuberosity 107, 174, 332
ischioanal fossa 327
ischiocavernosus muscle 174, 327
ischiococcygeus muscle 174
ischiopubic ramus 105
ischium 107, 327
isthmus of thyroid gland 254, 255
isthmus of uterine tube 324

J
jejunal arteries 281
jejunum 301, 307
joint (articular) capsule 132, 134, 135
joint cavity 128
jugular foramen 44, 49
jugular notch 54, 56
jugular or suprasternal notch 82
jugular process 54
jugular tubercle 54
jugum 58
junction of periosteum (removed) with fibrous membrane 128
junction of synovial membrane (removed) with articular cartilage 128
K
kidney 222, 256, 282, 301, 314, 320
L
labia majora 318, 323
labia minora 323
lacrimal bone 36, 38
lacrimal fossa 50

lacrimal groove 60, 70
lacrimal hamulus 70
lacrimal process 69
lacuna 12, 26
lambdoid border 54
lamella 26
lamellated corpuscle 230
lamina of vertebra 74, 76, 77, 126
lamina propria 302, 304, 305, 307, 310, 325, 331, 326
laryngopharynx 213, 291, 303
larynx 213
lateral antebrachial cutaneous nerve 221
lateral border of nasal bone 68
lateral border of scapula 88
lateral branch 215
lateral branches of interventricular artery 264
lateral cerebral sulcus 234
lateral condyle of femur 109
lateral condyle of tibia 111
lateral cord of brachial plexus 217
lateral crico-arytenoid 156
lateral cuneiform 114, 119
lateral cutaneous branch of subcostal nerve 221
lateral cutaneous branches (dorsal rami) 221
lateral cutaneous branches (ventral rami) 221
lateral epicondyle 90, 109
lateral femoral cutaneous nerve 218, 221
lateral funiculus of white matter 232
lateral horn of gray matter 232
lateral intercondylar tubercle 111
lateral malleolar facet 116
lateral malleolus 112, 206
lateral mass 74
lateral meniscus 136
lateral pectoral nerve 217
lateral plate of pterygoid process 58
lateral process of malleus 71
lateral process of talus 116
lateral pterygoid 148, 153, 229, 242, 271
lateral recess 241
lateral rectus 149, 155, 224
lateral sacral artery 282
lateral sacral crest 78
lateral supracondylar line 109
lateral sural cutaneous nerve 221
lateral surface of zygomatic bone 65
lateral surface of inferior nasal concha 69
lateral thalamic nucleus 242
lateral thoracic artery 275
lateral tubercle 116
lateral ventricle 236, 242, 245, 248, 250, 291
latissimus dorsi 142, 160, 163, 164, 173, 178, 180, 184, 217, 275
least occipital nerve 215
left atrium 63, 264, 270
left auricle 263,264
left axillary artery 269
left brachial artery 269
left colic (splenic) flexure 310
left colic artery 281, 282
left common carotid artery 223
left common carotid artery 223, 263, 264, 268, 269, 270, 278
left coronary artery 264, 268
left gastric artery 269, 281, 320
left gastro-omental artery 281
left inferior pulmonary vein 263
left inferior pulmonary vein 268
left lobe of liver 309
left lobe of thyroid gland 254, 255
left lung 253, 293
left main (primary) bronchus 268, 295, 298
left pulmonary artery 263, 264, 268
left pulmonary veins 264
left radial artery 269
left radial recurrent artery 269

left renal artery 269
left subclavian artery 223, 263, 264, 269, 270, 278
left superior pulmonary vein 263, 268
left suprarenal gland 256
left ulnar artery 269
left ventricle 263, 264, 270
lenticular process 71
lesser curvature of stomach 305
lesser horn of hyoid 72
lesser occipital nerve 216, 226
lesser omentum 301, 312
lesser palatine foramina 66
lesser palatine nerve 226
lesser sciatic notch 107
lesser splanchnic nerve 222
lesser trochanter 109
lesser tubercle 90, 186
lesser tympanic spine 56
lesser wing 58
leukocyte - monocyte (white blood cell) 262
leukocyte - neutrophil (white blood cell) 262
leukocyte or white blood cell (wbc) - monocyte 14
leukocyte or white blood cell (wbc) - neutrophil 14
levator anguli oris 145, 146, 149
levator ani muscle 174, 331, 332
levator labii superioris 145, 146, 149
levator labii superioris alaeque nasi 145, 146, 149
levator palpebrae superioris 149, 155, 224
levator scapulae 147, 149, 167, 168, 180, 184, 216, 217
levator veli palatini 151, 153
levatores costarum muscle 159, 160, 163, 164, 215
ligament of head of femur 135
ligamentum arteriosum 223, 263, 264, 269, 270
limen 238
linea alba 170, 177, 184, 201
linea aspera 109
lingual artery 271
lingual nerve 228
lingula 240
lipid storage area 10
lips 303
liver 257, 281, 301, 305, 307, 308, 312, 314
lobar (secondary) bronchus 223, 295, 298
locus ceruleus 241
long ciliary nerve 224
long gyrus 238
long limb of incus 71
long plantar ligament 209
long thoracic nerve 217
longissimus capitis muscle 160
longissimus cervicis muscle 160
longissimus muscle 159
longissimus thoracis muscle 160
longitudinal collagen fibers 21
longitudinal fissure 234
longus capitis 155
longus capitis muscle 167, 173
longus colli 147, 149, 167, 168, 173
loose connective tissue of stratum papillare 21
lower subscapular nerve 217
lumbar lordosis 73
lumbar vertebra 34, 73, 314
lumbar vertebral column 34
lumbosacral dorsal rootlets 214
lumbosacral trunk 218
lumbrical muscles 138, 188, 193, 194, 209
lunate 97, 99
lunate surface 105
lung 223, 264, 301, 312
lunula 22

M
major calyx 316
major duodenal papilla 308
malleolar articular facet 111

malleolar fossa 112
malleolar groove 111, 112
malleus 71
mammillary bodies 224, 236, 241, 250
mammillary process 77
mandible 36, 38, 40, 47, 124, 151, 155, 291
mandibular condyle 132, 242
mandibular foramen 62
mandibular fossa 56
mandibular notch 62
mandibular ramus 132
manubriosternal synchondrosis 126
manubrium 82
marginal artery 281, 282
masseter 132, 145, 146, 148, 153, 155, 177, 227, 291, 303
masseteric tuberosity 62
mast cell 10
mastoid air cells 132, 242
mastoid angle 52
mastoid border 54
mastoid canaliculus 56
mastoid notch 56
mastoid process 56, 132, 151
maturing t cells 253
maxilla 36, 38, 40, 42, 44, 47, 151
maxillary artery 271
maxillary branch 224
maxillary branch 226
maxillary process 69
maxillary sinus 47
maxillary sinus 60, 155, 227, 291
maxillary tuberosity 60
medial antebrachial cutaneous nerve 221
medial border of nasal bone 68
medial border of scapula 88, 182
medial condyle 109, 111
medial cord of brachial plexus 217
medial cuneiform 114, 119
medial cutaneous branches (dorsal rami) 221
medial eminence 241
medial epicondyle 90, 109
medial geniculate ganglion 241
medial geniculate nucleus 252
medial intercondylar tubercle 111
medial lumbar intertransversarii muscle 163, 164
medial malleolar facet 116
medial malleolus 111
medial meniscus 136
medial pectoral nerve 217
medial plate of pterygoid process 58
medial pterygoid 148, 153, 229, 242
medial rectus 155
medial supracondylar line 109
medial supracondylar ridge 90
medial surface of inferior nasal concha 69
medial thalamic nucleus 242
medial thigh muscles 177
medial tubercle of talus 116
median antebrachial vein 276
median aperture 240, 245
median cubital vein 276
median nerve 217
median sacral crest 78
mediastinal pleura 173
medulla oblongata 214, 224, 227, 234, 236, 240, 245, 250, 251, 252, 291
medullary cavity 19, 29
meniscus 128
mental foramen 62
mental nerve 226
mental protuberance 62
mental spines 62
mental tubercle 62
mentalis 145, 146, 149, 151, 155
mesenteric fat 314

mesentery 301, 307, 323
mesosalpinx 325
metacarpal bones 27, 30, 32, 84, 97
metaphysis 29
metatarsal bones 30, 32, 104, 114
microvilli 6
microvillus brush border 307
midbrain 227, 234, 236, 240, 245, 250, 251
middle cardiac vein 263, 270
middle cerebellar peduncle 241, 245
middle cerebral artery 242, 271, 272
middle circular muscle 325, 331
middle clinoid process 58
middle colic artery 281, 282
middle cranial fossa 49
middle cuneiform 119
middle facet for calcaneus 116
middle frontal gyrus 238
middle lobe of lung 297
middle meningeal artery and branches in dura mater 247
middle nasal concha 64, 155, 227, 291
middle nasal meatus 19, 40
middle phalanx of foot 114
middle phalanx of hand 97
middle pharyngeal constrictor 147, 149, 151, 153, 155
middle rectal artery 282
middle scalene 147, 149, 160, 167, 184, 216, 217, 275
middle superior alveolar nerve 228
middle talar articular surface 117
middle temporal gyrus 238
middle trunk of brachial plexus 217
minor calyx 316
mons pubis 323
mucosa of epididymis 329
mucosa of tongue 151
mucosa of uterine tube 325
mucous acini 303
mucous in goblet cell 6
mucous neck cell 305
multifidus cervicis muscle 163
multifidus lumborum muscle 163
multifidus muscle 159, 160, 164
multifidus thoracis muscle 163
muscle belly or body 140
muscle cell or fiber 140
muscles of facial expression 177, 178
muscles of mastication 178
muscular branches of femoral 284
muscularis mucosae 302, 304, 310
muscularis of uterine tube 325
musculocutaneous nerve 217
musculotubal canal 56
musculus uvulae 153, 155
myelin sheath 213
mylohyoid 145, 146, 151, 155, 168, 229
mylohyoid line 62
myocardium 264
myometrium 325

N

nail 22
nail bed 22
nasal bone 36, 38, 42, 47
nasal cavity 291
nasal foramina 68
nasal septum 251, 291
nasal spine 50
nasalis 145, 146, 148
nasociliary nerve 224
nasopalatine nerve 226
nasopharynx 252, 291, 303
navicular 114, 118
navicular articular surface 116
neck of rib 81
neck of scapula 88

neck of femur 109
neck of fibula 112
neck of malleus 71
nerve 266
nerve in perimysium 140
nerve of the pterygoid canal 226
nerve to geniohyoid muscle 216
nerve to inferior omohyoid muscle 216
nerve to mylohyoid muscle 228
nerve to sternohyoid muscle 216
nerve to sternothyroid muscle 216
nerve to superior omohyoid muscle 216
nerve to temporalis muscle 228
nerve to the obturator internus muscle 218
nerve to the subclavius muscle 217
nerve to thyrohyoid muscle 216
neurohypophysis 251, 252
neuron 230
nodulus 240
nonkeratinized stratified squamous epithelium of the mucosa 326
nuchal ligament 159, 160, 163
nuchal ligament (vertebral syndesmosis) 126
nucleus of adipose cell 10, 24
nucleus of fibroblast 10
nucleus of glial cell 16
nucleus of multipolar neuron 16
nucleus of osteocyte 26
nucleus of reticular cell 10
nucleus pulposus of intervertebral disc 126

O

obex 241
oblique arytenoid 153, 156
oblique fissure 297
oblique line 62
oblique popliteal ligament 136
oblique vein 270
obliquus capitis inferior muscle 163
obliquus capitis superior muscle 163
obliquus inferioris muscle 215
obliquus superioris muscle 215
obliterated umbilical artery 282
obturator artery 282
obturator externus 135, 174, 199, 201, 202, 218
obturator foramen 105, 107
obturator groove 107
obturator internus 135, 173, 199, 202, 203, 218, 331, 332
obturator nerve 174, 218, 282
occipital angle 52
occipital artery 271
occipital belly of occipitofrontalis 24, 147, 148
occipital bone 36, 38, 40, 42, 44, 47, 49, 155, 251, 252
occipital border 52
occipital condyle 54, 155, 242
occipital lobe 224, 234, 236, 245, 250, 291
occulomotor nerve 224, 241
olecranon 92, 134
olecranon fossa 90
olfactory bulb 224
olfactory nerve 224
olive 241, 242
omental or fatty appendices 310, 311
omohyoid 147, 149, 167, 168, 170, 180, 184, 216, 229
opening of straight sinus 214, 272
opponens digiti minimi 193, 194
opponens pollicis 193, 194
opthalmic branch 224
optic canal 58
optic chiasm 224, 241, 242, 248, 271
optic nerve 155, 224, 241
optic tract 224, 236, 241
oral cavity 291
orbicularis oculi 145, 146, 148, 151, 227, 228
orbicularis oris 145, 146, 149, 151, 155
orbit 36, 291

orbital plate 64
orbital process 66
oropharynx 291, 303
os coxae 30, 32, 135
osteocyte 12
osteon 26
outer lip of crest of ilium 106
outer longitudinal layer of tunica muscularis 326
ovarian ligament 323, 324
ovary 258, 323, 324
oxyphil cell 255

P

palatine bone 36, 40, 44, 47
palatine process 60
palatine tonsil 153
palatopharyngeus 153, 155
palmar aponeurosis 177, 188, 193
palmar interossei 193, 194
palmaris brevis 188, 193, 194
palmaris longus 188, 193
pancreas 257, 281, 301
pancreatic duct 257, 308
pancreatic ductule 308
pancreatic islet 257, 308
papillary muscle 264
parafollicular (c) cell 254
parietal bone 36, 38, 40, 42, 47, 49
parietal foramen 52
parietal lobe 224, 234, 236, 245, 250, 291
parietal pericardium 173, 264
parietal peritoneum 258
parietal tuber 52
parieto-occipital sulcus 238
parotid duct 227, 303
parotid gland 132, 145, 147, 149, 216, 303
patella 30, 32, 104, 128, 136
patellar ligament 124, 128, 136, 205, 206
patellar surface of femur 109
pecten pubis or pectineal line 107
pectinate muscle 264
pectineal or spiral line 109
pectineus 170, 197, 199, 201, 202
pectoral artery 275
pectoralis major 142, 170, 177, 178, 180, 184, 217, 275, 276
pectoralis minor 180, 184, 217, 275
pedicle 74, 76, 77
peduncle 252
peg cells 325
pelvic diaphragm 199, 202, 203
penis 174, 177, 199, 201, 202, 218, 284, 288, 311, 318, 331
perforating cutaneous nerve 218
perichondrium 12
perimetrium 325
perimysium 140
perineal body 174
perineal membrane 327
perineurium 213
periodontal membrane 124
periorbital fat 155
periosteum 19, 128, 136, 140
perirenal fat 314
perpendicular plate of palatine 66
perpendicular plate of ethmoid 64
petro-occipital fissure 49
petrosphenoidal fissure 49
petrotympanic fissure 56
petrous part of temporal bone 56
phalanges of hand 30, 32
phalanges of foot 84, 104
pharyngeal branch 226
pharyngeal tubercle 54
pharyngobasilar fascia 153
pharyngotympanic tube 153, 155
phrenic nerve 216, 217, 269, 275, 293

splenius cervicis muscle 159
spongy urethra 318, 328, 333
squamous part of temproal 50, 54, 56
stapes 71
stereocilia 329
sternal angle 82
sternal end 86
sternal facet 86
sternocleidomastoid 147, 149, 168, 170, 177, 178, 180, 184, 216,229, 242, 303
sternocostal (synchondrosis) 126
sternocostal (typically synovial but can be symphysial) 126
sternohyoid 147, 149, 167, 168, 170, 184, 216
sternothyroid 167, 168, 170, 184, 216
sternum 34, 173, 311
stomach 281, 282, 293, 301, 304, 305, 307, 310, 311, 312
straight sinus 224, 248
stratified squamous epithelium 19
stratum basale 8, 20
stratum basalis 325
stratum corneum 8, 20
stratum functionalis 325
stratum granulosum 8, 20
stratum lucidum 8, 20
stratum spinosum 8, 20
striated skeletal muscle 266
styloglossus 147, 149, 151, 153, 229
stylohyoid 147, 149, 151, 153, 168, 229
styloid process of temporal bone 56
styloid process of third metacarpal 101
stylomastoid foramen 56
stylopharyngeus 151, 153
subchondral bone 29
subclavian artery 168, 254, 275, 293
subclavian groove 86
subclavian vein 278
subclavius 180, 217
subcostal muscle 173, 222, 278
subcostal nerve 218, 220
subcutaneous layer 19, 24, 133, 135149, 286
submandibular fossa 62
submandibular ganglion 228
submandibular gland 147, 149, 227, 303
submucosal (brunner's) glands 302
subscapular artery 275
subscapular fossa 88
subscapularis 133, 182, 186, 217, 275
sulcus limitans 241
sulcus tali 116
superciliary arch 50
superficial circumflex iliac artery 286
superficial circumflex iliac vein 286
superficial epigastric artery 286
superficial epigastric vein 286
superficial external anal sphincter muscle 174
superficial inguinal lymph node 288
superficial inguinal ring 331
superficial middle cerebral vein and tributaries in subarachnoid space 247
superficial palmar arch 276
superficial temporal artery 271
superficial transverse metacarpal ligament 188
superficial transverse perinei muscle 174
superficial veins 24
superior angle of scapula 24, 88, 182, 186
superior articular process 78
superior articular process/facet 76, 77
superior cerebellar peduncle 241, 242
superior cervical ganglion 272
superior cluneal nerves 221
superior colliculus 236, 241, 245, 252
superior compartment of articular cavity 132
superior costal facet 76
superior frontal gyrus 238
superior gemellus 173, 199, 202, 203, 218
superior gluteal artery 282, 284

superior gluteal nerve 218
superior laryngeal nerve 228
superior lateral brachial cutaneous nerves 221
superior lateral genicular artery 286
superior lobe of lung 297
superior longitudinal muscle 155
superior medullary vellum 240, 241
superior mesenteric artery 222, 256, 269, 281, 282, 320,
superior mesenteric artery 256
superior mesenteric ganglion 222
superior mesenteric vein 281, 282
superior mesenteric vein and tributaries 311
superior nasal concha 64, 155, 291
superior nuchal line 54
superior oblique 149, 155, 224
superior orbital fissure 58
superior pancreaticoduodenal artery 281
superior parathyroid gland 255
superior pharyngeal constrictor 151, 153, 155, 229
superior posterior lateral nasal branch 226
superior pubic ramus 107
superior rectal artery 281
superior rectus 149, 155, 224
superior sagittal sinus 214, 224, 242, 248, 272
superior temporal gyrus 234, 238
superior temporal line 52
superior thoracic artery 275
superior thyroid artery 271
superior trunk of brachial plexus 216, 217
superior vena cava 222, 263, 264, 268, 270, 278
superior vermis 240, 242
superior vertebral notch 74, 76, 77
superior vesical artery 282
supinator 188, 190
supinator crest 92
supporting cell 230
supra-acetabular groove 106
supra-orbital notch or foramen 50
supraclavicular nerve 216, 221
supraglenoid tubercle 88
supraorbital nerve 226
suprapatellar bursa 128, 136, 138
suprarenal gland 222, 281
suprarenal vein 281
suprascapular artery 275
suprascapular nerve 217, 275
suprascapular notch 88
supraspinatus 133, 180, 182, 184, 186, 187
supraspinous fossa 88
supraspinous ligament 159, 164
suprastyloid crest 94
surface mucous cell 305
surgical neck 90
suspensory ligament of penis 333
sustentaculum tali 117
sutural bone 40
sweat glands in dermis 21
sympathetic trunk 222, 278, 282
sympathetic trunk ganglion 222
symphysial surface 107
synovial (tendon) sheath 138
synovial fold 128
synovial membrane 133
synovial membrane of joint capsule 128, 136

T

taeniae coli 310
tail of pancreas 308
talus 114
tarsal bones 104
tarsal sinus 117
tarsals 30, 32
taste bud 230
taste pore 230
teeth 303

uterine artery 282
uterine tube 258, 323, 325
uterus 258, 282, 318, 324
uvula 227, 240

V

vagina 258, 318, 323, 325, 326
vaginal artery 282
vaginal process 58
vagus nerve 168, 216, 222, 224, 241, 245, 254, 268, 269, 272, 293, 304
valve of inferior vena cava 264
vastus intermedius 199, 201, 284
vastus lateralis 135, 284, 197, 201, 202, 205
vastus medialis 142, 197, 201, 284
vein 303
vein with red blood cells (rbc) 295
venous valves 266
ventral horn of gray matter 232
ventral ramus 213
ventral root of spinal nerve 213, 232
ventral rootlets 213
venule 255
vermiform appendix 310
vertebral artery 213, 242, 271, 272, 275
vertebral body 74, 76, 77, 272, 291
vertebral column 30, 32, 303, 311
vertebral foramen 74
vertical muscle 155
vesicouterine pouch 323, 326
vestibular area 241
vestibule 327

vestibulocochlear nerve 224, 241, 242, 245
villi 302
visceral pericardium 264
vocalis muscle 213
vomer 36, 40, 44, 47
vomerine crest of choana 67
vomerine groove 67

W

white blood cells 266
white communicating ramus 222

X

xiphoid process 82

Z

zona fasciculata of cortex 256
zona glomerulosa of cortex 256
zona pellucida 324
zona reticularis of cortex 256
zone of calcified cartilage 27
zone of hypertrophied cartilage 27
zone of proliferating cartilage 27
zone of resting cartilage 27
zygomatic arch 38, 44, 151, 271
zygomatic bone 36, 38, 40, 42, 44
zygomatic process 56, 60
zygomatico-orbital foramen 65
zygomaticofacial foramen 65
zygomaticotemporal foramen 65
zygomaticus major 145, 146, 149, 227
zygomaticus minor 145, 149